PHYSICS AND TECHNOLOGY OF CRYSTALLINE OXIDE SEMICONDUCTOR CAAC-IGZO

Wiley-SID Series in Display Technology

Series Editors:
Anthony C. Lowe and Ian Sage

PHYSICS AND TECHNOLOGY OF CRYSTALLINE OXIDE SEMICONDUCTOR CAAC-IGZO

APPLICATION TO LSI

Edited by

Shunpei Yamazaki
Semiconductor Energy Laboratory Co., Ltd, Japan

Masahiro Fujita
University of Tokyo, Japan

This edition first published 2017
© 2017 John Wiley & Sons, Ltd

Registered Office
John Wiley & Sons, Ltd, The Atrium, Southern Gate, Chichester, West Sussex, PO19 8SQ, United Kingdom

For details of our global editorial offices, for customer services and for information about how to apply for permission to reuse the copyright material in this book please see our website at www.wiley.com.

Library of Congress Cataloging-in-Publication Data

Names: Yamazaki, Shunpei, 1942– author. | Fujita, Masahiro, 1956– author.
Title: Physics and technology of crystalline oxide semiconductor CAAC-IGZO.
    Application to LSI / Shunpei Yamazaki, Masahiro Fujita.
Description: Chichester, West Sussex, United Kingdom : John Wiley & Sons, Ltd
    Registered office John Wiley & Sons Ltd, [2017] | Includes bibliographical references and index.
Identifiers: LCCN 2016025860 | ISBN 9781119247340 (cloth) | ISBN 9781119247432 (epub) |
    ISBN 9781119247425 (Adobe PDF)
Subjects: LCSH: Semiconductors–Materials. | Semiconductors–Characterization. |
    Gallium compounds. | Zinc compounds.
Classification: LCC TK7871.85 .Y357598 2016 | DDC 621.39/5–dc23
LC record available at https://lccn.loc.gov/2016025860

A catalogue record for this book is available from the British Library.

Set in 10/12pt Times by SPi Global, Pondicherry, India

Printed and bound in Malaysia by Vivar Printing Sdn Bhd

10 9 8 7 6 5 4 3 2 1

Contents

About the Editors

Shunpei Yamazaki received his Ph.D., ME, BE, and honorary degrees from Doshisha University, Japan, in 1971, 1967, 1965, and 2011, respectively, and is the founder and president of Semiconductor Energy Laboratory Co., Ltd. He invented a basic device structure of non-volatile memory known as "flash memory" in 1970 during his Ph.D. program. Yamazaki is a distinguished foreign member of the Royal Swedish Academy of Engineering Sciences and a founder of Kato & Yamazaki Educational Foundation. Yamazaki has published or co-published over 400 papers and conference presentations and is the inventor or co-inventor of over 6314 patents (Guinness World Record in 2011).

1967	Completed Master's Degree Program at Doshisha University Graduate School of Engineering
1970	Invented a basic device of flash memory (Japanese Patent No. 886343; Japanese Examined Patent Application Publication No. Sho50-36955)
1971	Received Ph.D. in Engineering from Doshisha University Graduate School Doctoral Program
	Joined TDK Corporation (formerly TDK Electronics Co., Ltd.)
1980	Established Semiconductor Energy Laboratory Co., Ltd. and assumed position as president
1984	Awarded the Richard M. Fulrath Award by the American Ceramic Society (for research on MIS structure)
1995	Awarded the Medal with Dark Blue Ribbon from the Cabinet Office of the Japanese government (proceeds given to Japanese Red Cross Society) (awarded 6 times since 2015)
1997	Awarded the Medal with Purple Ribbon from the Cabinet Office of the Japanese government (for development of MOS LSI element technology)
2009	IVA (Royal Swedish Academy of Engineering Science) Foreign Member
2010	Awarded Okochi Memorial Technology Award from Okochi Memorial Foundation
2011	IEEE Life Fellow
	Received Honorary Doctor Degree of Culture from Doshisha University

Renewed his first Guinness World Record in 2004 (man holding the most patents in the world)

2015   Granted the title of "Friend of Doshisha" by Doshisha University

2015   SID Special Recognition Award for "discovering CAAC-IGZO semiconductors, leading their practical application, and paving the way to next-generation displays by developing new information-display devices such as foldable or 8K × 4K displays"

Masahiro Fujita received his Ph.D. in Information Engineering from the University of Tokyo in 1985 on his work on model checking of hardware designs by using logic programming languages. In 1985, he joined Fujitsu as a researcher and started to work on hardware automatic synthesis as well as formal verification methods and tools, including enhancements of BDD/ SAT-based techniques. From 1993 to 2000, he was director at Fujitsu Laboratories of America and headed a hardware formal verification group which was developing a formal verifier for real-life designs having more than several millions of gates. The developed tool has been used in production internally at Fujitsu and externally as well. Since March 2000, he has been a professor at VLSI Design and Education Center in the University of Tokyo. He has done innovative works in the areas of hardware verification, synthesis, testing, and software verification mostly targeting embedded software and web-based programs. He has been involved in a Japanese governmental research project for dependable system designs and has developed a formal verifier for C programs that could be used for both hardware and embedded software designs. The tool is now under evaluation jointly with industry with governmental support. He has authored and co-authored 10 books, and has more than 300 publications and has been given several awards from scientific societies. He has been involved as program and steering committee members in many prestigious conferences on CAD, VLSI designs, software engineering, and more. His current research interests include synthesis and verification of SoC (System on Chip), hardware/software co-designs targeting embedded systems and cyber physical systems, digital/analog co-designs, and formal analysis, verification, and synthesis of embedded programs.

List of Contributors

Shunpei Yamazaki (editor)   Semiconductor Energy Laboratory Co., Ltd.
Masahiro Fujita (editor)   The University of Tokyo

In alphabetical order:

Takeshi Aoki   Semiconductor Energy Laboratory Co., Ltd.
Masami Endo   Semiconductor Energy Laboratory Co., Ltd.
Hiroki Inoue   Semiconductor Energy Laboratory Co., Ltd.
Takahiko Ishizu   Semiconductor Energy Laboratory Co., Ltd.
Masayuki Kimura   Semiconductor Energy Laboratory Co., Ltd.
Munehiro Kozuma   Semiconductor Energy Laboratory Co., Ltd.
Yoshiyuki Kurokawa   Semiconductor Energy Laboratory Co., Ltd.
Shuhei Maeda   Semiconductor Energy Laboratory Co., Ltd.
Daisuke Matsubayashi   Semiconductor Energy Laboratory Co., Ltd.
Shinpei Matsuda   Semiconductor Energy Laboratory Co., Ltd.
Takanori Matsuzaki   Semiconductor Energy Laboratory Co., Ltd.
Shuhei Nagatsuka   Semiconductor Energy Laboratory Co., Ltd.
Satoru Okamoto   Semiconductor Energy Laboratory Co., Ltd.
Yuki Okamoto   Semiconductor Energy Laboratory Co., Ltd.
Tatsuya Onuki   Semiconductor Energy Laboratory Co., Ltd.
Takashi Shingu   Semiconductor Energy Laboratory Co., Ltd.
Yutaka Shionoiri   Semiconductor Energy Laboratory Co., Ltd.
Kei Takahashi   Semiconductor Energy Laboratory Co., Ltd.
Toshihiko Takeuchi   Semiconductor Energy Laboratory Co., Ltd.
Masashi Tsubuku   Semiconductor Energy Laboratory Co., Ltd.
Naoaki Tsutsui   Semiconductor Energy Laboratory Co., Ltd.
Seiichi Yoneda   Semiconductor Energy Laboratory Co., Ltd.

Series Editor's Foreword

The convergence of personal electronic devices towards small, powerful and multifunctional platforms throws into relief the conflict for resources between the display and other system electronics. On the one hand, high-quality, high-resolution and bright displays not only provide an essential human interface, but are one of the decisive factors in attracting users to purchase a device and differentiate between different models. On the other hand, the display has no purpose without the electronic systems which control and supply content to it – functions which now require powerful, fast data processing and information storage capabilities. Both display and system electronics must share the limited energy stored in a small, lightweight battery, and achieving excellent performance from the whole device, combined with adequate battery life from a small package, is a central challenge.

In this volume, Dr. Yamazaki and Professor Fujita bring together a comprehensive account of how CAAC oxide semiconductors can contribute to the ecosystem of large-scale integrated electronics. Many of the developments presented here provide routes to making major power-consumption savings in the operation of electronic systems. In other cases, performance improvements or new capabilities arise from the use of these CAAC oxide components: this is the case, for example, in the imaging sensors presented in Chapter 7 of this book.

The book you are holding is one of the three volumes planned by Dr. Yamazaki and his colleagues, to give a comprehensive overview of CAAC-IGZO technology. The first volume presents the basic science and technology of the materials: deposition conditions, structure, physical properties, and the physics and performance of the semiconductor devices using them. The origin of high carrier mobility and the exceptional low leakage current in CAAC-IGZO TFTs, as well as techniques for measuring it, are presented.

The third volume will describe in detail the application of CAAC oxides to display devices – LCD and OLED active matrix circuits, driver circuits and new technologies which apply particularly to flexible displays. Issues of stability and light sensitivity, which are of particular importance in displays, are thoroughly explored and routes to their solution are presented.

In the present volume, the application of CAAC-IGZO to LSI is presented. The book includes a thorough account of the TFT structures exploited and their fabrication, threshold control and switching characteristics, including their extremely low off-state current and relative immunity to short-channel effects. Then, the application of these components to the most important and relevant electronic subsystems is described: memory, CPUs and FPGAs. The benefits available from CAAC devices in these systems are described – long-term data storage without refresh, higher memory densities and power reduction through adoption of normally-off logic. The design changes which can realise these benefits and the actual performance of circuits are described. In imaging sensors, the low leakage current of CAAC devices allows high-performance global shuttering and on-sensor image processing to be realised, bringing new capabilities to the devices. The volume concludes with an overview of further application fields, including RF tags, X-ray imaging and CODEC systems.

The authors and editors bring to their subject an outstanding breadth of expertise in the research and development of CAAC-IGZO materials, devices and systems, and their account of the subject should provide a definitive source for those seeking to understand and exploit the impact of this developing technology on modern electronics.

Ian Sage
Malvern, UK, 2016

Preface

Entering the 21st century, it seems that the growth of the electronics industry is hitting saturation level, even though it is the largest industry in the world. This is because the amount of energy used by people, which has already become enormous – as reflected in the abrupt climate change in recent years – is going to increase even more with its growth. Especially, the energy consumptions of cloud computing and electronic devices such as smartphones and supercomputers will continue to increase. Therefore, it is not an exaggeration to say that the development of new energy-saving devices has a direct influence on the continued existence of all mankind.

For this reason, we started extensive research on crystalline oxide semiconductors (OS), especially on a c-axis-aligned crystalline indium – gallium – zinc oxide (CAAC-IGZO) semiconductor. Due to the economic downturn in the aftermath of the Lehman Brothers' bankruptcy in the autumn of 2008, many companies withdrew from research on this subject, but I never gave up and our research in this area has continued to the present day. One of the most important characteristics of a field-effect transistor (FET) using this wide-gap semiconductor is that the off-state current is on the order of yoctoampère per centimeter (10^{-24}A/cm) (yocto is the smallest SI prefix), which is smaller than that of any other device measured so far. This characteristic effectively reduces the energy consumption, and thus we believe that it coincides with society's need to save energy.

It has been less than 10 years since I started researching and developing oxide semiconductors, but I think that proposing their effectiveness without delay is the first step toward a contribution to humanity. That is why I would like to introduce this book series *Physics and Technology of Crystalline Oxide Semiconductor*, consisting of *Fundamentals*, *Application to LSI*, and *Application to Displays*, even though I know that it cannot be said that every detail is completely covered in the book series.

The book series contains the discovery of CAAC-IGZO by me, Shunpei Yamazaki, one of the editors and authors thereof, as well as the research results on its application obtained at Semiconductor Energy Laboratory Co., Ltd. (SEL), where I serve as president. We have decided to write the experimental facts down in as much detail as possible, and publish models whose principles have not yet been verified. The reason is that I would like to give a couple of hints to readers – graduate students, on-site researchers, and developers – so that they can conduct further R&D as soon as possible. For these reasons, as well as the limited number of pages, I would like you to accept my deepest apologies for not being able to publish all of the data in these books. Even after the publication of these three books about crystalline oxide semiconductors, I would like to continue making our CAAC-IGZO technology known to the public by conducting further research on it from both engineering and academic points of view.

This book covers a wide range of topics, such as the device physics of FETs using CAAC-IGZO and their applications to LSI.

In the past, Bell Laboratories published a set of books called *The Bell Telephone Laboratories series* about the invention of transistors and research results thereof, which accordingly spread the current concept of transistors throughout the world. We sincerely hope that our books will help to spread the CAAC-IGZO technology just as *The Bell Telephone Laboratories series* helped to popularize the concept of transistors. I think that CAAC-OS, especially CAAC-IGZO, still has many unexplored possibilities and thus more institutions and scientists should research it in cooperation with each other. I am expecting that the CAAC-IGZO which we discovered will flourish in the 21st century by publishing its physical properties and principles, as well as by applying it in the display and LSI fields, especially in energy-saving devices.

So far, we have made some efforts by submitting papers and giving presentations at various conferences about crystalline oxide semiconductors and OS FETs. However, we have never heard of another case where a ceramic was used for an active element on a mass-production basis in Si LSI or displays; thus, many companies (with the exception of Sharp Corporation) will face a lot of difficulties in terms of mass production. Note that a ceramic with an amorphous structure has been proposed before, but it was not put into practical use due to reliability problems. Especially, the great depression following 2008 made many companies quit their R&D of ceramics with an amorphous structure, which was deemed to be fruitless because a FET utilizing amorphous ceramic lacks reliability.

I, Shunpei Yamazaki, observed a TEM image of an IGZO film in front of a TEM screen to find a solution for the reliability issue. At that time, I discovered that a CAAC structure existed in the IGZO film. I thought that the problem of reliability could be solved by using this kind of material, and thus shifted the focus of our R&D to CAAC-IGZO. A FET using this CAAC-IGZO has a high level of reliability, which cannot be said of a FET which uses amorphous IGZO. Thus, a FET with CAAC-IGZO is excellent from a repeatability point of view in that it can be measured and evaluated stably, both on the material and device level. As a result of the stable measurement and evaluation, we discovered that the off-state current is on the order of yoctoamps per centimeter (10^{-24} A/cm), as mentioned above. Additionally, since IGZO has a wide solid-solution phase, we succeeded in fabricating FETs using CAAC-IGZOs having high mobilities of $30 - 70$ cm^2/V-s, thus exceeding 50 cm^2/V-s, by changing the composition ratio and the device structure. A mobility equaling that of an LTPS-FET means that the CAAC-IGZO might be able to not only fight evenly with an LTPS-FET, but also outperform it in the industry. Furthermore, we tried to apply CAAC-IGZO FETs to LSI, something which has never been done before, and discovered that such a FET can operate with a channel length of just $20 - 60$ nm.

Our data has been reviewed by many specialists, but it seems that to help people understand *the true value of the crystalline oxide semiconductor*, there is still a need to further explain the numerous issues concerning fundamental properties, which have not yet been fully understood. Moreover, a lot of people gave us the same advice: to help intellectuals grasp the whole picture of the technology by publishing a series of at least three books (*Fundamentals*, *Application to LSI*, and *Application to Displays*). Accordingly, I decided to publish them. Note that almost the whole content of these books is based on our experimental data. Hence, please acknowledge SEL and Advanced Film Device Inc. (AFD Inc.), a subsidiary of SEL, as the sources of these books, unless otherwise specified.

During the creation of this book, many people helped and guided us. I would like to express my deepest appreciation especially to Dr. Masahiro Fujita, who has improved the research environment in the field of OS LSI, for being a co-editor of this book, *Application to LSI*, and for training the employees of SEL.

Moreover, during the research and development on which these books are based, as well as during the writing process, many young researchers at SEL also contributed. The names of all the authors involved can be found in the List of Contributors.

We would also like to extend our heartfelt thanks to Dr. Johan Bergquist, Dr. Michio Tajima, Mr. Yukio Maehashi, Mr. Takashi Okuda, and Mr. Jun Koyama for helping us with the writing of this book – by checking for errors and giving us a great deal of advice on how to improve the text.

I was blessed with support and cooperation from many outstanding individuals. I would like to add that I could not have finished these books in such a short period of time without the efforts of Dr. Ian Sage, a Wiley-SID book series editor, who suggested the publication of the books within this time, as well as Ms. Alexandra Jackson and Ms. Nithya Sechin of John Wiley & Sons, Ltd. Last but not least, I would like to express my sincere gratitude to those publishers and authors who allowed us to use their figures as references in these books.

Shunpei Yamazaki
President of Semiconductor Energy Laboratory Co., Ltd.

Acknowledgments

First of all, we would like to thank **Dr. Johan Bergquist**, **Dr. Michio Tajima**, **Mr. Yukio Maehashi**, **Mr. Takashi Okuda**, and **Mr. Jun Koyama**, who encouraged us and gave valuable advice on writing manuscripts.

Furthermore, we would also like to thank many people in a variety of fields for providing data, cooperation in writing manuscripts and English translation, and all other aspects of this book.

Our heartfelt thanks to (in alphabetical order):

Ms. Mayumi Adachi
Ms. Yukari Amano
Mr. Yoshinobu Asami
Mr. Yuji Egi
Mr. Toshiya Endo
Ms. Nana Fujii
Ms. Ai Hattori
Mr. Shinji Hayakawa
Mr. Atsushi Hirose
Mr. Ryota Hodo
Mr. Mitsuhiro Ichijo
Ms. Yasuko Iharakumi
Dr. Kiyoshi Kato
Dr. Shuichi Katsui
Mr. Hajime Kimura
Ms. Kyoko Kitada

Ms. Yuko Konno
Mr. Masaki Koyama
Mr. Motomu Kurata
Mr. Tetsunori Maruyama
Ms. Michiyo Mashiyama
Mr. Hidekazu Miyairi
Ms. Tomoko Nakagawa
Mr. Hasumi Nomaguchi
Mr. Takuro Ohmaru
Mr. Naoki Okuno
Ms. Yoko Otake
Ms. Shiori Saga
Mr. Masayuki Sakakura
Mr. Naoya Sakamoto
Mr. Yujiro Sakurada
Mr. Shinya Sasagawa

Ms. Yu Sasaki
Mr. Yuichi Sato
Mr. Akihisa Shimomura
Ms. Erika Takahashi
Ms. Tamae Takano
Mr. Yasuhiko Takemura
Mr. Hikaru Tamura
Mr. Tetsuhiro Tanaka
Ms. Mika Tatsumi
Ms. Yukiko Tojo
Mr. Ryo Tokumaru
Ms. Hitomi Tsurui
Mr. Yuto Yakubo
Mr. Naoto Yamade
Ms. Chiaki Yamura
and many others

Shunpei Yamazaki
Masahiro Fujita

1

Introduction

1.1 Overview of this Book

The three books in this series deal with *c*-axis-aligned crystalline indium–gallium–zinc oxide (CAAC-IGZO), an oxide semiconductor (see Figure 1.1): *Physics and Technology of Crystalline Oxide Semiconductor CAAC-IGZO: Fundamentals* (hereinafter referred to as *Fundamentals*) [1], *Physics and Technology of Crystalline Oxide Semiconductor CAAC-IGZO: Application to LSI* (this book, hereinafter referred to as *Application to LSI*), and *Physics and Technology of Crystalline Oxide Semiconductor CAAC-IGZO: Application to Displays* (hereinafter referred to as *Application to Displays*) [2]. *Fundamentals* describes, for example, the material properties of oxide semiconductors, the formation mechanism and crystal structure analysis of IGZO, the fundamental physical properties of CAAC-IGZO, the electrical characteristics of field-effect transistors (FETs) with CAAC-IGZO active layer (hereinafter referred to as CAAC-IGZO FETs), and comparisons between CAAC-IGZO and silicon (Si) FETs. *Application to Displays* introduces applications of the CAAC-IGZO FET technology to liquid crystal and organic light-emitting diode displays, describing the process flows and characteristics of the FETs, the driver circuits for displays, the technologies for high-definition, low-power, flexible displays, and so on.

This volume, *Application to LSI*, aims to introduce the applications of CAAC-IGZO FET technology to large-scale integration (LSI) and broadly and concisely review the device physics of CAAC-IGZO FETs. On the basis of the distinct material features of these FETs disclosed in *Fundamentals*, such FETs have an attractive application field in LSIs, in addition to the display applications described in *Application to Displays*. Not only focusing on oxide semiconductor material aspects, this book will also describe device design and fabrication using such materials, combination with other technologies, and specific applications (see Figure 1.2).

Physics and Technology of Crystalline Oxide Semiconductor CAAC-IGZO: Application to LSI, First Edition.
Edited by Shunpei Yamazaki and Masahiro Fujita.
© 2017 John Wiley & Sons, Ltd. Published 2017 by John Wiley & Sons, Ltd.

Application to Displays

Application to LSI

Liquid Crystal Display

OLED Display

Flexible Display

Memory
- NOSRAM
- DOSRAM

CPU   FPGA   Image Sensor

Applications

OLED material

Liquid crystal material

Transfer technology

Process Technology

- high resolution
- large size substrate (high productivity and low cost)
- combination with other components such as wiring, insulating film, and capacitor

Process Technology

- microfabrication
- combination with Si Process (hybrid process)
- combination with other components such as wiring, insulating film, and capacitor

Elemental Technology

CAAC-IGZO Transistor

Defect Levels
Electrical Properties
Diversity of Transistor Structures

Fundamentals

Material

Film Deposition

Crystal Structure
- CAAC-IGZO
- nc-IGZO

Oxide Semiconductor

Fundamentals

Figure 1.1   Framework and summary of the book series

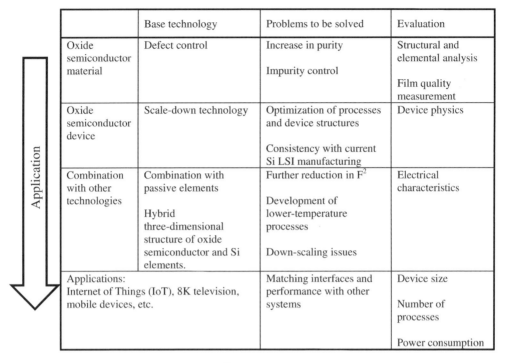

	Base technology	Problems to be solved	Evaluation
Oxide semiconductor material	Defect control	Increase in purity Impurity control	Structural and elemental analysis Film quality measurement
Oxide semiconductor device	Scale-down technology	Optimization of processes and device structures Consistency with current Si LSI manufacturing	Device physics
Combination with other technologies	Combination with passive elements Hybrid three-dimensional structure of oxide semiconductor and Si elements.	Further reduction in F^2 Development of lower-temperature processes Down-scaling issues	Electrical characteristics
Applications: Internet of Things (IoT), 8K television, mobile devices, etc.		Matching interfaces and performance with other systems	Device size Number of processes Power consumption

(left margin, vertical: Application)

Figure 1.2   Scope of this book. The symbol F^2 means the square of the feature size F, used as an index of the memory cell size

Application examples of CAAC-IGZO FET technologies to LSIs are specifically described in the subsequent chapters.

1.2   Background

The integrated circuit (IC) has a huge market [3]. As shown in Figure 1.3, the total market size, including analog, micro, logic, and memory applications, is worth approximately 278 billion US dollars. Here, "micro" applications are microprocessor units (MPUs), microcontroller units (MCUs), and digital signal processors (DSPs); "logic" applications include specified logic and custom logic, such as field-programmable gate arrays (FPGAs) and application-specific integrated circuits (ASICs). CAAC-IGZO FETs address this vast IC market.

1.2.1   Typical Characteristics of CAAC-IGZO FETs

In the LSI field, reduction of power consumption has so far been achieved mainly by scaling down the FETs, employing advanced power management schemes, and more recently, sub-threshold driving. Si FETs are currently scaled down to very small technology nodes, for example, gate lengths as small as 14 and 16 nm [4]. Such aggressive downscaling causes

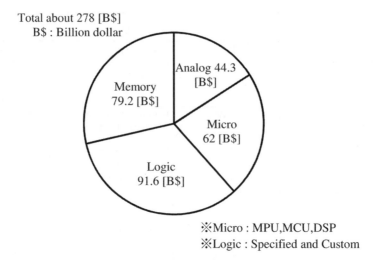

Total about 278 [B$]
B$: Billion dollar

※Micro : MPU,MCU,DSP
※Logic : Specified and Custom

Figure 1.3   Market size of ICs in 2014. *Source*: Adapted from [3]

an increase in the FET off-state current (leakage current in the FET in the off state), which poses new obstacles to further reduction of system power [5].

As reported by Kato *et al.* [6], CAAC-IGZO FETs exhibit extremely low off-state current, for example, 1.35×10^{-22} A/μm (135 yA/μm, where y stands for yocto) for a FET with channel length/width of 3/50 μm. In contrast, the off-state current in a single-crystal Si (sc-Si) FET of the same structure and dimensions has an off-state current of 1×10^{-12} A/μm (1 pA/μm), i.e., 10 orders of magnitude larger. When CAAC-IGZO FETs are used in LSI devices, such as dynamic random access memory (DRAM), non-volatile memories, and central processing units (CPUs), their extremely low off-state current will therefore reduce the system power consumption tremendously.

As reported by Matsubayashi *et al.* [7], CAAC-IGZO FETs with a channel node of 20 nm maintain the extremely low off-state current, despite the aggressive downscaling. Figure 1.4 shows the miniaturization progress of CAAC-IGZO and Si FETs during the past four to five years [8]. In the graph, the upper gray band corresponds to the achieved scaling values of CAAC-IGZO FETs, whereas the lower solid line shows the target scaling values of Si FETs disclosed by International Technology Roadmap for Semiconductors [9]. CAAC-IGZO FETs for processors, memories, and devices are denoted by diamond shapes, squares, and triangles, respectively. The number next to each mark corresponds to the conference shown below the graph where the device was disclosed. As shown, the scaling of CAAC-IGZO FETs gradually approaches that of Si FETs in recent years, so if the scaling continues to progress at this speed, it will catch up with that of Si FETs later in 2016 or 2017.

1.2.2   Possible Applications of CAAC-IGZO FETs

CAAC-IGZO FETs can be used in various LSIs (hereinafter called CAAC-IGZO LSIs), for example, in non-volatile memories [10–13], DRAMs [14], normally-off CPUs [15–17], FPGAs [18,19], and image sensors [20,21]. Non-volatile memories and DRAMs employing

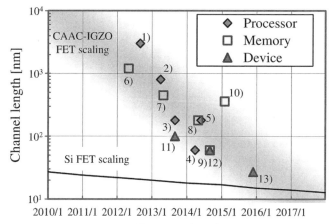

Processor:
1) SSDM 2012
2) COOL Chips 2013
3) SSDM 2013
4) COOL Chips 2014
5) VLSI 2014

Memory:
6) IMW 2012
7) IMW 2013
8) IMW 2014
9) SSDM 2014
10) ISSCC 2015

FET:
11) SSDM 2013
12) SSDM 2014
13) IEDM 2015

IEDM   : International Electron Devices Meeting
IMW   : International Memory Workshop
ISSCC   : International Solid-State Circuits Conference
SSDM   : International Conference on Solid State Devices and Materials
VLSI   : Symposia on VLSI Technology and Circuits

Figure 1.4   Comparison of scaling between CAAC-IGZO FETs and Si FET. *Source*: Adapted from [8].

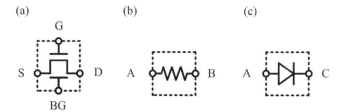

Figure 1.5   (a) CAAC-IGZO FET, an active element with four terminals (source S, drain D, gate G, back gate BG); (b) resistive element, a passive element with two terminals; and (c) diode, a two-terminal passive element with non-linear characteristics

CAAC-IGZO FETs are called non-volatile oxide semiconductor random access memory (NOSRAM) and dynamic oxide semiconductor random access memory (DOSRAM), respectively.

A CAAC-IGZO FET is an active element with four terminals: source, drain, gate, and back gate, as shown in Figure 1.5. New memory technologies that have recently attracted attention

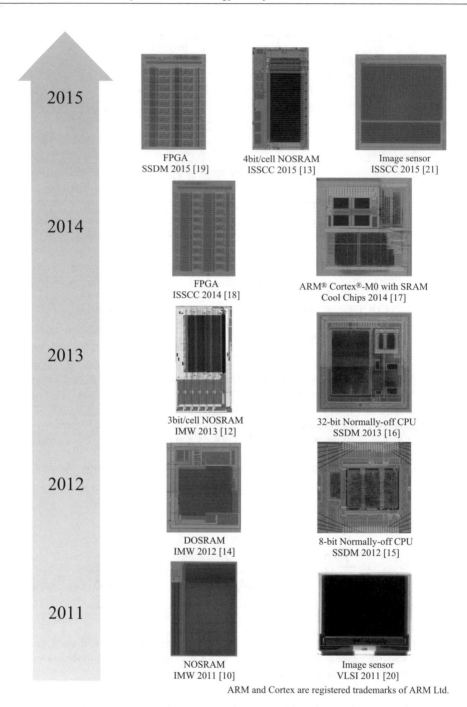

Figure 1.6 Examples of CAAC-IGZO LSIs fabricated between 2011 and 2015. *(For color detail, please see color plate section)*

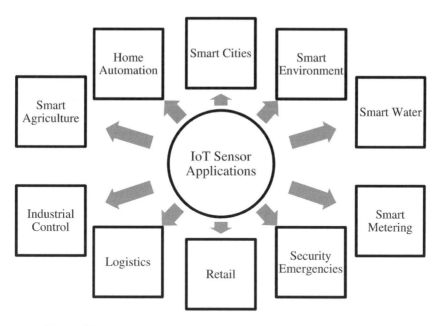

Figure 1.7 Application examples of IoT. *Source*: Adapted from [22,23]

include magnetoresistive random access memory (MRAM), resistive random access memory (ReRAM), phase change random access memory (PCRAM), and ferroelectric random access memory (FeRAM). These are all passive elements with two terminals, whereas CAAC-IGZO FETs with their four terminals may lead to new applications.

Figure 1.6 shows photographs of LSIs with CAAC-IGZO FETs that have been fabricated so far. Below each photograph, the type of LSI and the name of the conference where it was presented are written.

The concept of Internet of Things (IoT) is likely to be realized in the near future. In IoT, LSIs are used in various things to control them and collect and process information through the Internet. LSIs used for IoT should be inexpensive and autonomously powered, i.e., small in size and driven with low power, particularly in the idling state. It is therefore expected that CAAC-IGZO FETs, with their extremely low off-state current, will meet those requirements.

Janusz Bryzek, an advocate of IoT, proposes possible applications of IoT such as logistics, retail, security emergencies, and others (see Figure 1.7) [22,23]. He predicts that a total of one trillion sensors will be used in 2023 (i.e., he forecasts a huge possible market not only for the sensors themselves, but also for the necessary peripheral semiconductor circuits for preprocessing, temporary storage, and wireless transmission).

1.3 Summary of Each Chapter

The device physics, structure, and fabrication process of CAAC-IGZO FETs are briefly explained in Chapter 2, whereas application examples in LSIs and the like are described in and after Chapter 3.

Chapter 2 also includes a review of *Fundamentals*, followed by a description of various CAAC-IGZO FET structures and their basic electrical characteristics, both in general and with emphasis on the low off-state current. CAAC-IGZO FETs are resistant to the downsizing-induced reduction in field-effect mobility or short-channel effect, and unlike sc-Si FETs, the off-state current of CAAC-IGZO FETs does not increase at high temperatures. The possibility of downscaling is illustrated, with results of a CAAC-IGZO FET with channel node of 20 nm [7]. The fabrication process flow of an actual CAAC-IGZO FET with a typical structure is also explained. A hybrid structure that vertically combines Si and CAAC-IGZO FETs is also introduced.

Chapter 3 deals with NOSRAM, where the CAAC-IGZO FET technology is applied to non-volatile memories [10–13]. This non-volatile memory relies on the extremely low off-state current, and can operate at approximately 5 V (i.e., a quarter of the voltage of conventional flash memories). NOSRAM also exhibits an excellent write endurance. While conventional flash memory has a write endurance of approximately 10,000 cycles, NOSRAM endures one trillion writes, achieving a ten-million-fold increase. In addition, the electric potentials can be applied directly to the memory cell during data writing, thus providing accurate control over the accumulated electric charge. Therefore, NOSRAM enables multiple bits in one cell.

Chapter 4 presents DOSRAM in which the CAAC-IGZO FET technology is applied to the DRAM memory cell [14]. Compared with DRAM involving Si, DOSRAM features a long data retention period because the charge stored in the capacitor is hardly lost owing to the extremely low off-state current. Consequently, it requires less frequent refresh operations and therefore consumes less power than its Si FET-based equivalent. For the same reason, electric charges in a capacitor can be stored for a long time even at low capacitance. Accordingly, the capacitance required for data retention may be reduced, which is advantageous in miniaturization.

Chapter 5 describes a normally-off CPU deploying CAAC-IGZO FETs [15–17]. Similar to power gating, the power supply to a circuit stops when unused (the circuit switches to sleep mode) in a normally-off CPU, resulting in low power consumption. When a CPU circuit comprising Si FETs is subject to a power gating operation, there is an overhead in power consumption and performance due to saving and restoring of storage elements in the circuit. Consequently, power gating in short intervals has been problematic because the average CPU power consumption would increase instead. In contrast, a normally-off CPU implemented with CAAC-IGZO FETs reduces the overhead power consumption dramatically by exploiting the extremely low off-state current characteristics of CAAC-IGZO FETs, and shortens the time required for backup and recovery.

Chapter 6 provides an example wherein CAAC-IGZO FETs are applied to FPGAs [18,19]. An FPGA is an LSI that a user can configure after manufacture. In conventional FPGAs, the circuit configuration information is stored in a static random access memory (SRAM) that is used as configuration memory, but SRAM data are generally lost when the power is turned off. Consequently, setting information needs to be stored in the configuration memory every time the power is back on. If a non-volatile memory using a CAAC-IGZO FET replaces this SRAM, setting information is retained even when the power is turned off; thus, the memory does not need restoring in the configuration memory. Moreover, the area and power consumption compared with SRAM may be reduced. Therefore, the incorporation of a CAAC-IGZO FET is expected to produce an FPGA with higher density and lower power consumption. A power gating function can easily be implemented in FPGAs; consequently, turning off unused circuits may further reduce power consumption. Normally-off operation suitable for fine-grained multi-context structures is also possible by developing the above-mentioned features. Chapter 6 also introduces FPGAs with subthreshold operation, further reducing the power consumption via the lower operating voltage. Finally, the potential development of high-performance computing by combining an FPGA and a CPU, which has recently attracted extensive interest, is discussed.

Chapter 7 presents an example of an image sensor that uses CAAC-IGZO FETs [20,21]. Many of the existing complementary metal–oxide semiconductor (CMOS) image sensors use a rolling shutter mode whereby sensor pixels sequentially capture imaging data row by row. However, this mode exhibits a delay between first and last capturing sensor pixels. Therefore, a fast-moving object yields a distorted image. This delay occurs because captured data get leaked over time and are required to be read out immediately after their capture. When CAAC-IGZO FETs are introduced in an image sensor, the extremely low off-state current of the FETs enables the implementation of a global shutter mode whereby all sensor pixels simultaneously capture data. This off-state current also allows sensor pixels to retain captured data until readout, regardless of any difference in readout timing. Using multiple retention nodes in each sensor pixel allows multiple capture with very short shutter times, an attractive feature in machine vision. Adding an image difference detection function to the sensor pixel gives a motion sensor that performs detection of changes with respect to a reference frame in addition to normal imaging.

Chapter 8 presents other examples of CAAC-IGZO FET applications, demonstrating the versatility of this device. These examples include radio-frequency devices, X-ray detectors, encoder–decoders (CODECs), DC–DC converters (DC denotes direct current), analog programmable devices, and neural networks that may find use in various environments. Further, memory-based computing and an ultra-efficient power gating mechanism are presented.

LSIs with CAAC-IGZO have characteristics of very low off-state current and the associated reduction in system power consumption suggests that CAAC-IGZO LSIs may entirely replace Si LSIs in some applications.

References

[1] Yamazaki, S. and Kimizuka, N. (in press) *Physics and Technology of Crystalline Oxide Semiconductor CAAC-IGZO: Fundamentals.* New York: John Wiley.

[2] Yamazaki, S. and Tsutsui, T. (in press) *Physics and Technology of Crystalline Oxide Semiconductor CAAC-IGZO: Application to Displays.* New York: John Wiley.

[3] WSTS (2015) *The Final Semiconductor Market Figures for 2014.* World Semiconductor Trade Statistics.

[4] ITRS (2013) *Overall Roadmap Technology Characteristics (ORTC) Table.* International Technology Roadmap for Semiconductors.

[5] ITRS (2013) *Process Integration, Devices, and Structures Summary.* International Technology Roadmap for Semiconductors.

[6] Kato, K., Shionoiri, Y., Sekine, Y., Furutani, K., Hatano, T., Aoki, T., *et al.* (2012) "Evaluation of off-state current characteristics of transistor using oxide semiconductor material, indium–gallium–zinc oxide," *Jpn. J. Appl. Phys.*, **51**, 021201.

[7] Matsubayashi, D., Asami, Y., Okazaki, Y., Kurata, M., Sasagawa, S., Okamoto, S., *et al.* (2015) "20-nm-Node trench-gate-self-aligned crystalline In–Ga–Zn-oxide FET with high frequency and low off-state current," *IEEE IEDM Tech. Dig.*, 141.

[8] Yamazaki, S. (2016) "Unique technology from Japan to the world – super low power LSI using CAAC-OS." Available at: www.umc.com/2015_japan_forum/pdf/20150527_shunpei_yamazaki_eng.pdf [accessed February 11, 2016].

[9] ITRS (2009) *Table FEP2: High Performance Device Technical Requirements.* Available at: www.dropbox.com/sh/ia1jkem3v708hx1/AAB6fSsJmdHaQNEu538i9gKNa/2009%20Tables%20%26%20Graphs/FEP/2009Tables_FEP2.xls?dl=0 [accessed February 19, 2016].

[10] Matsuzaki, T., Inoue, H., Nagatsuka, S., Okazaki, Y., Sasaki, T., Noda, K., *et al.* (2011) "1Mb Non-volatile random access memory using oxide semiconductor," *Proc. IEEE Int. Memory Workshop*, 185.

[11] Inoue, H., Matsuzaki, T., Nagatsuka, S., Okazaki, Y., Sasaki, T., Noda, K., *et al.* (2012) "Nonvolatile memory with extremely low-leakage indium–gallium–zinc-oxide thin-film transistor," *IEEE J. Solid-State Circuits.*, **47**, 2258.

[12] Nagatsuka, S., Matsuzaki, T., Inoue, H., Ishizu, T., Onuki, T., Ando, Y., *et al.* (2013) "A 3bit/cell nonvolatile memory with crystalline In–Ga–Zn–O TFT," *Proc. IEEE Int. Memory Workshop*, 188.

[13] Matsuzaki, T., Onuki, T., Nagatsuka, S., Inoue, H., Ishizu, T., Ieda, Y., *et al.* (2015) "A 128kb 4b/cell nonvolatile memory with crystalline In–Ga–Zn Oxide FET using Vt cancel write method," *Int. Solid-State Circuits Conf. Dig. Tech. Pap.*, 306.

[14] Atsumi, T., Nagatsuka, S., Inoue, H., Onuki, T., Saito, T., Ieda, Y., *et al.* (2012) "DRAM using crystalline oxide semiconductor for access transistors and not requiring refresh for more than ten days," *Proc. IEEE Int. Memory Workshop*, 99.

[15] Ohmaru, T., Yoneda, S., Nishijima, T., Endo, M., Dembo, H., Fujita, M., *et al.* (2012) "Eight-bit CPU with non-volatile registers capable of holding data for 40 days at 85°C using crystalline In–Ga–Zn oxide thin film transistors," *Ext. Abstr. Solid. State. Dev. Mater.*, 1144.

[16] Sjökvist, N., Ohmaru, T., Furutani, K., Isobe, A., Tsutsui, N., Tamura, H., *et al.* (2013) "Zero area overhead state retention flip flop utilizing crystalline In–Ga–Zn oxide thin film transistor with simple power control implemented in a 32-bit CPU," *Ext. Abstr. Solid. State. Dev. Mater.*, 1088.

[17] Tamura, H., Kato, K., Ishizu, T., Uesugi, W., Isobe, A., Tsutsui, N., *et al.* (2014) "Embedded SRAM and Cortex-M0 core using a 60-nm crystalline oxide semiconductor," *IEEE Micro*, **34**, 42.

[18] Aoki, T., Okamoto, Y., Nakagawa, T., Ikeda, M., Kozuma, M., Osada, T., *et al.* (2014) "Normally-off computing with crystalline InGaZnO-based FPGA," *IEEE Int. Solid-State Circuits Conf. Dig. Tech. Pap.*, 502.

[19] Kozuma, M., Okamoto, Y., Nakagawa, T., Aoki, T., Kurokawa, Y., Ikeda, T., *et al.* (2015) "180-mV Subthreshold operation of crystalline oxide semiconductor FPGA realized by overdrive programmable power switch and programmable routing switch," *Ext. Abstr. Solid State Dev. Mater.*, 1174.

[20] Aoki, T., Ikeda, M., Kozuma, M., Tamura, H., Kurokawa, Y., Ikeda, T., *et al.* (2011) "Electronic global shutter CMOS image sensor using oxide semiconductor FET with extremely low off-state current," *Symp. IEEE Symp. VLSI Technol. Dig. Tech. Pap.*, 175.

[21] Ohmaru, T., Nakagawa, T., Maeda, S., Okamoto, Y., Kozuma, M., Yoneda, S., *et al.* (2015) "25.3 µW at 60 fps 240 × 160-Pixel vision sensor for motion capturing with in-pixel non-volatile analog memory using crystalline oxide semiconductor FET," *IEEE Int. Solid-State Circuits Conf. Dig. Tech. Pap.*, 118.

[22] TSenser Summit (2016) Genesis of TSensor. Available at: www.tsensorssummit.org/genesisoftsensor.html [accessed February 11, 2016].

[23] Libelium Comunicaciones Distribuidas S.L. (2016) 50 Sensor applications for a smarter world. Available at: www.libelium.com/top_50_iot_sensor_applications_ranking [accessed February 11, 2016].

2

Device Physics of CAAC-IGZO FET

2.1 Introduction

Kimizuka and Mohri [1] first synthesized an indium–gallium–zinc oxide (InGaZn oxide, hereinafter referred to as IGZO) in the 1980s, and revealed its crystal structure. In an IGZO crystal, repeat units, each having an InO_2 layer and a $(GaZn)O$ layer, are periodically stacked to form a layered structure in a phase-equilibrium state (see Figure 2.1) [2]. In 1995, Orita *et al.* [3] examined the conduction characteristics and bandgap of a bulk $InGaZnO_4$ crystal with an ytterbium iron oxide $(YbFe_2O_4)$ structure and reported that $InGaZnO_4$ is preferable as a transparent conductive material. Furthermore, Nomura *et al.* [4–6] made a single-crystal IGZO film on an yttria-stabilized zirconia (YSZ) substrate using reactive solid-phase epitaxy with heat treatment at 1400°C; they used the film as an active layer of a field-effect transistor (FET) and reported the FET characteristics. However, a FET with single-crystal IGZO as an active layer has not been put into practical use as of 2016.

Yamazaki *et al.* [7] reported a unique IGZO film with a crystal structure different from that of a single-crystal or polycrystalline IGZO, called a *c*-axis-aligned crystalline IGZO (CAAC-IGZO) film. An $InGaO_3(ZnO)$ crystal in the CAAC-IGZO film has the $YbFe_2O_4$ structure (see *Physics and Technology of Crystalline Oxide Semiconductor CAAC-IGZO: Fundamentals* [8] (hereinafter referred to as *Fundamentals*) for the details). In CAAC-IGZO, the *c*-axes of the crystals are aligned almost perpendicular to the CAAC-IGZO film surface, while the *a–b* planes are randomly oriented. In addition, there are no clear grain boundaries between the crystals. Formation of CAAC-IGZO does not require the high temperature of single-crystalline IGZO synthesis (1400°C, see Nomura *et al.* [4]). In addition, instead of epitaxial growth that slowly forms a film, high-speed sputtering can be used for the formation of CAAC-IGZO. A FET having CAAC-IGZO as an active layer (hereinafter referred to as a CAAC-IGZO FET) is characterized by: (1) a low density of defect states due to its crystallinity [9], which offers stable

Physics and Technology of Crystalline Oxide Semiconductor CAAC-IGZO: Application to LSI, First Edition.
Edited by Shunpei Yamazaki and Masahiro Fujita.
© 2017 John Wiley & Sons, Ltd. Published 2017 by John Wiley & Sons, Ltd.

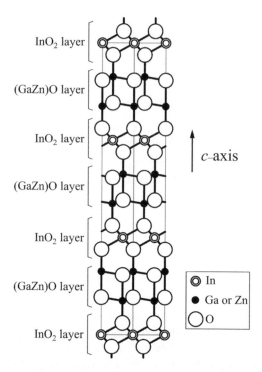

InO$_2$ layer

(GaZn)O layer

InO$_2$ layer

c-axis

(GaZn)O layer

InO$_2$ layer

(GaZn)O layer

◎ In
● Ga or Zn
○ O

InO$_2$ layer

Figure 2.1   Crystal structure of IGZO [InGaO$_3$(ZnO)$_m$]. InO$_2$ layers and (GaZn)O layers are periodically stacked to form a layered structure

characteristics; (2) an extremely low off-state current of yoctoamp order (yA/µm, y $=$ 10^{-24}) [10]; and (3) strength against the short-channel effect [11] (here, "off-state current" means the leakage current in the off state). Liquid crystal displays using CAAC-IGZO FETs in the backplanes are already being mass produced [12].

Figure 2.2 shows X-ray diffraction (XRD) spectra of a CAAC-IGZO film. The XRD spectrum in Figure 2.2(a) shows a peak of (009) indicating alignment of the c-axes of InGaO$_3$(ZnO) crystals almost perpendicular to the film surface. In Figure 2.2(b), no diffraction peak is observed in the a–b plane, suggesting no orientation of the crystal a–b planes with respect to the film surface (see *Fundamentals* [8] for further details on XRD).

The use of CAAC-IGZO FETs in LSIs proceeds similarly in the display field [13–16]. Miniaturized CAAC-IGZO FETs in an LSI also have the above-mentioned characteristics. The off-state current of a CAAC-IGZO switching FET can be extremely low, which leads to extremely low power consumption by the LSI.

This chapter explains the device physics of a CAAC-IGZO FET. In Section 2.2, its extremely low off-state current will be described. In addition to the basic electrical characteristics, a comparison with a silicon (Si) FET is shown to demonstrate how low the off-state current of the CAAC-IGZO FET is. In fact, it is lower than the detection limit of a normal current-measurement instrument (10^{-13} A), so an original measurement method had to be developed to measure this ultra-low off-state current.

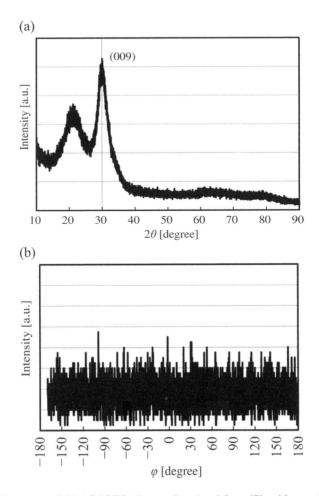

(a)

(009)

Intensity [a.u.]

10   20   30   40   50   60   70   80   90

2θ [degree]

(b)

Intensity [a.u.]

−180 −150 −120 −90 −60 −30 0 30 60 90 120 150 180

φ [degree]

Figure 2.2   XRD spectra of CAAC-IGZO. *Source*: Reprinted from [7], with permission from Wiley

In Section 2.3, a calculation method for estimating off-state currents lower than the detection limit of conventional devices, on the basis of transfer characteristics, is described. This method enables the estimation of I_{cut} (a drain current value I_d at a gate voltage V_g of 0 V) below the detection limit (10^{-13} A). Here, shifting the threshold voltage V_{th} of a CAAC-IGZO FET allows the acquisition of a lower I_{cut} value. The calculation method can be used to find out how much V_{th} should be shifted to obtain a desired I_{cut} value (i.e., the value required for a particular CAAC-IGZO FET application).

In Section 2.4, a technique for controlling the threshold voltage (V_{th}) of a CAAC-IGZO FET is described. The dynamic control of the FET's V_{th} is important not only for a reduction in variation of FET characteristics, but also for use of the extremely low off-state current of the CAAC-IGZO FET in LSI.

Section 2.5 explains the on-state current characteristics of a CAAC-IGZO FET. Although the electron mobility of a CAAC-IGZO FET is much lower than that of a Si FET, the difference in field-effect mobility is reduced for downscaled FETs. While a CAAC-IGZO FET with a channel length (L) of 1 μm has a field-effect mobility of approximately 10 cm^2/V-s (i.e., 100 times lower than that of a Si FET), miniaturized Si FETs exhibit an increased drift field strength, which accelerates electrons, hence turning them into hot electrons. Hot electrons, in turn, generate phonons, and their drift velocity becomes saturated. That is, the speed of the electrons becomes saturated in a Si FET as the channel gets shorter. In CAAC-IGZO FETs, on the contrary, electrons are not easily accelerated, and do not easily become hot electrons; therefore, a saturation in drift velocity does not occur to the same extent as in a Si FET. As a result, the difference in field-effect mobility versus Si decreases by downscaling. This suggests that CAAC-IGZO FET, if sufficiently small, can replace Si FETs in LSIs.

In Section 2.6, we present a measure to be used against the short-channel effect of a CAAC-IGZO FET. A surrounded-channel (S-ch) structure gives small characteristic degradation. For example, even when a FET has a channel as short as 30 nm and its gate insulator has a thickness as large as 11 nm in equivalent oxide thickness (EOT), characteristic degradation hardly occurs with the S-ch structure.

Section 2.7 introduces a recent scaled-down CAAC-IGZO FET. The CAAC-IGZO FET with a 20-nm node has a cut-off frequency of 34 GHz. In addition, even in the scaled-down FET, the off-state current is extremely low.

The process technology to fabricate LSI devices is introduced in Section 2.8. First, the methods used to fabricate CAAC-IGZO FETs with top-gate top-contact (TGTC) and trench-gate self-aligned (TGSA) structures are explained; then, a hybrid structure in which a CAAC-IGZO FET is placed over a Si FET is described.

2.2 Off-State Current

This section discusses how low the off-state current (leakage current in an off state) of the CAAC-IGZO FET is, compared with that of a conventional Si FET. A CAAC-IGZO FET has an off-state current on the order of yoctoamps (10^{-24} A), below the detection limit of common current measurement (0.1 pA). As possibilities for measuring the off-state current, methods utilizing a FET with increased channel width (W) and the voltage drop of a capacitor are discussed. The reason for the extremely low off-state current of the CAAC-IGZO FET is also discussed theoretically with reference to an energy band diagram.

2.2.1 Off-State Current Comparison between Si and CAAC-IGZO FETs

The off-state currents of Si and CAAC-IGZO FETs are compared with each other (see Figure 2.3). Their drain current–gate voltage (I_d–V_g) characteristics are obtained under −25°C, room temperature (R.T.), and 150°C at a drain voltage V_d of 1 V. As shown in Figure 2.3(a), the off-state current of the Si FET is not lower than the detection limit of 0.1 pA (10^{-13} A), and increases with the measurement temperature. Possible reasons for the off-state current in the Si FET are p–n junction leakage and current generated by interband thermal transition. In contrast, the CAAC-IGZO FET has an off-state current lower than the detection limit regardless of the temperature, as shown in Figure 2.3(b). Because the CAAC-IGZO FET operates in an n-channel accumulation mode, p–n junction leakage does not occur. In addition, CAAC-IGZO has a wide bandgap of approximately 3 eV and few mid-gap levels, so the conduction band carrier generation by interband transitions or excitations from deep levels is also negligible. The detailed mechanisms of the off-state leakage current of CAAC-IGZO FETs are theoretically discussed in Subsection 2.2.3. Measurement methods developed to detect the yoctoamp-order off-state current of the CAAC-IGZO FET are described in Subsection 2.2.2.

Such an extremely low off-state current, which cannot be obtained in Si FETs, leads to an ultra-low-power device. Various device applications of CAAC-IGZO FET technology have been reported (see Figure 2.4) [18].

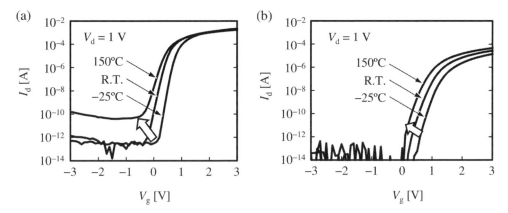

Figure 2.3 Comparison of off-state current between (a) Si FET with $W/L = 0.35\,\mu m/10\,\mu m$ and (b) CAAC-IGZO FET with $W/L = 0.45\,\mu m/10\,\mu m$. *Source*: Adapted from [17]

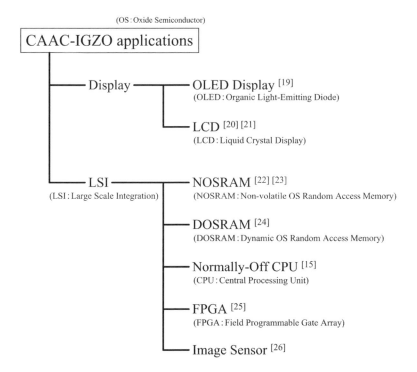

Figure 2.4 Applications of CAAC-IGZO FET technology to various devices. *Source*: Adapted from [18]

2.2.2   Measurement of Extremely Low Off-State Current

The extremely low off-state current of the CAAC-IGZO FET is on the order of 10^{-24} A, which is lower than the detection limit (0.1 pA) of common current measurement methods. The off-state current, flowing between a source and a drain when the FET is off, increases in proportion to the channel width. To allow an off-state current measurement, the channel width of the CAAC-IGZO FET is increased to be as wide as 1 m.

Figure 2.5 shows micrographs of a CAAC-IGZO FET with channel length L of 3 μm and channel width W of 1 m [10]. In the left photograph, 20,000 (200 × 100) CAAC-IGZO FETs, each having W = 50 μm, are aligned in the 6806 μm × 6878 μm region. The right photograph is an enlarged view of the region enclosed in a square in the left photograph. These CAAC-IGZO FETs are parallel-connected, forming a CAAC-IGZO FET with a total channel width of 1 m.

The cross-sectional structure of one CAAC-IGZO FET is shown in Figure 2.6. This FET has a TGTC structure and employs an overlap structure whereby the gate electrode overlaps with

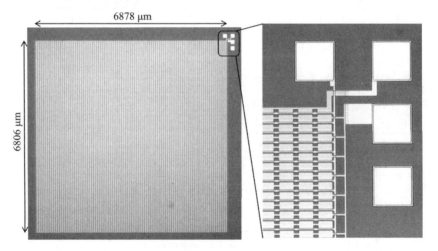

Figure 2.5   Micrographs of the CAAC-IGZO FET with W = 1 m. *Source*: Reproduced from [10], with permission of *Japanese Journal of Applied Physics*

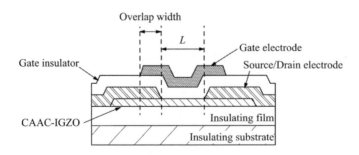

Figure 2.6   Cross-sectional view of the CAAC-IGZO FET

the source and drain electrodes (the overlap width is 2 μm). On a glass substrate, the following films are formed in the order they are listed.

- Base insulating film: 300-nm-thick silicon oxide (amorphous)
- Active layer: 30-nm-thick CAAC-IGZO film
- Source and drain electrodes: 100-nm-thick tungsten
- Gate insulator (sometimes referred to as GI): 100-nm-thick silicon oxide
- Gate electrode: stack of 15-nm-thick tantalum nitride and 135-nm-thick tungsten
- Passivation layer: 300-nm-thick silicon oxide (not shown)

The I_d–V_g characteristics of the CAAC-IGZO FET with $W = 1$ m at $V_d = 3$ V are shown in Figure 2.7. For the measurement, an Agilent 4156C Precision Semiconductor Parameter Analyzer is used. According to Figure 2.7, even though the channel width is as wide as 1 m, the off-state current of the fabricated CAAC-IGZO FET is lower than the detection limit of 0.1 pA.

In order to obtain the off-state current, another method has therefore been developed. To measure the minute current, the small amount of charge moved by the current should be increased to a detectable level. Then, a method of estimating the current by measuring the change in the charge over a long time is developed. Figure 2.8 shows the conceptual diagram of this measurement method. A device under test (here, a CAAC-IGZO FET), which is the element to be measured, is denoted by "DUT." Figure 2.8(a) shows a configuration used to measure the current of the DUT, and Figure 2.8(b) shows the time change of the potential

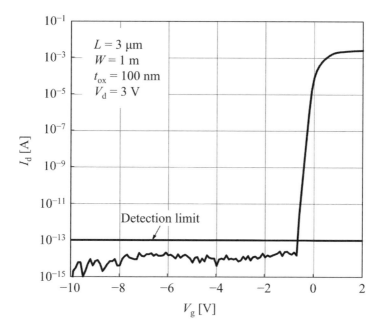

Figure 2.7 I_d–V_g characteristics of the CAAC-IGZO FET with $W = 1$ m. T_{ox} denotes the thickness of the gate insulator. *Source*: Reproduced from [10], with permission of *Japanese Journal of Applied Physics*

V_F of node F connected to the DUT. The leakage current I of the DUT is expressed by the following equation:

$$I = \frac{C_F \Delta V_F}{\Delta t} \tag{2.1}$$

where t and C_F denote the time and capacitance of node F, respectively. The measurement of the time change in potential V_F enables estimation of the off-state current of the DUT.

Equation (2.1) shows the following three important factors for obtaining high measurement accuracy: reduction in C_F, suppression of measurement noise for V_F, and long-term stable measurement. In light of these factors, the circuit configuration, device structure, and measurement environment are constructed.

The constructed circuit configuration comprises the DUT, a reading circuit (source follower), and a programming circuit (see Figure 2.9). In the drawing, G_W and D_W denote the gate and drain electrodes of the programming circuit, respectively; G and S denote the gate and source electrodes of the DUT, respectively; F denotes the floating node; G_R, D_R, and S_R denote the gate, drain, and source electrodes of the reading circuit, respectively; and V_{out} denotes the output potential.

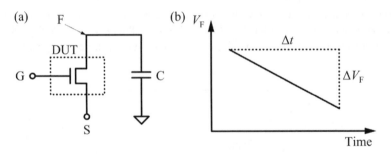

Figure 2.8 Conceptual diagrams of measurement method utilizing voltage drop: (a) circuit diagram; (b) behavior of V_F. *Source*: Adapted from [10]

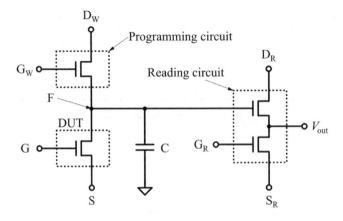

Figure 2.9 Circuit configuration used for measurement. *Source*: Adapted from [10]

To reduce the leakage current or the contribution to the capacitance C_F from sources other than the DUT, W is set as high as 1 m. The DUT structure itself also has some features. The overlap structure in Figure 2.6 has a large capacitance between the gate and the drain, causing C_F to be large. Thus, the CAAC-IGZO FET for the DUT adopts an offset structure as shown in Figure 2.10. The offset structure lacks the overlap of the gate and source/drain electrodes. The distance between the gate electrode and the source/drain electrode shown in the figure is referred to as an offset width. When the structure is changed from the overlap structure (overlap width 2 μm) to the offset structure (offset width 2 μm), the C_F per micrometer of channel width decreases greatly from 1 fF/μm to 0.07 fF/μm.

The timing diagrams for reading and programming are shown in Figure 2.11. To ensure that the CAAC-IGZO FET (DUT) is turned off, the gate voltage (V_G) and source voltage (V_S) of the DUT are set to −3 V and 0 V, respectively. In the diagrams, V_{DR}, V_{GR}, and V_{SR} denote the potentials applied to the drain electrode (D_R), gate electrode (G_R), and source electrode (S_R) of the reading circuit, respectively; V_{DW} and V_{GW} denote the potentials applied to the drain electrode (D_W) and gate electrode (G_W) of the programming circuit, respectively.

In reading, as shown in Figure 2.11(a), a V_{GR} of 0.5 V, V_{DR} of 5 V, and V_{SR} of 0 V are applied to the reading circuit for 10 s every 5 min. In this period, the output potential V_{out} is read out. During other periods, the voltages applied to the source follower are suppressed with $V_{GR} = V_{DR} = V_{SR} = 2.5$ V.

In programming, as shown in Figure 2.11(b), the programming circuit is supplied with V_{DW} of 3 V for 20 s and V_{GW} of 5 V for 10 s. During those 10 s, the output potential V_{out} is read out as in the reading operation. During other periods, V_{GW} is set to −3 V so as to turn off the programming CAAC-IGZO FET.

A set of off-state current measurements contains 2-time programming and 72-time reading operations conducted over approximately 6 h. The programming operation is conducted twice continuously to observe the saturation of the output potential V_{out} and confirm that 3 V is written in node F. To confirm the reproducibility, this set is repeated three or more times.

It is important to suppress various types of noise and to conduct stable measurements over a long period of time. Since the electric characteristics of the CAAC-IGZO FET are influenced by temperature, humidity, and light, these factors are controlled. In particular, the measurement is conducted at 85°C, and in dry air with a dew point of −60°C or lower in a dark room. In addition, an uninterruptible power supply is used to reduce power-supply noise, and an anti-vibration table is provided for the stage to reduce vibration. Accordingly, the measurement variation in V_{out} is suppressed to within ±1 mV over the long-term measurement period.

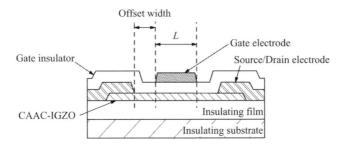

Figure 2.10   Cross-sectional view of the CAAC-IGZO FET with offset structure

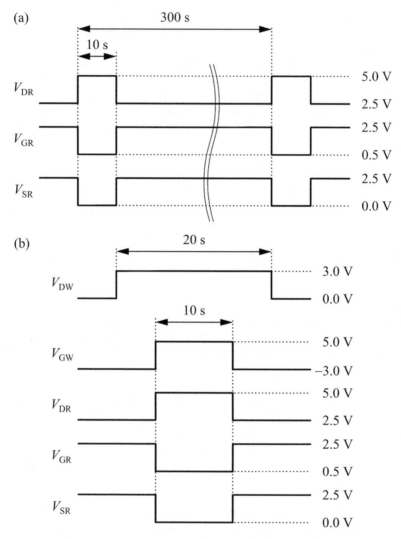

Figure 2.11 Timing diagram: (a) read and (b) program. *Source*: Reproduced from [10], with permission of *Japanese Journal of Applied Physics*

The measurement results are shown in Figure 2.12. The channel width of the DUT is 1 m in Figure 2.12(a), 100 mm in Figure 2.12(b), and 10 mm in Figure 2.12(c). The first vertical axis in the graph represents the output potential V_{out}. The graph shows that V_{out} decreases linearly in each set and that the leakage current flows at a constant rate.

The leakage current is calculated as follows. First, the slope of the output potential ($\Delta V_{out}/\Delta t$) is calculated from least-squares fitting of the measurement data over the last 3 h of each set. The second vertical axis in each graph of Figure 2.12 represents the deviation between the fitted line and the measurement data. The plotted differences show that the output potential V_{out} exhibits high linearity and the measurement noise falls within a range of ±1 mV.

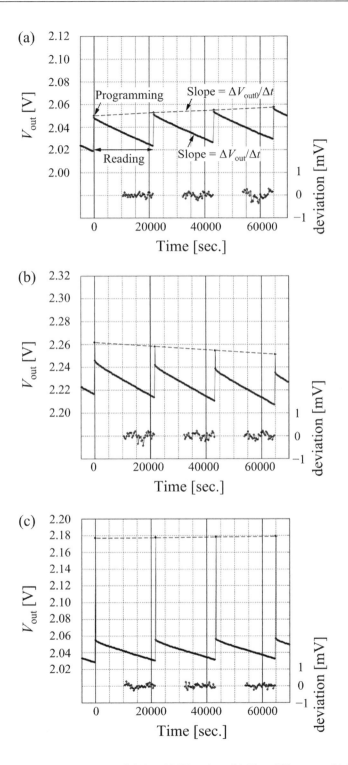

Figure 2.12   Change in the output potential V_{out}: (a) $W = 1$ m, (b) $W = 100$ mm, and (c) $W = 10$ mm.
Source: Reproduced from [10], with permission of *Japanese Journal of Applied Physics*

Next, the time change of the node-F potential is calculated from $\Delta V_F/\Delta t = (\Delta V_{out}/\Delta t - \Delta V_{out0}/\Delta t)/S_{SF}$, where S_{SF} is the output potential slope of the source follower with respect to the input potential (here, $S_{SF} = 0.93$). $\Delta V_{out0}/\Delta t$ is the slope of the output potential during programming (V_{out0}), shown by the dotted line in Figure 2.12. Since the voltage V_F is fixed at 3 V during programming, it is assumed that the change in the output voltage ΔV_{out0} is affected by the source follower and is not related to the leakage current of the DUT. Thus, a slope $\Delta V_{out0}/\Delta t$ is reasonably subtracted from a slope $\Delta V_{out}/\Delta t$.

The leakage current in each case is calculated from Equation (2.1) ($I = C_F \Delta V_F/\Delta t$), and the results are shown in Table 2.1. Focusing on the channel width dependence, the leakage current per micrometer of channel width (i_{leak}) is obtained in the following manner: When the channel width is denoted by W_j, the measured value of the leakage current (x_j) is estimated to be the sum of the leakage component of the DUT ($i_{leak} \times W_j$) and the other leakage component I_{cut} (here, j indicates an arbitrary number): in other words, $x_j = i_{leak} \times W_j + I_{cut}$. The i_{leak} and I_{cut} values that minimize $\sum_{j=1-9}(x_j/W_j - i_{leak} - I_{cut}/W_j)^2$ are thus obtained, with the results in Table 2.1. Figure 2.13 shows the measurement results and the fitted function $f(W)$. The off-state current per micrometer of channel width is calculated to be 135 yA/μm at 85°C, being extremely low [10].

Table 2.1 Measurement result of leakage current. *Source*: Reproduced from [10], with permission of *Japanese Journal of Applied Physics*

Time [s]	$I_{leak}(W = 1\ m)$ [A]	$I_{leak}(W = 100\ mm)$ [A]	$I_{leak}(W = 10\ mm)$ [A]
10,800–21,600	1.36×10^{-16}	1.38×10^{-17}	1.55×10^{-18}
32,400–43,200	1.30×10^{-16}	1.40×10^{-17}	1.52×10^{-18}
54,000–64,800	1.30×10^{-16}	1.41×10^{-17}	1.54×10^{-18}

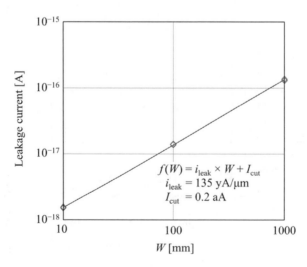

Figure 2.13 Measurement result of leakage current and plot of the fitted function $f(W)$. *Source*: Reproduced from [10], with permission of *Japanese Journal of Applied Physics*

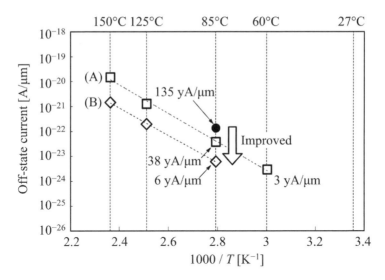

Figure 2.14 Arrhenius plot of off-state currents of CAAC-IGZO FETs. *Source*: Adapted from [28]

With use of the above measurement method, the off-state current of the CAAC-IGZO FET was successfully measured in 2011 [27]. The off-state currents of various CAAC-IGZO FETs have been measured based on an estimation of the amount of decrease in the charge stored in the capacitor over a long period. Figure 2.14 is an Arrhenius plot of the off-state current of two CAAC-IGZO FETs. As shown in the graph, the CAAC-IGZO FET (B) with an off-state current of 6 yA/μm at 85°C was obtained in the spring of 2014 [28].

This subsection has described the extremely low off-state current of the CAAC-IGZO FET and the method for measuring it. The long-term measurement of a change in the charge amount shows that the CAAC-IGZO FET has an extremely low off-state current – on the order of yoctoamps. In 2014, an off-state current of 6 yA/μm at 85°C, which was the lowest value recorded among CAAC-IGZO FETs tested so far, was obtained. Such a low off-state current is promising when the CAAC-IGZO FET is to be used in a memory. The application to memory is described in Chapters 3 and 4. Other applications are described in later chapters.

2.2.3 Theoretical Discussion with Energy Band Diagram

What contributes to the CAAC-IGZO FET's extremely low off-state current? This low off-state current is probably because of the wide bandgap. Revealing this mechanism will facilitate research on possible low-power technologies and new materials with low off-state current. In this subsection, the reason for the very low off-state current is theoretically discussed compared with a Si FET. In particular, the tunneling of holes from the drain to the channel is discussed as the main mechanism for the off-state current of the CAAC-IGZO FET.

In a CAAC-IGZO FET, the channel is connected directly to the source and drain electrodes. The junction characteristics are determined by the work functions of the source and drain electrodes, and the electron affinity and bandgap of CAAC-IGZO. Table 2.2 shows the physical properties of CAAC-IGZO and tungsten, and their measurement methods. CAAC-IGZO

Table 2.2   Physical properties of CAAC-IGZO and tungsten, and their respective measurement method. UPS stands for ultraviolet photoelectron spectroscopy. *Source*: Adapted from [29]. Copyright 2012 The Japan Society of Applied Physics

	Measured value [eV]	Measurement method
Ionization potential of CAAC-IGZO	7.8	UPS
Bandgap of CAAC-IGZO	3.2	Ellipsometer
Work function of tungsten	5.0	UPS

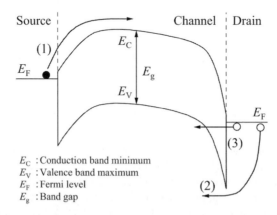

E_C : Conduction band minimum
E_V : Valence band maximum
E_F : Fermi level
E_g : Band gap

Figure 2.15   Schematic band diagram of the CAAC-IGZO FET in an off state. The black and white circles denote an electron and a hole, respectively. *Source*: Reproduced from [30], with permission of *Japanese Journal of Applied Physics*

has a bandgap of 3.2 eV, which is three times wider than that of Si (1.1 eV). From the difference between the ionization potential and the bandgap, the electron affinity of CAAC-IGZO is calculated to be 4.6 eV. Tungsten, a metal with a work function close to that value, is used for calculation as the source and drain electrodes.

The schematic band diagram of the CAAC-IGZO FET in an off state ($V_g < 0$) supplied with a positive drain voltage ($V_d > 0$) is shown in Figure 2.15. The Fermi level of the source electrode is denoted by E_F, and the energies of the conduction band minimum and valence band maximum around the channel center are denoted by E_C and E_V, respectively. In addition, the bandgap represented by a difference between E_C and E_V is denoted by E_g. As shown in Figure 2.15, the following three factors may contribute to the off-state current:

(1) injection of thermally excited electrons from the source to the channel;
(2) injection of thermally excited holes from the drain to the channel;
(3) tunneling of holes from the drain to the channel.

In the discussion below, IGZO is assumed to be a perfect crystal and an intrinsic semiconductor. This is because the off-state current is determined not by the mobility but by the potential barrier, as indicated by the above three factors. While impurities significantly affect the mobility, they do not significantly affect the potential barrier. In that regard, the discussion of the off-state current is different from that of the on-state current which relies on mobility.

The leakage current by (1) thermally excited electrons is proportional to $\exp(-E_{ele}/k_BT)$. Here, E_{ele} ($= E_C - E_F$) is the electron potential barrier, which becomes approximately equivalent to the bandgap as the gate voltage decreases. k_B is a Boltzmann constant. In that case, the ratio of exponential factors between an IGZO FET and a Si FET is estimated to be $\exp\left[-\left(E_g(\text{IGZO}) - E_g(\text{Si})\right)/k_BT\right] \sim 10^{-35}$. This indicates that the leakage current due to the thermally excited electrons is practically negligible in the IGZO FET.

The leakage current by (2) thermally excited holes is proportional to $\exp(-E_{hole}/k_BT)$. Here, E_{hole} is the potential barrier at the junction in case of the IGZO FET, calculated to be 2.8 eV from Table 2.2. The Si FET has $E_{hole} \sim E_g(\text{Si})$. The ratio of the exponential factors between the IGZO and Si FETs is estimated to be $\exp\left[-\left(2.8\,\text{eV} - E_g(\text{Si})\right)/k_BT\right] \sim 10^{-27}$. The leakage current due to the thermally excited holes is also negligible in the IGZO FET.

The leakage current by (3) tunneling of holes from the drain to the channel is discussed below. To simplify the tunneling situation, the case where particles tunnel across a one-dimensional triangular potential barrier with a height of V_0 and width of a is assumed. The effective mass of the particle in case of $x < 0$ and that in case of $x \geq 0$ are denoted by m and m^*, respectively. The tunneling current density J when the particle enters the barrier from the region of $x < 0$ and tunnels through the barrier in the region of $0 \leq x \leq a$ to the region of $x > a$ is obtained by the Fowler–Nordheim equation under the assumptions of the Wentzel–Kramers–Brillouin (WKB) approximation and absolute zero [31]:

$$J = \frac{q^3 F^2}{8h\pi V_0}\sqrt{\frac{m}{m^*}}\exp\left(-\frac{8\pi\sqrt{2m^*}}{3qhF}V_0^{3/2}\right) \qquad (2.2)$$

In this equation, F ($= V_0/a$), h, and q represent the electric field strength, Planck's constant, and elementary charge, respectively. In this case, the particle, the region $x < 0$, and the region $x \geq 0$ correspond to a hole, a drain, and a channel, respectively. Thus, m^* appearing in the exponential part corresponds to the hole effective mass. The effective mass m^* is calculated from the first-principles calculation based on density functional theory (DFT), and V_0 and a are calculated by device simulation, to find the tunneling current density J.

First, the effective mass of a hole is calculated. The energy band structure of the crystalline IGZO is obtained by the first-principles calculation, and the hole effective mass is estimated from the structure of the valence band maximum. The IGZO used in the calculation is a perfect crystal.

The IGZO crystal with the $YbFe_2O_4$ structure is shown in Figure 2.16. The crystal has a layered structure, and the number of atoms in the unit cell is 84. The norm-conserving pseudo-potential DFT introduced in OpenMX [32] is applied to the unit cell, and the Perdew–Burke–Ernzerhof (PBE)-type generalized gradient approximation (GGA) is applied to the exchange interaction potential of the electrons. The cut-off energy of the localized basis function is set to 200 Ry, and the k point is sampled with a $5 \times 5 \times 3$ mesh.

Figure 2.17 shows the calculated energy band diagram of the crystalline IGZO together with the Brillouin zone. Compared with the dispersion of the conduction band, the dispersion of the valence band is very flat. This means that the hole effective mass is larger than the electron effective mass.

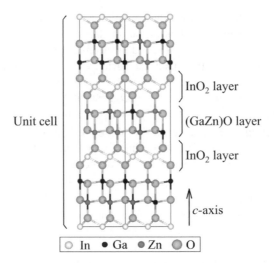

Figure 2.16   Crystal structure of IGZO

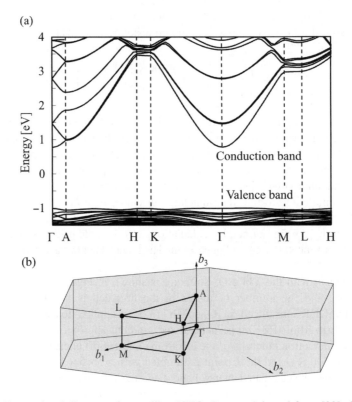

Figure 2.17   Energy band diagram of crystalline IGZO. *Source*: Adapted from [29]. Copyright 2012 The Japan Society of Applied Physics

Table 2.3 Effective mass of holes and electrons in IGZO and Si. *Source*: Adapted from [29]. Copyright 2012 The Japan Society of Applied Physics

Material	IGZO	Si
Effective mass of electrons (m_e^*/m_e)	0.25(b_1)	0.19 (transverse)
	0.25(b_2)	0.98 (longitudinal)
	0.23(b_3)	
Effective mass of holes (m_h^*/m_e)	21(b_1)	0.16 (light)
	41(b_2)	0.49 (heavy)
	11(b_3)	

Table 2.4 Parameters of the IGZO FET in device simulation. *Source*: Adapted from [18]. Copyright 2012 The Japan Society of Applied Physics

Channel length [μm]	3
Gate insulator thickness [nm]	200
Dielectric constant	15
Donor concentration [cm^{-3}]	10^{-10}

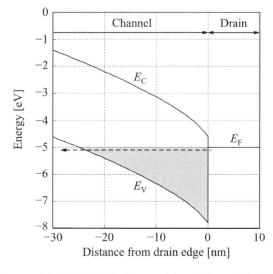

Figure 2.18 Band diagram of the IGZO FET. *Source*: Adapted from [29]. Copyright 2012 The Japan Society of Applied Physics

The effective masses of holes and electrons in crystalline IGZO and those in Si [30] are shown in Table 2.3. The hole effective mass in the IGZO is above 10, which is more than 50 times higher than the effective mass of light holes in Si. This suggests that the heavy holes strongly suppress the tunneling current in crystalline IGZO.

Next, the band bending of the IGZO FET is calculated by device simulation to estimate the height and width of the potential barrier. The Synopsys Sentaurus Device [33] is used as the

Table 2.5   Parameters in Table 2.2 for tunneling current density J. *Source*: Adapted from [29]. Copyright 2012 The Japan Society of Applied Physics

	IGZO FET	Si FET
V_0 [eV]	2.8	1.1
a [nm]	25	7
$F = V_0/a$ [MV/m]	112	157

Table 2.6   Hole tunneling current densities of the IGZO and Si FET. *Source*: Adapted from [29]. Copyright 2012 The Japan Society of Applied Physics

	IGZO FET	Si FET
Tunneling current density J [yA/μm^2]	~ exp (−900)	1.8×10^{14}

simulator. The parameters used for the calculation are shown in Table 2.4. The assumed IGZO is practically an intrinsic semiconductor. The voltages V_g and V_d are set to −10 V and 7.5 V, respectively. The calculated band diagram is shown in Figure 2.18. The graph shows that the above-assumed one-dimensional triangular potential barrier is appropriate.

The calculated height V_0, width a of the potential barrier, and electric field F of the IGZO FET and of the Si FET are shown in Table 2.5. In the calculation for the Si FET, the acceptor concentration of the channel and the dielectric constant are set to 10^{16} cm^{-3} and 11.9, respectively. The potential barrier of the IGZO FET reflects the bandgap difference, being three times as high as that of the Si FET.

From the above calculation result, the hole tunneling current density is obtained (see Table 2.6). In the case of the Si FET, $m_h{}^*$ is set to 0.16 as the light holes affect the hole effective mass. As shown in the table, the hole tunneling current density in the IGZO FET is almost zero, which is significantly smaller than that of the Si FET.

2.2.4   Conclusion

A change in the charge amount in a node connected to a CAAC-IGZO FET was measured for a long period, revealing that this FET has an extremely low off-state current (on the order of yoctoamps). Possible reasons for such a low off-state current are given below. Since a crystalline IGZO FET has a wide bandgap, leakage current due to thermally excited electrons and holes practically does not flow. Furthermore, the hole effective mass in the channel is as high as approximately 10, and thus the hole tunneling current does not flow either. In addition to the above explanations based on band theory, the small number of recombination centers in the bandgap of a CAAC-IGZO film is also an important factor. Such features should be basically maintained, even when the CAAC-IGZO FET is scaled down. CAAC-IGZO FETs with such low off-state current can be used for low-power LSIs, as described in Chapter 3 and after.

2.3 Subthreshold Characteristics

As shown in Section 2.2, a CAAC-IGZO FET has an extremely low off-state current (I_{off}), on the order of yoctoamps per micrometer [10]. The application of CAAC-IGZO FETs to low-power LSI, taking advantage of the low- I_{off} property, has been proposed in [13–16]. For example, if a cut-off current, I_{cut} (a drain current I_d at a gate voltage V_g of 0 V) is lowered to this extremely low off-state current, a capacitor connected to the CAAC-IGZO FET can retain charges for a very long time, such as approximately 10 years, enabling a non-volatile memory.

Figure 2.19 shows the conceptual graph of the I_d–V_g characteristics of a CAAC-IGZO FET. In the I_d–V_g curve, the V_g voltage region below the threshold voltage V_{th} is called the subthreshold region, and I_d flowing in this region is called the subthreshold leakage. Subthreshold leakage current is inevitable in FETs. In light of the extremely small I_{off} of the CAAC-IGZO FET, the subthreshold region of the FET should extend into an extremely small I_d (on the contrary, the subthreshold region of a Si FET does not extend into such a small level, and even the lowest I_d in the region is 0.01 pA [34]). While the reason for the very small off-state current measured in Section 2.2 has not been clearly revealed, the subthreshold leakage seems to be the major source of the leakage current at least down to 1 yA. Thus, I_{cut} in the normally-off I_d–V_g characteristic can be considered to be generally determined by the subthreshold leakage.

To lower I_{cut} to its required value, which differs depending on the applications of low-power LSI, V_{th} should be shifted in the positive direction. If we can find the I_{cut} value before shifting, we can understand how much we should shift V_{th}. However, the subthreshold leakage of a CAAC-IGZO FET is probably lower than the lower limit (0.1 pA = 1×10^{-13} A) of a normal measurement instrument [10]. Although Section 2.2 explained the method of estimating a very small current below the detection lower limit in detail, the method is not practical for estimating

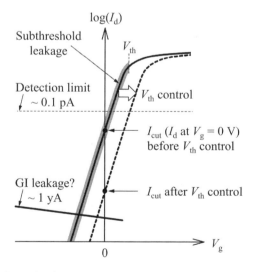

Figure 2.19 Conceptual graph of I_d–V_g characteristics of a CAAC-IGZO FET. The subthreshold leakage is the main contributor in a CAAC-IGZO FET down to an extremely small drain current of 1 yA. Shifting V_{th} in a positive direction can lower the I_{cut} determined by subthreshold leakage

I_{cut} in arbitrary I_d–V_g characteristics because it requires special circuits and long-term measurement.

This section introduces a simple method to estimate a subthreshold leakage at $I_d < 0.1$ pA from the I_d–V_g characteristics ($I_d > 0.1$ pA) of a CAAC-IGZO FET [35]. This method reveals that a subthreshold swing (*SS*) at $I_d > 0.1$ pA is degraded by the presence of electron trap levels, while the *SS* converges on the ideal value in a deep subthreshold region of $I_d < 0.1$ pA. The estimated I_{cut} (< 0.1 pA) is at the equivalent order to the value obtained from the memory-retention experiment with a memory cell having a CAAC-IGZO FET, which indicates that the estimation is appropriate. This method enables easy estimation of the V_{th} shift amount to obtain a desired I_{cut} without long-term measurement.

2.3.1   Estimation of I_{cut} by SS

One method of estimating I_{cut} below the lower detection limit is extrapolation using the *SS* of measured I_d–V_g characteristics. Let us estimate I_{cut} by this method before discussing the method newly proposed in this section.

The CAAC-IGZO FET used for estimating I_{cut} has a double-gate, top-contact structure (see Figure 2.20). The active layer is a 20-nm-thick CAAC-IGZO film with a relative dielectric constant ε_r of 15. As shown in the cross-sectional TEM image of Figure 2.21(a), this film has a layered structure in the direction perpendicular to the substrate, as in the case of a single-crystal InGaZnO$_4$ [Figure 2.21(b)] [36]. The film was formed by DC sputtering with a polycrystalline target with a composition of In : Ga : Zn = 1 : 1 : 1 under an argon–oxygen (Ar–O$_2$) atmosphere at a substrate temperature of 300°C. The active layer is sandwiched between buffer layers (with $\varepsilon_r \sim 15$) with thicknesses of 5 and 40 nm that have wider bandgaps than InGaZnO$_4$.

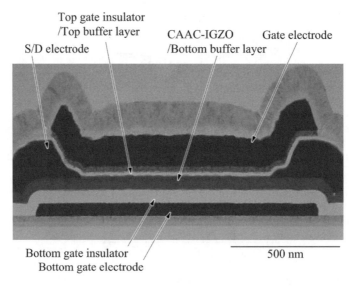

Figure 2.20   Cross-sectional STEM image of a CAAC-IGZO FET. *Source*: Adapted from [35]. Copyright 2015 The Japan Society of Applied Physics

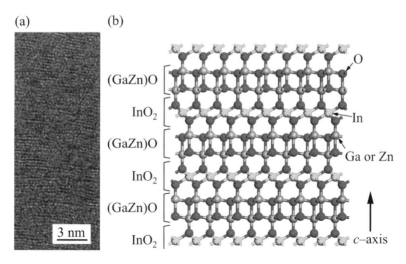

(a)   (b)

Figure 2.21 (a) Cross-sectional TEM image of CAAC-IGZO film; (b) schematic diagram of InGaZnO$_4$ crystal structure. *Source*: Reprinted from [35], with permission of IEEE, © 2015

A bottom-gate insulator and a top-gate insulator are oxides ($\varepsilon_r \sim 4.1$) with thicknesses of 60 and 20 nm, respectively. The channel width W and length L of the FET are 0.80 and 0.85 μm, respectively. The I_d–V_g characteristics are measured at room temperature by a semiconductor parameter analyzer.

The measured I_d–V_g characteristics of the fabricated CAAC-IGZO FETs [the number of FETs is 13 ($n = 13$)] are shown in Figure 2.22. The drain voltage V_d is 0.1 and 1.8 V. The graph shows that the normally-off characteristics are obtained, and I_d at $V_g = 0$ V, i.e., I_{cut} is lower than the lower detection limit ($I_d < 0.1$ pA). When a shift voltage V_{sh} is defined as the value of V_g at $I_d = 1$ pA, the value of V_{sh} at $V_d = 1.8$ V is 0.539 V on average. The value of $SS\ (= dV_g/d\log(I_d))$ at $I_d = 1$ pA is 126.4 mV/decade on average. The value of I_{cut} obtained from extrapolation with these values is

$$I_{cut} = 10^{-12-V_{sh}/SS} = 10^{-12-0.539/126.4} = 5.4 \times 10^{-17} [\text{A}]. \tag{2.3}$$

Is this value reliable?

To examine the reliability of the I_{cut} obtained by the extrapolation using SS at $I_d = 1$ pA and V_{sh}, the value is compared with another I_{cut} value estimated from the measured retention characteristics of a memory with a CAAC-IGZO FET. Figure 2.23(a) and (b) shows a circuit diagram and a timing diagram, respectively, of a memory cell that includes a CAAC-IGZO FET with a channel width W of 0.80 μm and a channel length L of 0.85 μm ($W/L = 0.80$ μm$/0.85$ μm). The CAAC-IGZO FET allows data writing and data retention at a floating node (F). A p-channel Si FET is used for data readout. The threshold voltage of the Si FET is denoted by $V_{th(Si)}$. A potential of 3.3 V is applied to the wiring WWL to turn on the CAAC-IGZO FET, and a potential of 1.8 V is applied to the wiring BL to write 1.8 V on the F. Subsequently, WWL is set to 0 V so that the potential of F (V_F) is retained. When the potential of SL is increased from 0 V to monitor a change in V_F, the readout Si FET is turned on at a certain potential. The potential of SL at which the readout Si FET is turned on is called V_{RM}.

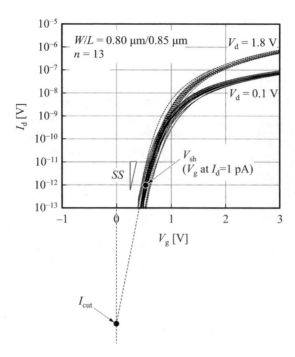

Figure 2.22 Measured I_d–V_g characteristics of CAAC-IGZO FETs with $W/L = 0.80$ μm/0.85 μm. The broken line is an extrapolation of a current below the lower detection limit by SS

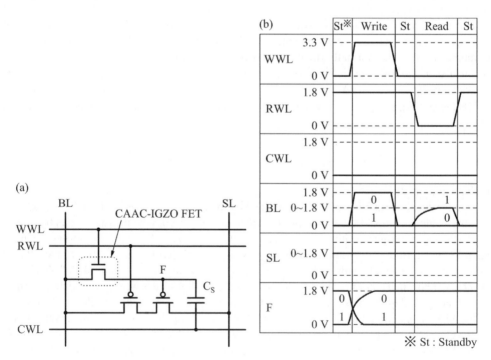

Figure 2.23 (a) Circuit diagram and (b) timing diagram of a memory cell having a CAAC-IGZO FET with $W/L = 0.80$ μm/0.85 μm

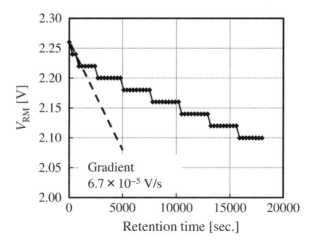

Figure 2.24   Time change in V_{RM} in memory retention

Here, the equation $V_{RM} = V_F - V_{th(Si)}$ is satisfied, and thus the time change of V_{RM} corresponds to the time change of V_F.

Figure 2.24 shows the time change of V_{RM} during retention after writing of 1.8 V to F. The monotonic V_{RM} decrease indicates the loss of charges from F. From the I_d-V_g characteristics in Figure 2.22, this loss of charges is considered to be derived from the subthreshold leakage of the CAAC-IGZO FET. The value of I_{cut} at V_d = 1.8 V corresponds to the initial time change of V_{RM}, and is thus calculated to be

$$I_{cut} = C_S \frac{dV_{RM}}{dt} = C_S \frac{dV_F}{dt} = 20\,[\text{fF}] \times 6.7 \times 10^{-5}\,[\text{V/s}] = 1.3 \times 10^{-18}\,[\text{A}]. \qquad (2.4)$$

Here, C_S is the retention capacitance of 20 fF in F. Compared with the value of I_{cut} estimated from the measured retention characteristics with Equation (2.4), the value of I_{cut} from Equation (2.3) with extrapolation using the SS is several tens of times overestimated. This means that the estimation of I_{cut} by the extrapolation with V_{sh} and SS is not very reliable.

2.3.2   Extraction Method of Interface Levels

Why is I_{cut} overestimated when extrapolated with SS obtained from the I_d-V_g characteristics? One possible reason is that the SS in the I_d region apparent from the I_d-V_g characteristics is higher in value than the SS in the I_d region below the lower detection limit, $I_d < 0.1$ pA. Increased SS suggests the existence of degrading factors. It is generally known that carrier (electron) trap levels in a channel or channel interface degrades SS. Thus, the density of states (DOS) of the trap levels, especially shallow ones, in the range of $I_d > 0.1$ pA is expected to be higher than that in the range of $I_d < 0.1$ pA. If the DOS of the trap levels is extracted from the measured I_d-V_g characteristics of a CAAC-IGZO FET, a device simulation that takes the extracted DOS into consideration can reproduce the measured I_d-V_g curve and can even estimate a minute current region below the lower detection limit.

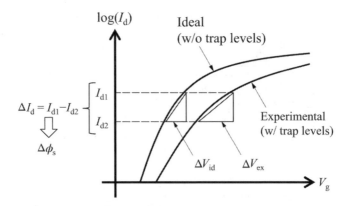

Figure 2.25 Schematic diagram of the measured I_d–V_g characteristics (with electron trap levels) and ideal I_d–V_g characteristics (without electron trap levels). *Source*: Adapted from [35]. Copyright 2015 The Japan Society of Applied Physics

From this motivation, a method for extracting the DOS of the interface trap levels (N_{it}) has been developed. In the method, N_{it} is extracted from the comparison between the measured I_d–V_g characteristics and the simulated ideal I_d–V_g characteristics. The degrading factor of the I_d–V_g characteristics is considered to be N_{it} in this method; however, whether or not the interface trap levels are actually present requires further discussion.

The N_{it} extraction method is described with reference to Figure 2.25. When the drain current changes from I_{d1} to I_{d2}, the measured and ideal I_d–V_g characteristics are assumed to change in the gate voltage by ΔV_{ex} and ΔV_{id}, respectively. The surface potential ϕ_s is assumed to change by $\Delta\phi_s$ in both cases. At this time, the number of electrons N_{trap} trapped by the interface levels per unit area and per unit energy is estimated from the following equation:

$$N_{trap} = \frac{C_{tg}}{q} \lim_{\Delta\phi_s \to 0} \left(\frac{\Delta V_{ex}}{\Delta\phi_s} - \frac{\Delta V_{id}}{\Delta\phi_s} \right) = C_{tg} \left(\frac{\partial V_{ex}}{\partial\phi_s} - \frac{\partial V_{id}}{\partial\phi_s} \right) \tag{2.5}$$

where C_{tg} and q denote top-gate capacitance and elementary charge, respectively. N_{trap} and N_{it} have the following relationship:

$$N_{trap} = \frac{\partial}{\partial\phi_s} \int_{-\infty}^{\infty} N_{it}(E)f(E)dE \tag{2.6}$$

where $f(E)$ is the Fermi–Dirac distribution function – i.e., the probability density of energy E at a certain temperature and the Fermi energy. This means that N_{trap} is expected to have a distribution modulated by the thermal distribution of $f(E)$. N_{it} is determined by the fitting N_{trap} calculated by Equation (2.5) with Equation (2.6). Device simulation using this N_{it} provides the I_d–V_g characteristics, including $I_d < 0.1$ pA. There are many references for other extraction methods of the carrier trap levels, including those discussing their physical significance [36–44].

2.3.3   Reproduction of Measured Value and Estimation of I_{cut}

The actual fitting result with the measured I_d–V_g characteristics obtained by the above method is described below. Figure 2.26 and Table 2.7 show the schematic cross-sectional diagram and calculation conditions used for device simulation of the ideal I_d–V_g characteristics. For the device simulation, a two-dimensional calculation is conducted with the use of ATLAS by Silvaco Inc. [45]. The assumed active layer of the CAAC-IGZO FET model has n^+ regions under source and drain electrodes and has an intrinsic channel region. To heighten the fitting accuracy, the electron mobility μ_n, which has a large influence on the on-state characteristics, is also treated as a fitting parameter.

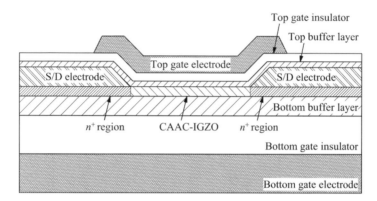

Figure 2.26   Schematic structure used for device simulation

Table 2.7   Device simulation conditions

FET size	Channel length L [μm]	0.85
	Channel width W [μm]	0.8
CAAC-IGZO	Electron affinity [eV]	4.6
	E_g [eV]	3.2
	Dielectric constant	15
	N_d [cm^{-3}]	6.6×10^{-9}
	N_d (below S/D electrodes)[cm^{-3}]	5×10^{18}
	Electron mobility μ_n [cm^2/Vs]	parameter
	Hole mobility μ_h [cm^2/Vs]	0.01
	N_c [cm^{-3}]	5×10^{18}
	N_v [cm^{-3}]	5×10^{18}
	Thickness [nm]	20
Buffer layers	Dielectric constant	15
	Top/bottom thickness [nm]	5/40
Gate insulators	Dielectric constant	4.1
	Top/bottom thickness [nm]	20/60

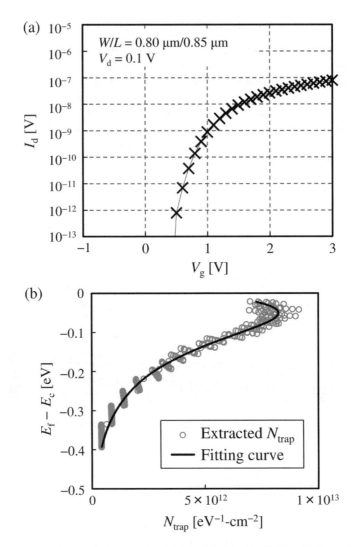

Figure 2.27 (a) I_d–V_g characteristics of a CAAC-IGZO FET with $W/L = 0.80\ \mu m/0.85\ \mu m$. This is one of the I_d–V_g curves at $V_d = 0.1\,V$ in Figure 2.22. Crosses denote measured data points at a V_g step of 0.1 V. (b) The number of electrons N_{trap} trapped by interface levels per unit area and unit energy (circles) and fitted curve (solid line). N_{trap} is extracted from the measured data points shown by the crosses using Equation (2.5). The vertical axis is the Fermi energy E_f from the conduction band minimum E_c of the CAAC-IGZO. *Source*: Adapted from [35]

Figure 2.27(a) shows one of the I_d–V_g curves at $V_d = 0.1\,V$ in Figure 2.22 ($n = 13$). The crosses denote measured data points with a V_g step of 0.1 V. The circles in Figure 2.27(b) denote the N_{trap} values extracted from the measured data points [crosses in Figure 2.27(a)] with Equation (2.5). The vertical axis is the Fermi energy E_f at the interface between IGZO and gate-insulating films, from the conduction band minimum E_c of CAAC-IGZO. N_{trap} has, regardless

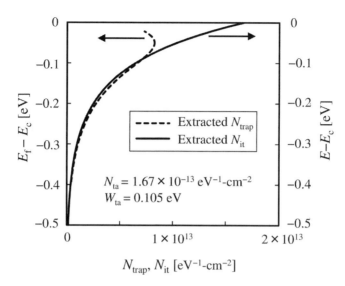

Figure 2.28 Relationship between the DOS of the interface trap levels, N_{it}, and the electron number, N_{trap}, trapped by interface levels per unit area and unit energy. N_{it} has a tail distribution with a peak value N_{ta} of 1.67×10^{13} eV^{-1}-cm^{-2} and a characteristic width W_{ta} of 0.105 eV.

of its large variation, its maximum value immediately below E_c and decreases as it moves away from E_c. Here, as N_{it} of Equation (2.6), the following tail distribution is assumed:

$$N_{it}(E) = N_{ta}\exp\left[\frac{E-E_c}{W_{ta}}\right]. \tag{2.7}$$

Then, N_{trap} is well fitted as shown by the solid line in Figure 2.27(b). As fitting parameters, $\mu_n = 5.5$ cm^2/V-s, a peak value $N_{ta} = 1.67 \times 10^{13}$ eV^{-1}-cm^{-2}, and a characteristic width $W_{ta} = 0.105$ eV are obtained. The relationship between N_{it} and N_{trap} is shown in Figure 2.28.

The device simulation with the obtained tail levels provides the I_d–V_g characteristics shown in Figure 2.29. The measured points are superposed for comparison. The calculated result well reproduces the measured result. In addition, the subthreshold leakage below the lower detection limit ($I_d < 0.1$ pA) is estimated without problem. While SS at $I_d = 1$ pA is 126 mV/decade, SS converges to 82 mV/decade for I_d much smaller than 0.1 pA due to the exponential decrease in tail levels. This value is coincident with the following ideal SS_{id} of the CAAC-IGZO FET:

$$SS_{id} = \frac{k_B T}{q}\left(1 + \frac{C_{tg}^{-1} + C_{act}^{-1}}{C_{bg}^{-1}}\right) = 82 \text{ mV/decade} \tag{2.8}$$

where C_{act} is the active-layer capacitance and C_{tg} (C_{bg}) is the series capacitance of the top (bottom)-gate insulator and the top (bottom) buffer layer. From the fitted curve in Figure 2.29, I_{cut} at $V_d = 1.8$ V is estimated to be 6.7×10^{-18} A. Figure 2.29 shows the extent to which I_{cut} extrapolated with SS at $I_d = 1$ pA is overestimated.

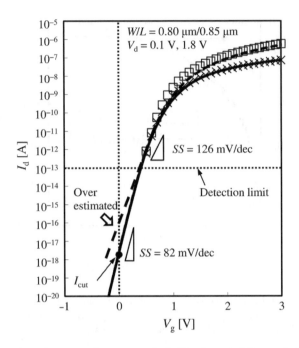

Figure 2.29 Comparison between the measured I_d–V_g characteristics and the calculated I_d–V_g characteristics by device simulation with extracted tail level. SS at $I_d = 1$ pA is 126 mV/decade. At $I_d < 1$ pA, SS is 82 mV/decade, which is coincident with the ideal value of Equation (2.8). *Source:* Adapted from [35]

When a similar analysis is performed with Figure 2.22 ($n = 13$), the average value of I_{cut} is calculated to be 2.0×10^{-18} A. From the measured retention characteristics of the memory cell with CAAC-IGZO FETs that are formed by the same process and have the same size, I_{cut} is estimated to be around 1.0×10^{-18} A, as shown in Equation (2.4), which corresponds to the above-calculated value. On the basis of this calculated value, to reduce the subthreshold leakage to 1 yA, V_{th} should be shifted in the positive direction by the following amount:

$$\Delta V_{th} = SS_{id} \times \left[\log(I_{cut}) - \log\left(1.0 \times 10^{-24}\right)\right]$$
$$= 82\,[\text{mV/dec}] \times \left[\log\left(2.0 \times 10^{-18}\right) - \log\left(1.0 \times 10^{-24}\right)\right]$$
$$= 82\,[\text{mV/dec}] \times (-17.7 + 24)$$
$$= 0.52\,[\text{V}].$$

(2.9)

2.3.4 Conclusion

This section has introduced a calculation method for estimating subthreshold leakage at $I_d < 0.1$ pA by extracting electron trap levels from the measured I_d–V_g characteristics of the CAAC-IGZO FET. This method estimates that the SS of the FET should have its ideal value in a deep subthreshold region because the electron trap levels decrease exponentially in a gap. This method can estimate I_{cut} values below the normal lower detection limit, thereby providing the amount by which V_{th} should be shifted to obtain the desired I_{cut} for a certain application and enabling process development or circuit design with CAAC-IGZO FETs.

2.4 Technique for Controlling Threshold Voltage (V_{th})

In general, desirable V_{th} values of inversion-mode Si FETs can be achieved by impurity doping to the channel regions. In contrast, the CAAC-IGZO FET is an n–i–n accumulation mode FET, i.e., its channel is intrinsic, the source and drain are n-type, and electrons are the majority carrier. Unlike Si FETs, which are not accumulation-mode FETs, it is not possible to shift the threshold voltage (V_{th}) in the positive direction by impurity-doping in the channel. Therefore, the method of applying an additional electric field to the channel region is employed to control V_{th}. This dynamic control of the V_{th} of the FET is not only effective for reducing variation in the electrical characteristics but can also be used to shift V_{th} to a value corresponding to an extremely low off-state current (low leakage mode). Si FETs are often formed on bulk Si substrates and utilize epitaxial growth; as such, they cannot have gate electrodes under the channel. There are reports of Si FETs utilizing a substrate bias for V_{th} control [46], but this method uniformly applies a voltage to all the FETs globally over the entire substrate and cannot be used to adjust FETs locally. On the contrary, CAAC-IGZO FET fabrication allows a design with a back gate that can be used for V_{th} control of individual FETs.

In this section, three techniques for controlling V_{th} in CAAC-IGZO FETs are explained: (1) constant back-gate bias, (2) retention circuit applying back-gate bias only when needed, and (3) a charge trap layer that holds injected charges. Table 2.8 contains a list of the features of each method. Methods (1) and (2) utilize back-gate bias application with and without a retention circuit, respectively. The circuit structure in method (1) is simple, while that in method (2) is more complicated. Method (3) has a simple circuit structure. Turning the power off is inapplicable (i.e., V_{th} control is applicable only during power application) in method (1), whereas it is possible (i.e., V_{th} is kept controlled even after powering off) in methods (2) and (3). The details of methods (1), (2), and (3) are explained in Subsections 2.4.1, 2.4.2, and 2.4.3, respectively.

2.4.1 V_{th} Control by Application of Back-Gate Bias

The method (1) of constantly applying back-gate bias is explained below. Figure 2.30 shows a cross-sectional view of a CAAC-IGZO FET with a back gate in the channel-length direction.

Table 2.8 Features of V_{th} control methods

	Back-gate bias application		(3) Charge trap layer
	(1) Constant bias	(2) Retention circuit	
Circuit configuration	Simple	Complex	Simple
Turning power off (low power consumption)	Inapplicable	Possible	Possible

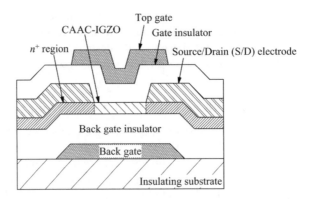

Figure 2.30   Cross-sectional schematic view of a CAAC-IGZO FET in a TGTC structure with a back gate (in the channel-length direction)

By applying voltage to the back gate, V_{th} can be controlled to a desirable value. The back gate only needs to be fixed at a potential to set V_{th} to the desirable voltage. When a negative voltage is applied to the back gate, V_{th} shifts in the positive direction. Conversely, the application of a positive voltage to the back gate shifts V_{th} in the negative direction. Depending on the application, another method, such as dynamical changing of the back-gate voltage, may also be applicable. For example, it is possible to switch between subthreshold driving and normal driving without changing the (top) gate voltage.

Here, experimental results of V_{th} control by back-gate bias application are introduced.

To fabricate a back-gate CAAC-IGZO FET, the back-gate electrode is first formed on an insulating substrate. For the back gate, tungsten (W) is used. Next, an alumina (AlO_x) is deposited as a back-gate insulator by atomic layer deposition (ALD) to a thickness of 20 nm. Next, a silicon oxide (SiO_x) film is deposited by plasma enhancement chemical vapor deposition (PE-CVD) to a thickness of 30 nm, followed by sputtering of the active layer (CAAC-IGZO). After this step, the source/drain electrodes, gate insulator, and top gate are formed. Finally, a passivation film is deposited.

Figure 2.31 shows I_d–V_{tg} curves with V_{bg} as parameter in the fabricated CAAC-IGZO FET with $W/L = 69$ nm $/52$ nm. Here, W denotes the channel width, L the channel length, V_{tg} the top-gate voltage, V_{bg} the back-gate voltage, and V_d the drain voltage. From Figure 2.31 it is seen that the I_d–V_{tg} curve shifts depending on the value and polarity of V_{bg}: a negative/positive V_{bg} shifts the curve in the positive/negative direction, respectively, which proves that V_{th} can be controlled by applying a back-gate voltage.

Note that in this section, an index of V_{sh} is used for the actual measurements instead of V_{th}. Here, V_{sh} is defined as the gate voltage (V_g) at which the drain current (I_d) is 1 pA (see the I_d–V_g characteristics in Figure 2.32). V_{sh} is used to indicate the polarity and degree of I_d–V_g curve shifting, in order to easily estimate the subthreshold leakage. The applications of the CAAC-IGZO FET technology described in and after Chapter 3 utilize the extremely low off-state current.

V_{bg} dependences of V_{th} and V_{sh} are exhibited in Figure 2.33, which indicates that both V_{th} and V_{sh} are proportional to V_{bg}, with negative slopes.

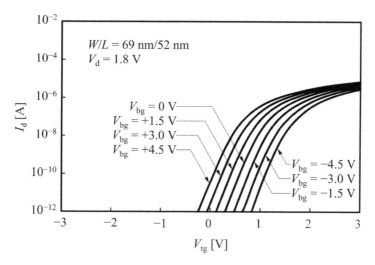

Figure 2.31 V_{bg}-dependence of the I_d–V_{tg} curve

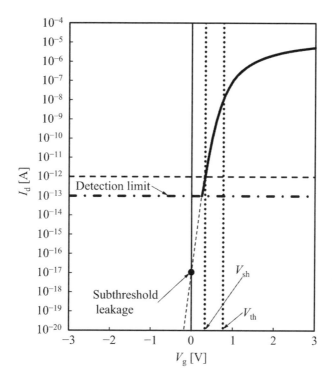

Figure 2.32   The graph for defining V_{sh} and the subthreshold leakage

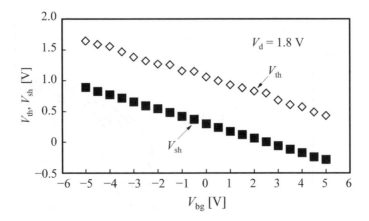

Figure 2.33 V_{bg}-dependence of V_{th} and V_{sh}

Next the FET reliability was evaluated under a state of V_{th} control, i.e., applied back-gate bias. To evaluate the reliability, the temporal drift of the I_d–V_g characteristics under the application of a constant back-gate voltage at 125°C was examined. This time, assuming an application utilizing the very low off-state current, the back-gate voltage was negatively applied to shift V_{th} in the positive direction. For the measurement, two CAAC-IGZO FETs with a channel size of $W/L = 290$ nm $/240$ nm were used. As back-gate insulators (BGI), two different films with EOTs of 12 and 48 nm were prepared, respectively. EOT is the thickness corresponding to a SiO_x thickness giving the same dielectric constant as the actual film.

I_d–V_g measurements were initially conducted at 125°C. A V_{bg} of −5 V was applied to the FET with EOT = 12 nm, while a V_{bg} of −11 V was applied to one with EOT = 48 nm. The measurement was conducted at V_d = 3.3 V, and the V_{sh} drift over 12 h was examined. The change in V_{sh} (ΔV_{sh}) from the initial value is shown in Figure 2.34. After 12 h, ΔV_{sh} in the case of EOT = 12 nm was 0.003 V, and that in the case of EOT = 48 nm was 0.04 V. It has been confirmed that ΔV_{sh} was small and the characteristics of the CAAC-IGZO FET were stable, even with continuous application of the back-gate voltage.

2.4.2 V_{th} Control by Formation of Circuit for Retaining Back-Gate Bias

The advantages of method (1), applying voltage by connecting the power to the back gate, are its simple circuit configuration and stable voltage application; however, method (1) has two disadvantages. One is that the voltage needs to be applied constantly, resulting in high power consumption. The other is that when the power is turned off, the voltage is no longer applied, so V_{th} shifts back to the original value. Especially when the CAAC-IGZO FET operates in a low-leakage mode to be used as a non-volatile memory, the voltage is required to be applied to the back gate even when the power is off.

One method to continue to apply voltage to the back gate even after power-off is to provide a circuit in which the back gate is connected to a capacitor so as to prevent release of a charge from the capacitor [method (2)]. The shifting direction of V_{th} can be controlled by adjusting the

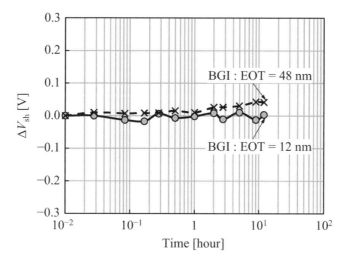

Figure 2.34 V_{sh} time-dependence at 125°C with back-gate bias

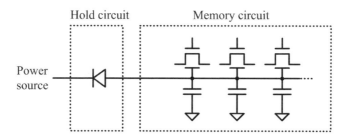

Figure 2.35   Back-gate bias-retention circuit

polarity of the charge stored in the capacitor. This method requires a technique to prevent discharge of the capacitor, resulting in a more complex circuit configuration compared with the method of directly applying the voltage to the back gate. However, after the charge is stored in the capacitor, power will not be consumed. Figure 2.35 shows such a circuit for retaining back-gate bias.

An example of a back-gate bias-retention circuit with a CAAC-IGZO FET is shown in Figure 2.36. When negative voltage is supplied to the back gate of the memory circuit, the node N_1 has negative voltage. N_1 is connected to the gate and back gate of the CAAC-IGZO FET for retention, and the negative voltage is applied to the gate and back gate even when the power source does not supply negative voltage. The condition of $V_{gs} < 0$ (where V_{gs} is the gate–source voltage of the CAAC-IGZO FET in the retention circuit) is satisfied, and the off-state current can be sufficiently small; thus, the retention circuit can shut off the charge flow.

Another example of a back-gate bias-retention circuit with a CAAC-IGZO FET is shown in Figure 2.37. This retention circuit has diodes with CAAC-IGZO FETs that are serially connected, having a lower off-state current than the circuit of Figure 2.36. When the number of serially connected diodes increases, the off-state current can be decreased, but the voltage drops

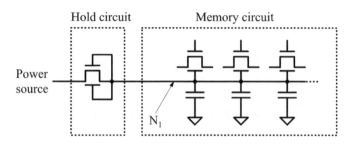

Figure 2.36 Back-gate bias-retention circuit with an IGZO FET

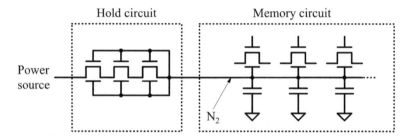

Figure 2.37 Back-gate bias-retention circuit with serially connected IGZO FETs

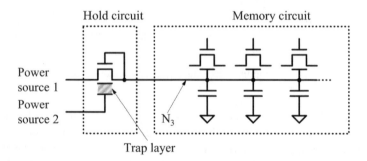

Figure 2.38 Back-gate bias-retention circuit with IGZO FET and trap layer

associated with the power source and N_2 increase and the transmission speed of the power sup-ply to N_2 slows down. These points should be taken into consideration at the time of circuit design.

Another example of a back-gate bias-retention circuit is shown in Figure 2.38. A charge trap layer is provided to the CAAC-IGZO FET, the detail of which is described in the next subsec-tion. The V_{th} of the CAAC-IGZO FET in the retention circuit can be shifted sufficiently in the positive direction with use of the charge trap layer, so that the off-state current will be suffi-ciently low even without application of negative voltage from the power source 2. This method offers a diode with sufficiently low off-state current with the CAAC-IGZO FET. The negative voltage for shifting V_{th} to the desired value is supplied from the power source. The CAAC-IGZO FET in this circuit can have a channel length, channel width, gate insulator thickness, and semiconductor material as long as the FET can be resistant to negative voltage application and have a low off-state current.

2.4.3 V_{th} Control by Charge Injection into the Charge Trap Layer

A third way to control V_{th} in the positive direction is to inject electrons into a charge trap layer (CT layer), provided on the back-gate side.

In this method, V_{th} control is performed by a negative electric field caused by the electrons trapped in the CT layer below the channel. The CT layer can be located on the side of the back or top gate. If electrons are trapped in the CT layer once, there will be no further need to apply the electric field and thus, in principle, the method does not require any power source.

Next, the experimental results of V_{th} control by charge injection into the CT layer are introduced. For the evaluation, a CAAC-IGZO FET in a TGTC structure with a back gate is fabricated. A CT layer is formed between the insulators sandwiched by the back gate and the CAAC-IGZO. Figure 2.39 shows the schematic cross-sectional view of the FET.

To trap electrons in the CT layer, the conduction band (E_C) of the CT layer should be lower than those of the insulators sandwiching the CT layer, and the CT layer material is selected accordingly. Figure 2.40 shows the band diagram from the back gate to the CAAC-IGZO. In this experiment, hafnium oxide, HfO_x, is employed for the CT layer. SiO_x films are adopted for Insulators 1 and 2, which are placed below and above the CT layer. Hence, E_C of the CT layer is lower than those of the sandwiching insulators by 1.2 eV. This E_C difference keeps the electrons injected into the CT layer stable. In addition, it has a large dielectric constant (four to six times that of SiO_x), hence resulting in a large bias voltage per unit charge. HfO_x is used in optical coatings (high refractive index) and as high-k dielectric material in DRAM capacitors. Note that the conditions for injecting charge into the CT layer can be changed by adjusting the thickness of the two insulators sandwiching the CT layer (Insulators 1 and 2) or changing the CT layer material.

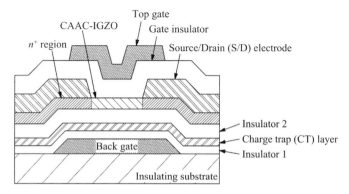

Figure 2.39 Schematic cross-sectional view of a CAAC-IGZO FET in a TGTC structure with a back gate and a CT layer (in the channel-length direction)

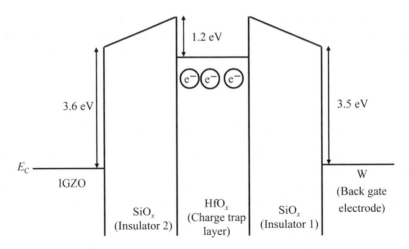

Figure 2.40 Energy band diagram for the CT layer

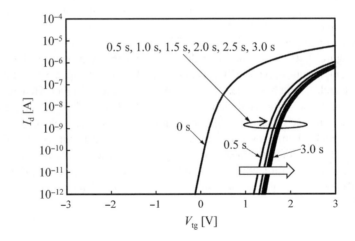

Figure 2.41 Charge injection time dependence of the I_d–V_{tg} curve

Next, the fabrication process is explained. First, a back-gate electrode (W) is formed on the insulating substrate. Next, insulators are deposited as shown in Figure 2.40. A SiO_x film, which will be Insulator 1, is deposited to a thickness of 10 nm by PE-CVD. A HfO_x film is deposited to a thickness of 20 nm by ALD to form the CT layer. Another SiO_x film, which will be Insulator 2, is deposited to 30 nm by PE-CVD. Then, a CAAC-IGZO serving as an active layer is deposited by sputtering. After that, the source/drain electrodes, gate insulator, and top gate are formed. Finally, a passivation film is deposited.

Figure 2.41 shows room-temperature I_d–V_{tg} curves at $V_{bg} = +38$ V with the charge injection time as parameter for a CAAC-IGZO FET with a channel W/L of 0.26 μm/0.19 μm. The measurement is conducted for 3 s at 0.5 s intervals. From Figure 2.41, it is confirmed that the I_d–V_{tg} curve shifts in the positive direction due to charge injection.

In addition, Figure 2.42 shows the charge injection time dependence of ΔV_{sh}, where V_{sh} is defined as V_{tg} at $I_d = 1 \times 10^{-12}$. From Figure 2.42 it is shown that ΔV_{sh} is proportional to the logarithm of the charge-injection time. This fact indicates that the amount of electrons injected into the CT layer is proportional to the logarithm of the injection time.

Next, variations in V_{th} control by charge injection into the CT layer are measured. The condition for charge injection is set at R.T. and $V_{bg} = +38$ V, with a charge injection time of 3 s. The I_d–V_{tg} curves of 56 test elements before and after charge injection are shown in Figure 2.43. The corresponding normal probability distribution plot of V_{sh} before and after charge injection

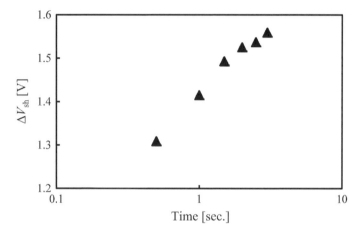

Figure 2.42   Charge injection time dependence of ΔV_{sh} ($V_{bg} = +38$ V, R.T.)

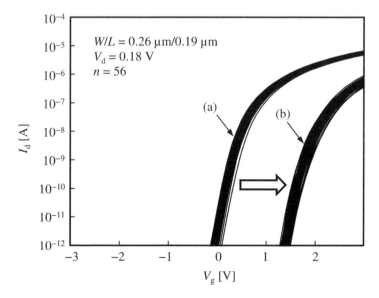

Figure 2.43 I_d–V_g characteristics of a CAAC-IGZO FET (a) before and (b) after charge injection into the CT layer ($V_{bg} = +38$ V, 3 s, R.T.)

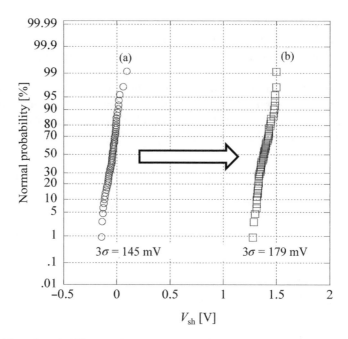

Figure 2.44 Normal probability distribution plot of V_{sh} (a) before and (b) after charge injection ($V_{bg} = +38$ V, 3 s, R.T.)

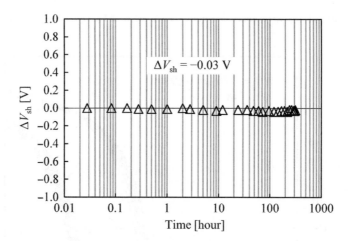

Figure 2.45 V_{sh} time dependence at 150°C after charge injection ($V_{bg} = +40$ V, 200 ms, R.T.)

is illustrated in Figure 2.44. No drastic increase in variations before and after charge injection is observed. 3σ of V_{sh} before charge injection is 145 mV, and that after injection is 179 mV.

Finally, the drift of V_{sh} after charge injection is measured at a temperature of 150°C (see Figure 2.45). The measurement procedure is shown in Figure 2.46 and also explained below.

A voltage V_{bg} of +40 V is first applied to the back gate for 200 ms at R.T. to inject charge into the CT layer. This causes V_{sh} to shift from 0 to 1.3 V. Next, the sample is heated to 150°C and the I_d–V_{tg} characteristics are measured by sweeping V_{tg} from −3 to 3 V while applying a V_d of

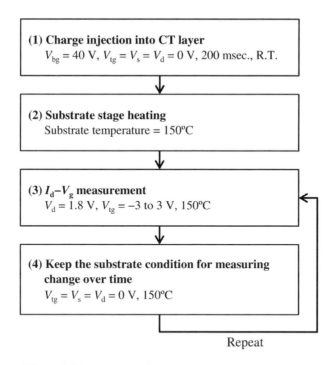

Figure 2.46   Procedure for measuring temperature stability

1.8 V. The time required to bring the sample from R.T. to 150°C is approximately 5 minutes. V_{sh} in the initial measurement, after elevating the temperature to 150°C, is 0.82 V. Figure 2.45 exhibits the change from the initial V_{sh}. During the measurement of V_{sh} drift, the I_d-V_g measurement is performed at 150°C and, during the retention state, V_{tg}, V_{bg}, V_S, and V_d are kept at 0 V. As can be seen in Figure 2.45, ΔV_{sh} is only −0.03 V even after 300 h at 150°C, indicating that the injected charge is maintained stably. From the aforementioned results, it is concluded that V_{th} control by charge injection into a CT layer is a reliable process.

2.4.4   Conclusion

In this section, the three methods for controlling V_{th} of the CAAC-IGZO FET have been described. The feasibililty of V_{th} control was confirmed for all methods, but each method has advantages and disadvantages, and thus an appropriate one should be selected depending on the required LSI performance. V_{th} control of the FET is an important technique for exploiting the very low off-state current of the CAAC-IGZO FET in LSIs.

2.5   On-State Characteristics

Although the most outstanding feature of CAAC-IGZO FETs is their extremely low off-state leakage current, high on-state current of FETs is required for high-speed LSI circuits. That is, high-mobility FETs are necessary for high-speed LSI operation. The electron mobility in CAAC-IGZO (In : Ga : Zn = 1 : 1 : 1) is reported to be approximately 10 cm^2/V-s [17], i.e., two orders of magnitude smaller than that of single-crystal Si (1450 cm^2/V-s), which is widely used in LSIs. Therefore, it may be argued that the low electron mobility of CAAC-IGZO is a bottleneck in CAAC-IGZO LSI.

The above discussion is an intuitive one, but it includes a critical misunderstanding: carrier mobility as a physical property of semiconductor materials is confused with field-effect mobility as an index of the current drivability of FETs. To compare the expected speed of CAAC-IGZO LSIs with that of LSIs with silicon requires a discussion based on a comparison of the field-effect mobilities of n-channel Si (Nch-Si) and CAAC-IGZO FETs, rather than on a comparison of their electron mobilities. In this section, we focus on the field-effect mobility and its dependence on channel length L, particularly when shortened to a deep submicron level.

Besides, to demonstrate the possibility of high-speed CAAC-IGZO LSIs, the cut-off frequency of CAAC-IGZO FETs was measured and the results are presented.

2.5.1   Channel-Length Dependence of Field-Effect Mobility

Matsuda *et al.* [47] reported the channel-length L dependence of the field-effect mobility μ_{FE} of CAAC-IGZO FETs and compared it with that of single-crystal Si FETs. In that paper, planar CAAC-IGZO FETs ($L = 0.45 - 100\,\mu m$), S-ch CAAC-IGZO FETs ($L = 0.055 - 0.515\,\mu m$), and planar Si FETs ($L = 0.30 - 7.95\,\mu m$) were fabricated, and their μ_{FE} values were measured and compared.

2.5.1.1   Sample Fabrication and Structure

S-ch CAAC-IGZO FETs have a three-dimensional (3D) gate structure to suppress the short-channel effects [11]. Figure 2.47 shows a schematic diagram and cross-sectional STEM images of the S-ch CAAC-IGZO FET. As discussed in detail in the next section, such FETs exhibit excellent characteristics, even for $L < 100\,nm$. Both types of CAAC-IGZO FET have a gate insulator with an EOT of 11 nm. The channel width W of the planar type is 10 μm, and the width of the fin-shaped channel island of the S-ch type is 47 nm. The active-layer thickness of the planar type is 15 nm, and that of the S-ch type is 40 nm. S-ch FETs have an effective channel width W_{eff} of 127 nm, equal to the sum of the active-layer width and twice the active-layer thickness.

For comparison, the L dependence of μ_{FE} of single-crystal Si (sc-Si) FETs was measured. Si FETs ($L = 0.30 - 7.95\,\mu m$) were fabricated using a thin film of single-crystal Si transferred on a glass substrate [48]. The EOT of gate-insulating films in these Si FETs was 20 nm, and their W was 6 μm.

2.5.1.2   Measurement Conditions

Field-effect mobility is defined by the following equation:

$$\mu_{FE} = \max \left[\frac{L}{WC_{OX}} \frac{1}{V_d} \left(\frac{\partial I_d}{\partial V_g} \right) \right] \tag{2.10}$$

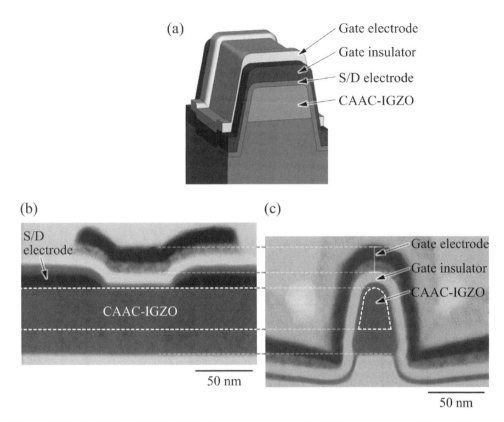

Figure 2.47   S-ch CAAC-IGZO FET: (a) schematic view, (b) STEM cross-sectional image (plane is parallel to the longitudinal axis of the channel), (c) STEM cross-sectional image (plane is normal relative to the longitudinal axis of the channel). *Source*: Reproduced from [47] with permission of *Japanese Journal of Applied Physics*

where W is the channel width, C_{OX} is the capacitance of the gate insulator, V_d is the drain voltage (here, $V_d = 1$ V), and I_d is the drain current. Here, μ_{FE} indicates the FET's normalized ability to drive current at a V_d of 1 V. Here, the effective channel width W_{eff} was used to calculate μ_{FE} of S-ch CAAC-IGZO FETs.

2.5.1.3   Measurement Results

The measured μ_{FE} values of CAAC-IGZO FETs and Si FETs are plotted in Figure 2.48(a) and (b), respectively. In Figure 2.48(a), the squares indicate the measurement results for the planar FETs and the triangles indicate those for the S-ch FETs.

For sc-Si FETs, μ_{FE} decreases from 503 to 113 cm^2/V-s as L decreases from 7.95 to 0.30 μm. In contrast, μ_{FE} of planar CAAC-IGZO FETs is approximately 7 cm^2/V-s and almost independent of L. For S-ch CAAC-IGZO FETs, μ_{FE} decreases from 7.8 to 5.1 cm^2/V-s, as L decreases from 0.515 to 0.055 μm, but the reduction rate of μ_{FE} by miniaturization is much smaller for S-ch CAAC-IGZO FETs than for sc-Si FETs.

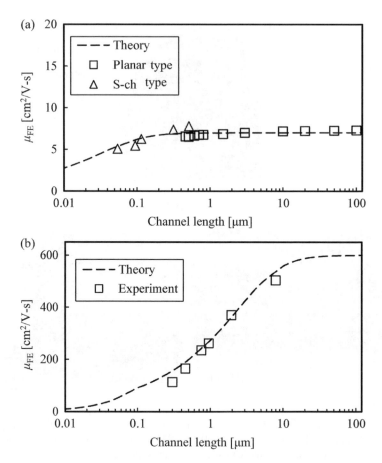

Figure 2.48 Channel-length dependence of field-effect mobility: (a) CAAC-IGZO FETs, (b) Si FETs. *Source*: Reproduced from [47] with permission of *Japanese Journal of Applied Physics*

2.5.1.4 Discussion

This drastic difference in the dependence of μ_{FE} on L between the two types of FET results from differences in phonon scattering susceptibilities. In other words, electrons in CAAC-IGZO are less likely to be affected by phonon scattering than those in sc-Si. When V_d is low or L is long, the drift velocity v_d of carriers in a semiconductor is a product of the electric field E and the "low-field mobility" μ_0:

$$v_d = \mu_0 E. \tag{2.11}$$

Increasing E by reducing L increases the v_d of the electrons to the sound velocity C_S in a solid. At this point, the electron temperature is higher than the lattice temperature (these electrons are referred to as "hot electrons"). Hot electrons are scattered by acoustic phonons and dissipate their energy. In that case, the relationship between v_d and E is modified as [49]

$$v_d = \mu_0 E \sqrt{\frac{T}{T_e}} = \mu_0 E \cdot \sqrt{2} \cdot \left\{ 1 + \left[1 + \frac{3\pi}{8} \left(\frac{\mu_0 E}{C_S} \right)^2 \right]^{1/2} \right\}^{-1/2} \qquad (2.12)$$

where T_e is the electron temperature and T is the lattice temperature.

When L is further reduced and the electron kinetic energy reaches the optical phonon energy E_P at the Γ point, the entire energy of the electrons proceeds to produce optical phonons; thus, v_d saturates. This phenomenon is called "velocity saturation," and the saturated electron drift velocity v_{sat} is expressed as [49]

$$v_{sat} = \sqrt{\frac{8E_P}{3\pi m_0}}. \qquad (2.13)$$

To obtain C_S and E_P for CAAC-IGZO, we calculated the phonon dispersion relationship using the CASTEP code [50, 51]. All calculations used the local-density approximation based on DFT, and norm-conserving pseudo-potentials were employed with a plane-wave cut-off of 800 eV.

The phonon dispersion relation was obtained by solving the dynamical matrix eigenvalue problem, and the dynamical matrix was calculated by the linear response method. For the $InGaO_3(ZnO)$ structure, a 28-atom cell obtained by converting the lattice vectors of a conventional $YbFe_2O_4$ unit cell was used.

Figure 2.49 shows the phonon dispersion relation obtained from these calculations. For the dispersion relation shown in Figure 2.49(a), the E_P of IGZO was 9.42 meV.

C_S was determined from the group velocity ($1/\hbar \cdot d\omega/dk$) at the Γ point, i.e., the slope of the acoustic phonon dispersion curve. Table 2.9 lists C_S values parallel (Γ-B) and perpendicular (Γ-F) to the c-axis. In a CAAC-IGZO FET, electrons flow in a plane perpendicular to the c-axis; thus, longitudinal acoustic (LA) phonons in the Γ-F direction scatter electrons. The C_S for LA phonons was calculated to be 5.84×10^5 cm/s.

We used published values of C_S and E_P for the sc-Si FET [49], and μ_0 values for sc-Si and CAAC-IGZO were derived by fitting the experimental curves of sc-Si and CAAC-IGZO, as shown in Figure 2.48. Table 2.10 lists the physical properties used for the calculations.

Figure 2.50 shows v_d as a function of E calculated from the parameters listed in Table 2.10. As E increases, the difference between the v_d in Si and IGZO decreases. The relation between v_d and E shown in Figure 2.50 can also be expressed in terms of a relation between μ_{FE} and L for a given drain voltage V_d. Assuming V_d of 1 V and that the electric field intensity E is uniform in the channel, L can be written as

$$L_{ch} = \frac{V_d}{E}. \qquad (2.14)$$

Under this assumption, the drift velocity v_d is also uniform in the channel, and μ_{FE} can be written as follows:

$$\mu_{FE} = \frac{v_d}{E}. \qquad (2.15)$$

The results of Equation (2.15) are shown as dashed lines in Figure 2.48, consistent with the experimental results for both the sc-Si and CAAC-IGZO FETs.

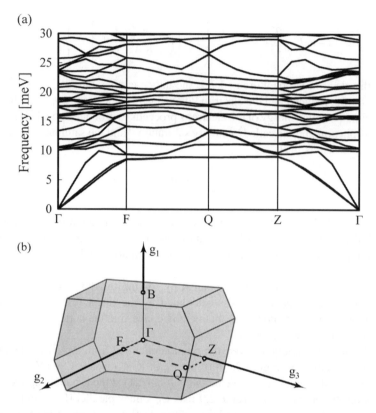

Figure 2.49   (a) Phonon-dispersion relation for InGaO$_3$(ZnO), (b) reciprocal lattice of InGaO$_3$(ZnO) in a 28-atom cell. *Source*: Reproduced from [47] with permission of *Japanese Journal of Applied Physics*

Table 2.9   Velocity of phonons in a single-crystalline IGZO. *Source*: Reproduced from [47], with permission of *Japanese Journal of Applied Physics*

Mode	Parallel to c-axis	Perpendicular to c-axis
LA	6.68	5.84
T$_1$A	2.74	3.09
T$_2$A	3.19	3.03

Table 2.10   Physical properties for calculating the drift velocities and field-effect mobilities of FETs. *Source*: Reproduced from [47], with permission of *Japanese Journal of Applied Physics*

	CAAC-IGZO	sc-Si
μ_0 [cm^2/V-s]	7	600
C_S [cm/s]	5.8×10^5	7×10^5
E_P [meV]	9.4	63

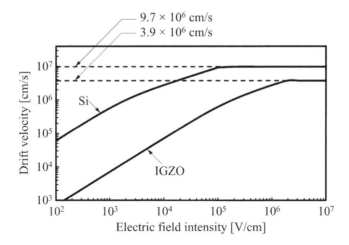

Figure 2.50 Calculated drift velocity v_d as a function of electric field intensity E. *Source*: Reproduced from [47] with permission of *Japanese Journal of Applied Physics*

In sc-Si FETs, the low-field mobility μ_0 is large and the drift velocity v_d of an electron easily reaches the speed of sound C_S; thus, hot electrons are easily generated. By contrast, hot electrons are less likely to be generated in the CAAC-IGZO FET because μ_0 is small and the v_d of an electron barely reaches C_S. The CAAC-IGZO FETs are thus less susceptible to phonon scattering compared with sc-Si FETs. Therefore, the μ_{FE} value of the sc-Si FET decreases significantly as L shortens, whereas it remains essentially constant for the CAAC-IGZO FET.

According to the comparison of v_{sat} values of both Si and IGZO, we expect that, by further reducing L, the μ_{FE} ratio of the CAAC-IGZO FET to that of the sc-Si FET should reach 1/2.6. However, this result is considered to be overestimated because the "velocity overshoot" phenomenon in sc-Si, which is discussed below, was ignored.

2.5.1.5 Effect of Velocity Overshoot for Electrons in Si FETs

In FETs with extremely short L, some carriers injected from the source region reach the drain region without any scattering in the channel. Such a mode of carrier transportation is referred to as "ballistic transport" [52]. When ballistic transport of carriers occurs, the average drift velocity v_d becomes faster than the saturation velocity v_{sat}. This phenomenon is referred to as "velocity overshoot" [53, 54]. According to the paper by Ruch [53], velocity overshoot occurs in a Si MOSFET when L is shorter than about 100 nm. Therefore, to estimate the mobility ratio of CAAC-IGZO FETs to Si FETs precisely, the velocity overshoot phenomenon needs to be taken into account.

Hereafter, the μ_{FE} of Si FETs derived by Monte Carlo simulation is discussed. Note that it is assumed that velocity overshoot can be ignored for CAAC-IGZO FETs, because the lower electron mobility of CAAC-IGZO means a shorter mean free path of electrons in the IGZO channel. Thus, ballistic transport hardly occurs in CAAC-IGZO FETs. The Monte Carlo simulation of electron transport in a crystalline IGZO is one of the challenges in research of the electrical conduction mechanisms of crystalline IGZO.

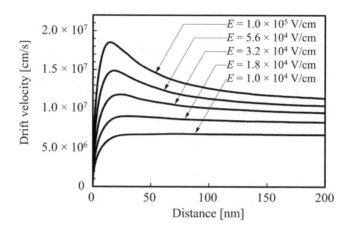

Figure 2.51   Simulation results of the velocity overshoot of electrons in bulk Si

First, the velocity overshoot phenomenon of electrons in *bulk* Si was simulated. In this calculation, the device simulator ATLAS (Silvaco, Inc.) and its module for Monte Carlo device simulation, MCDevice, were used [55]. Figure 2.51 shows the simulation results. When the electric field intensity is higher than 1.0×10^5 V/cm, we found that the velocity of electrons is faster than the v_{sat} of 1.0×10^7 cm/s, or that velocity overshoot occurs, especially with $L < 100$ nm for a drain voltage of 1 V.

In order to verify its validity, this Monte Carlo calculation result was compared with an analytical mobility model called the Schafetter and Gummel model (S&G model) [56]. According to the S&G model, the velocity of carriers is expressed as follows:

$$
v = \mu_0 E \left[1 + \cfrac{N}{\cfrac{N}{S} + N_{\text{ref}}} + \cfrac{\left(\cfrac{E}{A}\right)^2}{\cfrac{E}{A} + G} + \left(\cfrac{E}{B}\right)^2 \right]^{-\frac{1}{2}}. \tag{2.16}
$$

For carriers in Si, the best-fitting parameters listed in Table 2.11 were used. In Figure 2.52, the E dependences of the velocity of electrons in Si, calculated by the Monte Carlo simulation and the S&G model, were compared. For both calculations, no impurity doping was assumed. Note that the Monte Carlo simulation was carried out until the velocity of electrons converged to a steady state. The Monte Carlo simulation results were consistent with the analytical model, and their validity was confirmed.

Next, Monte Carlo simulation of carrier transport in FETs was carried out, in order to study the impact of the velocity overshoot on the μ_{FE} of the inversion-mode Nch-Si FETs. The simulation model is shown in Figure 2.53. In Figure 2.53, the gate length (L_{gate}) is 50 nm, the channel length L is 25 nm, and the gate insulator thickness is 1.5 nm. The density of acceptor-impurity doping in the bulk is 1×10^{15} cm^{-3}. In this calculation, interface scattering was ignored. The distribution of the electron velocity beneath the gate insulator was derived with this model on the condition that V_{d} is 1 V and V_{g} is 3 V, as shown in Figure 2.54. The velocity around the middle region of the channel was faster than v_{sat}, which means that velocity overshoot occurs.

Table 2.11   Fitting parameters of S&G models for both electrons and holes.
Source: Adapted from [57]

	Electron	Hole
μ_0 [cm²/V-s]	1400	480
S	350	81
N_{ref} [cm^{-3}]	3.0×10^{16}	4.0×10^{16}
G	8.8	1.6
A [V/cm]	3.5×10^3	6.1×10^3
B [V/cm]	7.4×10^3	2.5×10^4

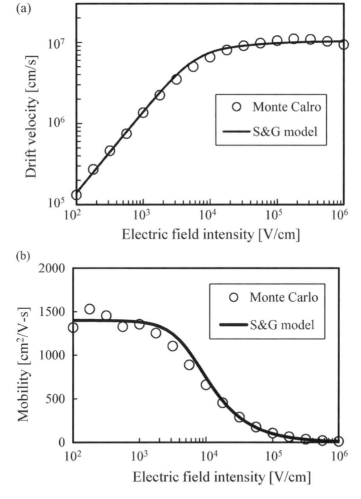

Figure 2.52   Comparison of the Monte Carlo simulation results and the S&G mobility model. (a) The dependence of the v_d of electrons on E; (b) the dependence of the electron mobility on E

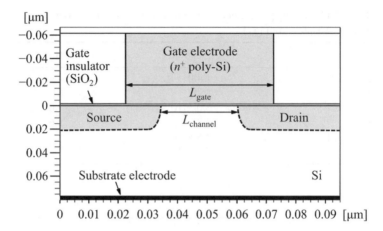

Figure 2.53   Monte Carlo simulation model

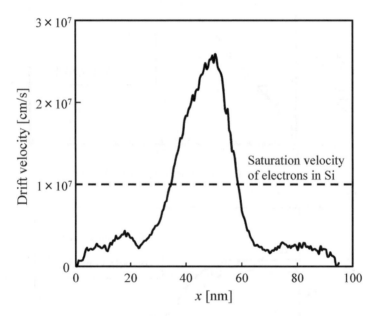

Figure 2.54   The distribution of electron drift velocity just beneath the GI film (L = 25 nm, L_{gate} = 50 nm, V_g = 3.0 V, V_d = 1.0V)

Simulation models of Nch-Si FETs with various values of L (L = 17.5 nm–1.6 µm) were made in accordance with the scaling rule [56, 58]. Here, the "voltage constant scaling rule" was adopted. In other words, as the FET size was equally scaled down by K^{-1} times, the doped impurity density increased by K^2 times. The μ_{FE} was defined as the quotient of the average velocity of electrons and the electric field intensity E in the channel. The L dependence of

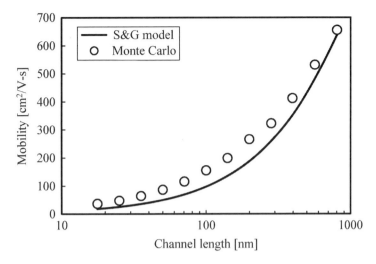

Figure 2.55 Channel-length dependence of electron mobility derived by Monte Carlo simulation and the S&G model

μ_{FE} of Nch-Si FETs was calculated by this Monte Carlo simulation and by the S&G model. Both calculation results are shown in Figure 2.55. The μ_{FE} derived from the Monte Carlo simulation is estimated to be larger than the analytical model for any values of L, because velocity overshoot is taken into account for the Monte Carlo simulation. According to the simulation, when a Nch-Si FET is miniaturized to be much shorter than 100 nm, the μ_{FE} is estimated to be two times higher than the analytical model in which the velocity overshoot is not taken into account. As discussed before, the μ_{FE} ratio of CAAC-IGZO FETs to that of Nch-Si FETs tends to decrease as L shortens. When L is reduced to 30 nm or shorter, the μ_{FE} ratio of CAAC-IGZO FETs to that of Nch-Si FETs would be approximately 1/10.

2.5.2 Measurement of Cut-off Frequency

Frequency parameters such as clock frequency are used for the speed index of LSIs. These depend not only on the current drivability of the FETs, but also on the load capacitance and resistance, and hence also on the purpose or design of the LSIs. Thus, the frequency parameter of a single FET is an essential factor, rather than the circuit speed being essential, and the cut-off frequency f_T is widely used for this purpose.

f_T denotes the frequency at which the current gain of a FET is equal to 1 [59]. The current gain of a FET obeys the following equation:

$$\text{Current Gain} = \frac{\Delta I_d}{\Delta I_g} = \frac{g_m \Delta V_g}{2\pi f C_{OX} WL \Delta V_g} = \frac{g_m}{2\pi f C_G} \qquad (2.17)$$

where C_{OX} is the gate capacitance per unit area, C_G ($C_G = C_{OX}WL$) is the gate capacitance, and g_m is the transconductance of the FET. Here, g_m is expressed as follows:

$$g_m = \frac{W}{L}\mu_{FE}C_{OX}V_d. \tag{2.18}$$

When the current gain is equal to 1, f_T is defined as

$$f_T = \frac{g_m}{2\pi C_G} = \frac{\mu_{FE}V_d}{2\pi L^2}. \tag{2.19}$$

To improve f_T, there are hence three options:

1. shorten the channel length;
2. strengthen the drain voltage;
3. use a high-mobility semiconductor.

f_T is proportional to the reciprocal of the time constant that is necessary to charge and discharge the gate capacitance. Indeed, the time constant τ for charging the gate capacitance when the gate voltage is changed by ΔV_g is expressed as follows:

$$\tau = \frac{C_{OX}WL\Delta V_g}{g_m\Delta V_g} = \frac{C_g}{g_m}. \tag{2.20}$$

With time constant τ, f_T can be expressed in the following manner:

$$f_T = \frac{1}{2\pi\tau}. \tag{2.21}$$

No LSI can be operated at a frequency higher than the f_T of its FETs in the LSI. Consequently, an f_T smaller than the required clock frequency means that the LSI cannot operate as intended. Yakubo et al. [60] reported that an f_T of 1.9 GHz was achieved with a S-ch CAAC-IGZO FET ($W/L = 300\ \mu m/60\ nm$) comprising 5000 parallel-connected FETs ($W/L = 60\ nm/60\ nm$). The measurement conditions were $V_d = 1.0\ V$ and $V_g = 2.2\ V$ (see Figure 2.56). Note that a V_g of 2.2 V is the condition at which g_m becomes maximal, as shown in Figure 2.57.

After Yakubo et al. published their paper, an f_T of 20.1 GHz was achieved with a V_d of 2.0 V and a V_g of 1.95 V by means of an improvement to the TEG (Test Element Group) structure and an In-rich oxide semiconductor (TYPE B) as a substitute for the oxide semiconductor used in Figure 2.56 (see Figure 2.58). The TYPE A FET uses the same oxide semiconductor as that of Figure 2.56.

Matsubayashi et al. [61] reported a f_T of 34.4 GHz using TGSA-structure CAAC-IGZO FETs with $L < 30$ nm. For the FET structure reported by Matsubayashi et al., both the miniaturization of L and the decrease in overlapped capacitance between the gate and the S/D electrodes contributed to the improved f_T. Detailed results and discussions are described in Section 2.7.

As discussed above, the f_T of CAAC-IGZO FETs with $L < 100$ nm is established to be on the order of 10 GHz. This high f_T suggests the possibility of fast operational speed of CAAC-IGZO LSIs.

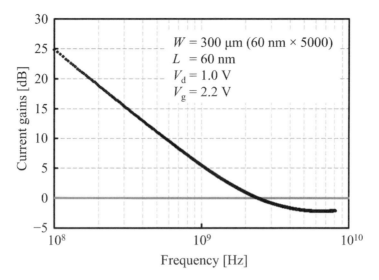

Figure 2.56 RF gains vs. the frequency of a S-ch CAAC-IGZO FET with $W/L = 300\,\mu m/60\,nm$. *Source*: Adapted from [60]

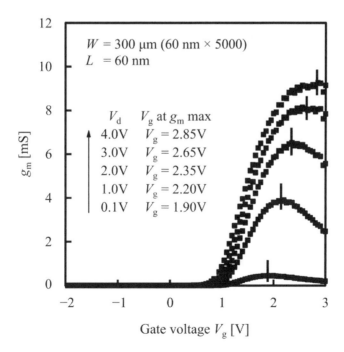

Figure 2.57 g_m Characteristics of a CAAC-IGZO FET with $W/L = 300\,\mu m/60\,nm$. *Source*: Adapted from [60]. Copyright 2014 The Japan Society of Applied Physics

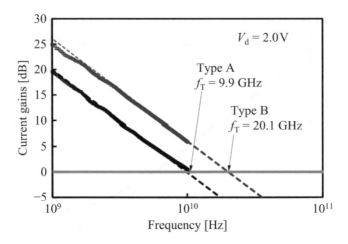

Figure 2.58   Measurement results of the cut-off frequency of IGZO FETs

2.5.3   Summary

Although the electron mobility in CAAC-IGZO is two orders of magnitude smaller than that of single-crystal Si, electrons in crystalline IGZO are less likely to be affected by phonon scattering under a high electric field compared with Si. This leads to a reduction in the difference in field-effect mobility μ_{FE} of CAAC-IGZO FETs and single-crystal Si FETs, when both FETs are miniaturized. According to a simple discussion of velocity saturation of electrons in crystalline IGZO and Si, the μ_{FE} ratio of CAAC-IGZO FETs to that of Si FETs should reach 1/2.6. A Monte Carlo simulation was carried out to estimate the precise μ_{FE} ratio between CAAC-IGZO and Si FETs, and it was found that this ratio should be limited to 1/10 because of the velocity overshoot phenomenon in single-crystal Si FETs.

Furthermore, the cut-off frequency of CAAC-IGZO FETs, which is an index of the switching speed of a single FET, was measured. The latest results achieved a cut-off frequency of more than 30 GHz.

In conclusion, the disadvantage of the low electron mobility of CAAC-IGZO compared with single-crystal Si can be offset by FET's miniaturization. In other words, its low electron mobility in CAAC-IGZO is *not* a critical bottleneck of operational speed of CAAC-IGZO LSIs, provided that the FETs are miniaturized to deep submicron level.

2.6   Short-Channel Effect

Degradation resulting from miniaturization, and especially from a shortened channel, is called a short-channel effect in general. As shown in the schematic diagram of Figure 2.59, the main factors of the short-channel effect are a negative shift of the threshold voltage (V_{th} roll-off), an increase in subthreshold swing (*SS*), a drain-induced barrier lowering (*DIBL*), and an increase in the off-state current (I_{off}).

The effective countermeasure against these short-channel effects is to increase the gate control over a channel region by thinning the gate insulator. In fact, the EOT of a gate insulator has been thinned to 1–2 nm in the state-of-the-art Si technology. However, with a thickness of

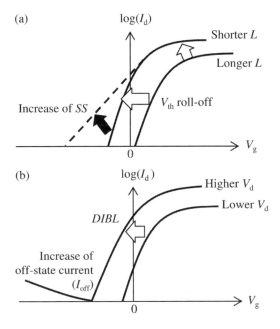

Figure 2.59   Graphs explaining the short-channel effect: (a) V_{th} roll-off, increase in SS and (b) *DIBL*, increase in I_{off}

1–2 nm, a tunnel current through the gate insulator is problematic. Thus, such countermeasures might spoil the advantages of CAAC-IGZO FETs, in particular, the extremely low I_{off} off-state current [10]. To avoid the tunnel current through a gate insulator, an EOT of approximately 10 nm is necessary. It has been considered that there is a trade-off between a decrease in the EOT of a gate insulator and extremely low I_{off}.

In this section, a miniaturized CAAC-IGZO FET with a small short-channel effect is introduced [62]. The EOT is about 11 nm and the channel length is about 30 nm. The FET has an S-ch structure in which the top surface and the side surfaces of an active layer are covered with a gate electrode (i.e., it is a 3D structure) [16].

As shown by the device simulation in Section 2.6.2, a narrower channel width can suppress the short-channel effect [11], as discussed with device simulation. Other factors, such as the physical properties of the CAAC-IGZO film and the S-ch structure, are also discussed.

2.6.1   Features of S-ch CAAC-IGZO FETs

Figure 2.60(a) and (b) shows a schematic view and a plane view, respectively, of the S-ch CAAC-IGZO FET using a CAAC-IGZO film as active layer [16]. Figure 2.60(c) and (d) shows STEM images of cross-sections in the channel-length and channel-width directions, respectively. The FET has a TGTC structure, where the gate electrode overlaps with the source and drain electrodes [see Figure 2.60(c)]. The gate electrode covers both the top face and side surfaces of the CAAC-IGZO film serving as a channel [see Figure 2.60(d)]. This is called the S-ch structure. As illustrated in Figure 2.60(b), the channel length L is defined to be the distance

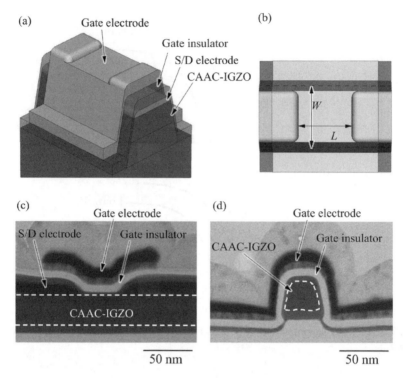

Figure 2.60   S-ch CAAC-IGZO FET using CAAC-IGZO film as active layer: (a) schematic view; (b) plane view; (c) cross-sectional STEM image in channel-length direction; and (d) cross-sectional STEM image in channel-width direction. *Source*: Reprinted from [16], with permission of IEEE, © 2014. *(For color detail, please see color plate section)*

between the source and the drain, and the channel width W is defined to be the width of the CAAC-IGZO film.

The CAAC-IGZO film was deposited to a thickness of 40 nm by DC sputtering under Ar/O$_2$ atmosphere at a substrate temperature of 300°C, with a polycrystalline target containing indium, gallium, and zinc, where the ratio of In:Ga:Zn was 1:1:1. Figure 2.61(a) shows the out-of-plane XRD spectrum of the CAAC-IGZO film deposited on a quartz substrate under the same conditions as the above S-ch CAAC-IGZO FET. A (009) diffraction peak is observed around $2\theta = 31°$, which is attributed to a CAAC structure (Figure 2.1). Figure 2.61(b) is a cross-sectional STEM image of the CAAC-IGZO film. The upper side is near the CAAC-IGZO surface. As shown in Figure 2.61(b), CAAC-IGZO has a stacked structure, so the *c*-axis of the IGZO crystal contained in the CAAC-IGZO film is oriented vertically with respect to the substrate.

Figure 2.62(a) shows the I_d–V_g characteristics of the S-ch CAAC-IGZO FET with drain voltage V_d as parameter. The EOT is 11 nm and $W/L = $ 47 nm /56 nm. As can be seen, the I_d–V_g characteristics are normally off and the off-state current is below the lower detection limit (below 0.1 pA) for all values of V_d. Neither any *DIBL* nor any increase in I_{off} as a result of increasing V_d could be observed.

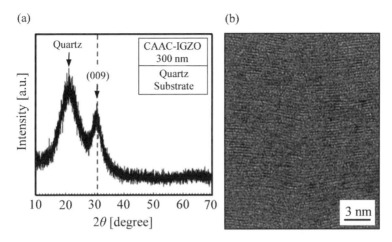

Figure 2.61   (a) Out-of-plane XRD spectrum of the CAAC-IGZO film; (b) cross-sectional TEM image of CAAC-IGZO film. *Source*: Reprinted from [16], with permission of IEEE, © 2014

Figure 2.62(b) shows V_d dependences of a shift voltage V_{sh} and SS. Nine samples were measured for each V_d. Again, the V_{sh} is defined as the gate voltage at which the drain current is 1 pA. The results indicate that V_{sh} and SS are independent of V_d, despite the short channel length of 56 nm. Specifically, the medians of $DIBL$ and SS are 67 mV/V and 92 mV/decade ($V_d = 1$ V), respectively. The results can be explained by the effect of the gate electric field from the side of a CAAC-IGZO island, which increases with decreasing channel width. Figure 2.63(a) shows a W dependence in the I_d–V_g characteristics of the S-ch CAAC-IGZO FET with $L = 56$ nm and drain voltage $V_d = 1$ V. Figure 2.63(b) shows the W dependences of V_{sh} and SS at $V_d = 1$ V. SS increased as W got wider, whereas V_{sh} shifted in the negative direction. When W was smaller than 100 nm ($W < 100$ mm), V_{sh} and SS gradually saturated. This proves that the narrower channel width can effectively suppress the short-channel effect.

Figure 2.64(a) shows the L dependence of the I_d–V_g characteristics of the S-ch CAAC-IGZO FET with $W = 47$ nm and $V_d = 1$ V, and Figure 2.64(b) shows the L dependence of V_{sh} and SS. No degradation resulting from the short-channel effect occurs even for $L = 56$ nm. To examine the characteristics of a more miniaturized FET, a 30-nm node S-ch CAAC-IGZO FET was fabricated (see Figure 2.65) [62]. Figure 2.66(a) shows the I_d–V_d characteristics where data is obtained by varying V_g from 0.0 to 2.0 V in 0.2 V steps. Figure 2.66(b) shows the I_d–V_g characteristics for $V_d = 0.1$ and 1.8 V, respectively. As shown, the miniaturized FET offers good saturation characteristics, low I_{off} below the detection limit (0.1 pA), and an SS of 90 mV/decade, even when its EOT is 11 nm. Figure 2.67(a) depicts the L dependence of the I_d–V_g characteristics of S-ch CAAC-IGZO FETs for $W = 33$ nm and $V_d = 1.8$ V. Figure 2.67(b) depicts the L dependence of SS ($V_d = 0.1$ V) and V_{th} ($V_d = 0.1$ V). Degradation resulting from the short-channel effect does not occur even when the FET is scaled down to $L = 32$ nm and the EOT of the gate insulator is 11 nm. To evaluate the retention characteristics, an S-ch CAAC-IGZO FET with $W/L = 40$ nm $/53$ nm and a 1-bit memory with a retention capacitance of 20 fF was fabricated, and the floating-node voltage was measured at 125°C (see Figure 2.68). The voltage

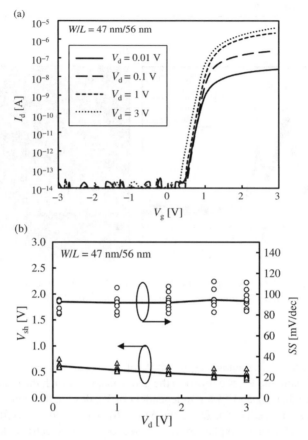

Figure 2.62 (a) Dependence on drain voltage (V_d) of the I_d–V_g characteristic and (b) dependence on V_d of the V_{sh} and SS of an S-ch CAAC-IGZO FET with a W/L of W/L of 47 nm/56 nm ($n = 9$). *Source:* Reprinted from [16], with permission of IEEE, © 2014

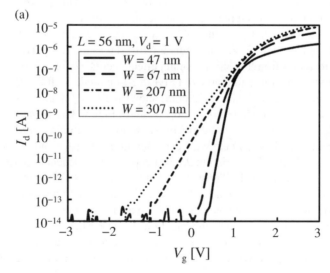

Figure 2.63 (a) Dependence on W of the I_d–V_g characteristic and (b) dependence on W of the V_{sh} and SS of an S-ch CAAC-IGZO FET with W/L of 47 nm/56 nm ($n = 9$), V_d of 1 V. *Source:* Reprinted from [16], with permission of IEEE, © 2014

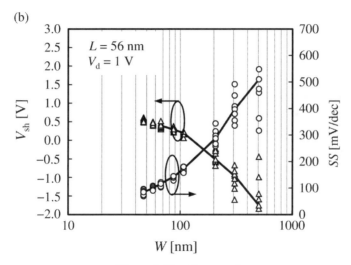

Figure 2.63 *(Continued)*

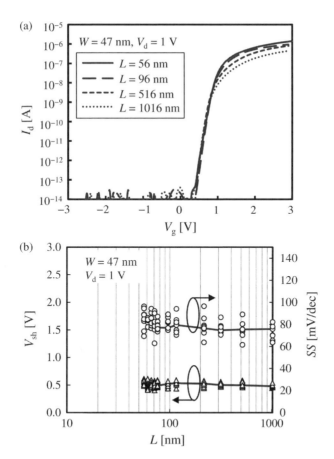

Figure 2.64 (a) Dependence on L of the I_d–V_g characteristic and (b) dependence on L of V_{sh} and SS of an S-ch CAAC-IGZO FET ($n = 9$) with a channel length of 56 nm, V_d of 1 V. *Source*: Reprinted from [16], with permission of IEEE, © 2014

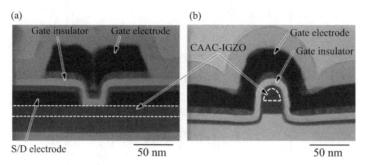

Figure 2.65 (a) Cross-sectional STEM image in channel-length direction and (b) cross-sectional STEM image in channel-width direction of an S-ch CAAC-IGZO FET with $W/L = 33\,\text{nm}/32\,\text{nm}$. *Source*: Reprinted from [62], with permission of IEEE, © 2015

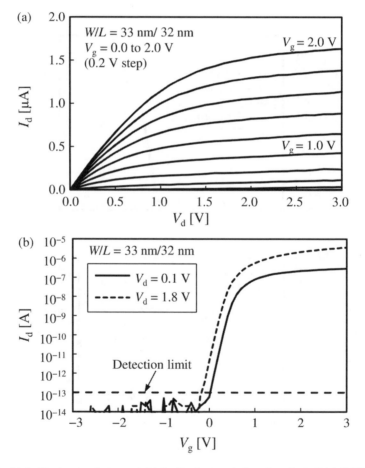

Figure 2.66 (a) I_d–V_d characteristics and (b) I_d–V_g characteristics of an S-ch CAAC-IGZO FET with $W/L = 33\,\text{nm}/32\,\text{nm}$. *Source*: Reprinted from [62], with permission of IEEE, © 2015

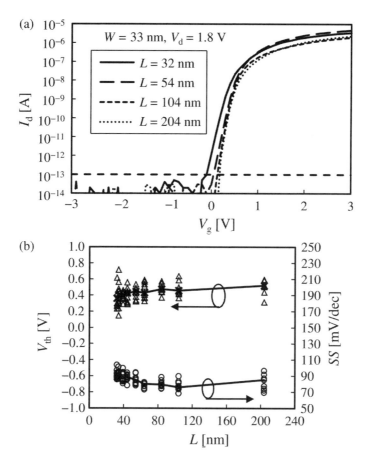

Figure 2.67 (a) Dependence on L of the I_d–V_d characteristics and (b) L dependence of the SS ($V_d = 0.1\,V$) and V_{th} ($V_d = 1.8\,V$) of an S-ch CAAC-IGZO FET with $W/L = 33\,nm/32\,nm$. *Source:* Reprinted from [62], with permission of IEEE, © 2015

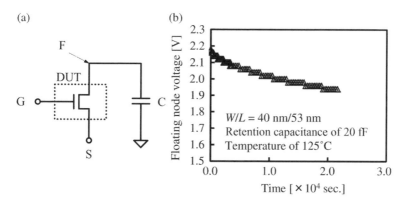

Figure 2.68 (a) Diagram of a circuit using the CAAC-IGZO FET with $W/L = 40\,nm/53\,nm$; (b) retention characteristics at 125°C. *Source:* Reprinted from [62], with permission of IEEE, © 2015

decreased by only 0.24 V, even after 20,000 s, which proves that even the miniaturized S-ch CAAC-IGZO FET has an extremely low I_{off}.

2.6.2   Effect of S-ch Structure

As explained in Subsection 2.6.1, the short-channel effect is prevented in S-ch CAAC-IGZO FETs by keeping the width W narrow. One reason why $DIBL$ is small, as shown in Figure 2.62, is discussed here [11,16]. The distance at which a drain electric field influences the potential in a channel is referred to as the natural length λ_n [63]. The natural length is represented by

$$\lambda_n = \sqrt{\frac{\varepsilon_S}{n\varepsilon_{OX}}\left(1 + \frac{\varepsilon_S t_S}{4\varepsilon_S t_{OX}}\right) t_S t_{OX}} \qquad (2.22)$$

where ε_S is the dielectric constant of the active layer, ε_{OX} is the dielectric constant of the gate insulator, t_S is the thickness of the active layer, and t_{OX} is the thickness of the gate insulator. Here, n represents the effective number of gates per channel. For instance, a single gate is one gate, a dual gate is two gates, a triple gate is three gates, and a quadruple gate is four gates. The drain electric field has less influence on the potential in the channel as λ_n becomes smaller. In the S-ch CAAC-IGZO FET, the gate electrode on the top side of the channel is dominant and n is close to 1 in Equation (2.22), when W is wide. On the contrary, when W is small, the gate electrode on the side surface of the channel is also dominant, n is close to 3 in Equation (2.22), and λ_n is shortened. For that reason, $DIBL$ is improved as W becomes smaller.

Next, the reason why SS is improved with narrower W in Figure 2.63 is discussed based on the result of the device simulation. Sentaurus, produced by Synopsys, Inc., was used for the simulation [33]. Figure 2.69 shows the electron current density distribution of the active layer

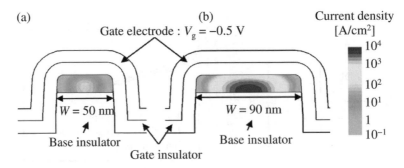

Figure 2.69   Electron current–density distribution of active layer in the W cross-section of a S-ch CAAC-IGZO FET at (a) $W = 50$ nm and (b) $W = 90$ nm

in the W cross-section of the S-ch CAAC-IGZO FET. Here, V_g is −0.5 V and V_d is 1 V. Figure 2.69(b) shows the S-ch CAAC-IGZO FET with W of 90 nm; it can be seen that the current density on the back-channel side of the active layer, which is far from the gate electrode, is increased. In contrast, as seen from Figure 2.69(a) showing the S-ch CAAC-IGZO FET with $W = 50$ nm, the current density on the back-channel side greatly decreases. Accordingly, with narrower W, the current density on the back-channel side can be controlled successfully, which leads to an improvement in SS.

2.6.3 Intrinsic Accumulation-Mode Device

One important factor allowing the CAAC-IGZO FET to suppress the short-channel effect is an accumulation-mode device with an intrinsic channel, in addition to the device structure described above. The potential rises more sharply in the intrinsic accumulation-mode device than in the inversion-mode FET; that is, the characteristic length is short [65]. This is explained analytically below [30].

Figure 2.70 shows a schematic view of a FET structure. Gauss's law is applied to a narrow zone from x to $(x + dx)$, which constitutes a channel portion:

$$-\varepsilon_S t_S \left[-\frac{d\phi(x)}{dx} + \frac{d\phi(x+dx)}{dx} \right] - \varepsilon_{OX} \frac{V_g - V_{FB} - \phi(x)}{t_{OX}} dx = -en_i \, \exp\left\{ \frac{e[\phi(x) - \phi_F]}{k_B T} \right\} t_S dx$$

$$(2.23)$$

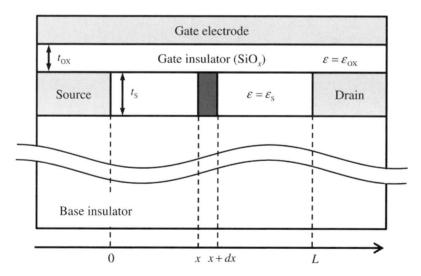

Figure 2.70 Schematic view of a FET structure. *Source*: Reproduced from [30], with permission of *Japanese Journal of Applied Physics*

where ε_S represents the dielectric constant of the semiconductor layer, ε_{OX} represents the dielectric constant of the gate insulator, t_S is the thickness of the semiconductor film, t_{OX} represents the thickness of the gate insulator, ϕ_F represents the Fermi potential, V_{FB} represents the flat band voltage, $\phi(x)$ represents the potential (a surface potential) at position x, and n_i represents the intrinsic carrier density. In addition, the base film is sufficiently thick.

The equation is rewritten as follows:

$$\frac{d^2\phi(x)}{dx^2} + \frac{1}{l^2}\left[V_g - V_{FB} - \phi(x)\right] = \frac{e}{\varepsilon_S}n_i \ \exp\left\{\frac{e[\phi(x) - \phi_F]}{k_B T}\right\}. \tag{2.24}$$

Here, $\dfrac{1}{l^2} = \dfrac{\varepsilon_{OX}}{\varepsilon_S t_S t_{OX}}$ is satisfied, where l is the characteristic length of the inversion-mode FET. The equation cannot be solved analytically, so an approximation is discussed below. Specifically, the right-hand side is expanded using $x = x_1 + x'$ and $\phi(x) = \phi_1 + \Delta\phi(x')$. Figure 2.71 shows the relationship between $x = x_1 + x'$ and $\phi(x) = \phi_1 + \Delta\phi(x')$ in the vicinity of the source electrode.

First, substitute $x = x_1 + x'$ and $\phi(x) = \phi_1 + \Delta\phi(x')$ in Equation (2.24) to obtain the following:

$$\frac{d^2\Delta\phi(x')}{dx'^2} + \frac{1}{l^2}\left[V_g - V_{FB} - \phi_1 - \Delta\phi(x')\right] = \frac{e}{\varepsilon_S}n_i \ \exp\left[\frac{e(\phi_1 - \phi_F)}{k_B T}\right]\exp\left[\frac{e\Delta\phi(x')}{k_B T}\right]$$

$$= \frac{e}{\varepsilon_S}n_1 \ \exp\left[\frac{e\Delta\phi(x')}{k_B T}\right] \tag{2.25}$$

$$\approx \frac{e}{\varepsilon_S}n_i\left[1 + \frac{e\Delta\phi(x')}{k_B T}\right].$$

Here, the electron density n_1 of $x = x_1$ is

$$n_1 = n_i \ \exp\left[\frac{e(\phi_1 - \phi_F)}{k_B T}\right]. \tag{2.26}$$

Thus, Equation (2.24) becomes

$$\frac{d^2\Delta\phi(x')}{dx'^2} + \frac{1}{l^2}\left[V_g - V_{FB} - \phi_1 - \Delta\phi(x')\right] = \frac{e}{\varepsilon_S}n_1 \ \exp\left[\frac{e\Delta\phi(x')}{k_B T}\right]. \tag{2.27}$$

The right-hand side is approximated by the first term of the Taylor expansion, and thus

$$\frac{d^2\Delta\phi(x')}{dx'^2} + \frac{1}{l^2}\left[V_g - V_{FB} - \phi_1 - \Delta\phi(x')\right] = \frac{e}{\varepsilon_S}n_1\left[1 + \frac{e\Delta\phi(x')}{k_B T}\right] \tag{2.28}$$

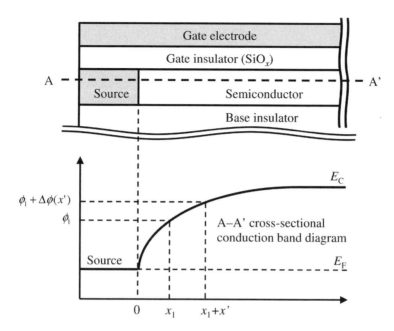

Figure 2.71 Schematic view of the FET structure and conduction band showing the relationship between $x = x_1 + x'$ and $\phi(x) = \phi_1 + \Delta\phi(x')$. *Source*: Reproduced from [30], with permission of *Japanese Journal of Applied Physics*

is obtained. The equation is rearranged by transposition:

$$\frac{d^2\Delta\phi(x')}{dx'^2} - \left(\frac{1}{l^2} + \frac{1}{\lambda_1{}^2}\right)\Delta\phi(x') + \frac{1}{l^2}\left(V_g - V_{FB} - \phi_1\right) - \frac{e}{\varepsilon_S}n_1 = 0. \qquad (2.29)$$

Here, $\dfrac{1}{\lambda_1{}^2} = \dfrac{e^2 n_1}{\varepsilon_S k_B T}$ is satisfied. Furthermore, the equation is rearranged with $\dfrac{1}{l'^2} = \dfrac{1}{l^2} + \dfrac{1}{\lambda_1{}^2}$ to obtain

$$\frac{d^2\Delta\phi(x')}{dx'^2} - \frac{1}{l'^2}\Delta\phi(x') + \frac{1}{l'^2}\left[\frac{l'^2}{l^2}\left(V_g - V_{FB} - \phi_1\right) - l'^2\frac{e}{\varepsilon_S}n_1\right] = 0. \qquad (2.30)$$

The characteristic length l of the inversion-mode FET is represented anew as $l_{(inv)}$, and the characteristic length l' of the intrinsic accumulation-mode FET used as the CAAC-IGZO FET is represented anew as $l_{(acc)}$. They both satisfy the following respective equations:

$$\frac{1}{l_{(\text{inv})}{}^2} = \frac{\varepsilon_{\text{OX}}}{\varepsilon_S t_S t_{\text{OX}}}$$ (2.31)

$$\frac{1}{l_{(\text{acc})}{}^2} = \frac{\varepsilon_{\text{OX}}}{\varepsilon_S t_S t_{\text{OX}}} + \frac{e^2 n_1}{\varepsilon_S k_B T}.$$ (2.32)

By comparison, $l_{(\text{inv})} > l_{(\text{acc})}$ can be seen.

Note that the obtained characteristic length of the intrinsic accumulation-mode FET varies depending on the term to be developed, but in any case, it is shorter than that of the inversion-mode FET. This means that intrinsic accumulation-mode FETs, including CAAC-IGZO FETs, have a larger immunity against the short-channel effect compared with the inversion-mode FETs.

2.6.4 Dielectric Anisotropy

CAAC-IGZO is supposed to exhibit different dielectric anisotropies between its a- and b-axes and its c-axis, because of its layered structure. The dielectric anisotropy of the crystalline IGZO was estimated by DFT calculation [62]. The model structure of InGaZnO$_4$ used for the calculation is shown in Figure 2.72.

The number of atomic layers per unit cell is 28. Periodic boundary conditions were used for the calculation. In the calculation model, Ga and Zn ions alternate, although they are randomly distributed in an actual CAAC-IGZO film. The calculated dielectric constants are shown in Table 2.12; ε^0 is the statics dielectric constant and ε^∞ is the optical dielectric constant of each axis. The ε^0 of the c-axis is larger than those of the a- and b-axes in crystalline IGZO.

The next discussion is on the influence of the dielectric anisotropy of CAAC-IGZO on FET characteristics. The c-axis of the crystal in the CAAC-IGZO film is oriented vertically, and the

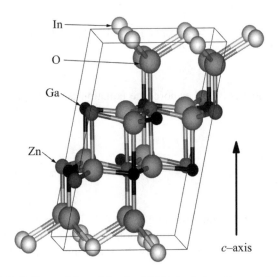

Figure 2.72   InGaZnO$_4$ model used for the DFT calculation. *Source*: Adapted from [62]

channel-length direction is along the *a*- or *b*-axis. As the dielectric constant along the *c*-axis becomes larger, the effective thickness of the active layer becomes smaller, and as the dielectric constant along the *a*- or *b*-axes becomes smaller, the effective channel length becomes longer. Thus, the anisotropic dielectric constant of CAAC-IGZO contributes to the resistance against the short-channel effect. Device simulation was conducted to examine the effect of the dielectric constant on FET characteristics. Figure 2.73(a) shows the I_d–V_g characteristics, where the dielectric constant in the vertical direction of the IGZO layer ε_c is 13.1 and those in the channel-length direction of the IGZO layer ε_{ab} are 5, 10, and 15. Figure 2.73(b) presents the I_d–V_g characteristics where $\varepsilon_{ab} = 8.2$ and $\varepsilon_c = 5$, 10, and 15.

The results of the device simulation elucidate that *SS* decreases when ε_{ab} is small and ε_c is large. This indicates that the dielectric anisotropy of CAAC-IGZO is one of the reasons for the high immunity to the short-channel effect of the CAAC-IGZO FETs.

Table 2.12   DFT calculation results of the relative dielectric constants of single-crystal InGaZnO$_4$. *Source*: Reprinted from [62], with permission of IEEE, © 2015

	a, b-axes		*c*-axis
	XX	YY	ZZ
ε^0	8.20	8.20	13.10
ε^∞	4.16	4.17	4.10
$\varepsilon^\infty - \varepsilon^0$	4.04	4.03	8.99

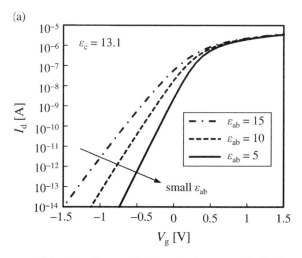

Figure 2.73   Influence of dielectric anisotropy in the active layer upon I_d–V_g characteristics (calculation results of two-dimensional device simulation): (a) dielectric constant along *a*- or *b*-axes; (b) dielectric constant along *c*-axis. *Source*: Adapted from [62]

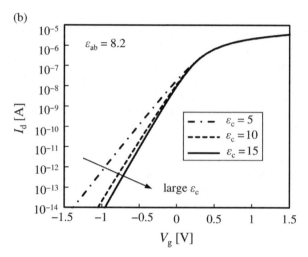

Figure 2.73 (*Continued*)

2.6.5 Numerical Calculation of the Band Diagrams in IGZO FETs

As discussed above, there are several reasons for the large immunity to the short-channel effect of the CAAC-IGZO FETs. By solving Poisson's equation numerically, the influences of these factors on short-channel effect immunity are discussed quantitatively. As discussed in Subsection 2.6.3, we divide the channel into narrow zones and apply Gauss's law to these zones from x to $(x + dx)$:

$$-\varepsilon_S t_S \left[-\frac{d\phi(x)}{dx} + \frac{d\phi(x + dx)}{dx} \right] - \varepsilon_{OX} \frac{V_g - V_{FB} - \phi(x)}{t_{OX}} dx = \rho(x) t_S dx. \tag{2.33}$$

In this equation, $\rho(x)$ represents the charge density of the channel length. In the inversion-mode FET:

$$\rho = -eN_A \tag{2.34}$$

is obtained. In the n-channel accumulation-mode FET:

$$\rho = -en_i \ \exp\left(\frac{e[\phi(x) - \phi_F]}{k_B T} \right). \tag{2.35}$$

Here, $\phi(x)$ represents the intrinsic Fermi level and ϕ_F means the Fermi level. Poisson's equation is converted into a differential equation, and numerically solved by an iterative method. The dispersed Poisson's equation is as follows:

$$\phi_i^{(k)} = \left(2 + \frac{\delta^2}{l^2} \right)^{-1} \left[\phi_{i-1}^{(k)} + \phi_{i+1}^{(k-1)} + \frac{\delta^2}{l^2} (V_g - V_{FB}) + \frac{\rho_i^{(k-1)}}{\varepsilon_S} \delta^2 \right] \tag{2.36}$$

where l is the characteristic length of the inversion-mode FET and δ is the length of the divided narrow zone:

Table 2.13 Definitions and values of physical quantities

Constant	Definition	Value
ε_S	Dielectric constant of the active layer [F/m]	$8.2\varepsilon_0$
ε_{OX}	Dielectric constant of the gate-insulating film [F/m]	$4.1\varepsilon_0$
ε_0	Dielectric constant of vacuum [F/m]	8.854×10^{-12}
t_S	Film thickness of active layer [nm]	15
t_{OX}	Film thickness of gate-insulating film [nm]	10
V_G	Gate voltage [V]	0
V_{FB}	Flat-band voltage [V]	−1.2
e	Elementary charge [C]	1.602×10^{-19}
N_A	Acceptor impurity density [cm^{-3}] (only for inversion-mode FET)	1×10^{15}
n_i	Intrinsic carrier density [cm^{-3}]	6.6×10^{-9}
ϕ_F	Fermi level [eV]	−1.6
k_B	Boltzmann constant [J/K]	1.381×10^{-23}
T	Temperature [K]	300

$$l = \sqrt{\frac{\varepsilon_S t_S t_{OX}}{\varepsilon_{OX}}} \qquad (2.37)$$

$$\delta = \frac{L}{N}. \qquad (2.38)$$

Here, i is a positive integer showing a certain position within the channel, k is a positive integer representing the iteration count of the numerical calculation, and N is the number of narrow divided zones. In addition, $\phi_i^{(k)}$ and $\rho_i^{(k)}$ represent the potential and the charge density at position $x = \delta \times i$, obtained in the kth iteration, respectively. Table 2.13 shows default values of the constants in Equation (2.36).

2.6.5.1 Comparison between Inversion-Mode and Accumulation-Mode FETs

The difference in band diagrams between the inversion-mode and accumulation-mode FETs described in Subsection 2.6.3 is discussed again based on the numerical calculations.

The structure is common to both the accumulation-mode and inversion-mode FETs. The acceptor impurity density in the channel portion of the inversion-mode FET is assumed to be 1×10^{15} cm^{-3} for any channel length. The dielectric constant of the semiconductor is set to 15.0. The band diagram of the conduction band minimum E_C in the channel direction for length L is obtained by plotting $-e\phi(x)$. In Figure 2.74, band diagrams of the accumulation-mode FET and the inversion-mode FET, the channel lengths of which are 500 nm, are compared.

As can be seen in Figure 2.74, the band minimum energy of the accumulation-mode FET changes more sharply close to the ends of the channel compared with the inversion-mode FET. In both FETs, the band barrier in the source and drain electrodes and the channel is as high as the difference between the work function of the gate electrode and the electron affinity

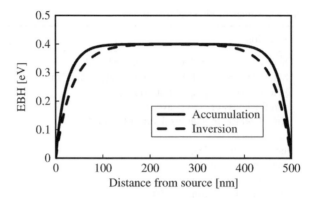

Figure 2.74   Comparison of band diagrams at $L = 500$ nm

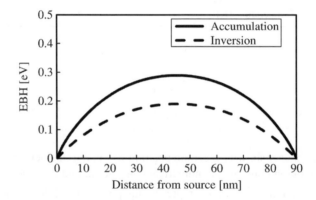

Figure 2.75   Minimum conduction band energy of a FET with $L = 90$ nm in the channel direction (comparison between an accumulation-mode FET and an inversion-mode FET)

of the semiconductor, when the channel region is weakly doped. Thus, the height of the potential barrier to electrons is suitable as an indicator for immunity to the short-channel effect. A band diagram of a FET with $L = 90$ nm in the channel-length direction is shown in Figure 2.75.

The energy barrier between the source region and the channel, which is a maximum of the band in the channel, is smaller in the inversion-mode FET than in the accumulation-mode FET. This means that the short-channel effect is smaller in the accumulation-mode FET than in the inversion-mode FET. The maximum in the band diagram at the middle of the channel is referred to as the energy barrier height (EBH) hereafter, and the index is used to examine the short-channel effect. EBH is given by the following equation:

$$EBH = \max[-e\phi(x)]. \tag{2.39}$$

The EBHs of the inversion-mode FET and the accumulation-mode FET are plotted as functions of channel length (see Figure 2.76).

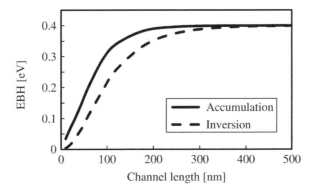

Figure 2.76 Dependence of EBH on channel length (comparison between inversion- and accumulation-mode FETs)

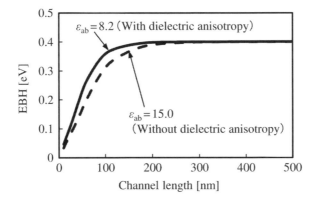

Figure 2.77 Dependence of EBH on channel length (impact of dielectric anisotropy)

The EBH decreases sharply below $L = 180$ nm in the inversion-mode FET, while the EBH starts to decrease below $L = 120$ nm in the accumulation-mode FET. This result indicates that the intrinsic accumulation-mode FET has a higher immunity to the short-channel effect compared with the inversion-mode FET.

2.6.5.2 Influence of the Dielectric Anisotropy

The ε_S in Equation (2.33) is a dielectric constant in the channel-length direction, and the effect of the dielectric anisotropy of CAAC-IGZO is naturally taken into account. The EBH at $\varepsilon_S = 8.2$ and the EBH at $\varepsilon_S = 15.0$ are compared in Figure 2.77. The EBH is prevented from decreasing due to the short-channel effect in consideration of the dielectric anisotropy. In other words, the dielectric anisotropy of CAAC-IGZO acts to increase the resistance to the short-channel effect.

In this calculation, when considering the dielectric anisotropy, the EBH decreases by only 1% at $L = 180$ nm when the gate thickness is kept at 10 nm. Thus, the short-channel effect hardly occurs. However, when L is smaller than 100 nm, the EBH decreases noticeably. Under short channel lengths of 100 nm or smaller, the short-channel effect cannot be prevented in the planar CAAC-IGZO FET with 10 nm thickness of the gate insulator. Therefore, an S-ch structure, as described in Subsection 2.6.1, is needed.

2.6.5.3 Impact of the Thickness of the Active Layer

The planar CAAC-IGZO FET can be scaled down to about 180 nm (in channel length) without scaling down the thickness of the gate insulator, which is far smaller than that which can be achieved with Si FETs.

One reason for its resistance to the short-channel effect of the CAAC-IGZO FET is that the active layer of the CAAC-IGZO film is very thin.

Also, in Si LSI technology, a wafer with a thin active layer using a buried oxide layer is used, which is known as silicon-on-insulator (SOI) technology. SOI is a known technology to prevent the short-channel effect, like the 3D gate structure of a Fin FET. For the CAAC-IGZO FET, a thin active layer of about 15 nm can be steadily deposited by sputtering, offering high immunity to the short-channel effect.

To examine how the thickness of the active layer impacts the immunity against the short-channel effect, a planar CAAC-IGZO FET with a 100-nm-thick active layer and a planar CAAC-IGZO FET with a 15-nm-thick active layer were assumed, and the dependences of the EBH on channel length were calculated for comparison.

As shown in Figure 2.78, the EBH decreased rapidly with scaling down of L when $t_S =$ 100 nm. Thus, the thinness of the active layer in the CAAC-IGZO FET greatly contributes to its high immunity against the short-channel effect.

2.6.5.4 Impact of 3D Gate Structure

A 3D gate structure is discussed below in view of the EBH. Poisson's equation for the accumulation-mode FET with the S-ch structure, considering channel width and the gate electrode on the side of the channel, is given as follows:

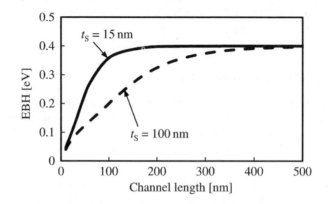

Figure 2.78 Dependence of EBH on channel length (impact of thickness of active layer)

$$-\varepsilon_S t_S W_{ch}\left[-\frac{d\phi(x)}{dx}+\frac{d\phi(x+dx)}{dx}\right]-(W_{ch}+2t_S)\varepsilon_{OX}\frac{V_g-V_{FB}-\phi(x)}{t_{OX}}dx$$
$$=-en_i\exp\left(\frac{e[\phi(x)-\phi_F]}{k_B T}\right)t_S W_{ch}dx.$$

(2.40)

This can be rearranged to

$$\frac{d^2\phi(x)}{dx^2}+\left(1+\frac{2t_S}{W_{ch}}\right)\frac{\varepsilon_{OX}}{\varepsilon_S t_S t_{OX}}\left[V_g-V_{FB}-\phi(x)\right]-\frac{en_i}{\varepsilon_S}\exp\left(\frac{e[\phi(x)-\phi_F]}{k_B T}\right)=0.$$

(2.41)

The factor describing the shape of the channel is defined as

$$G=1+\frac{2t_S}{W_{ch}}.$$

(2.42)

As the channel width W decreases, this factor G becomes larger. A large G may offer immunity against the short-channel effect. The impact of G on the immunity to the short-channel effect can also be discussed based on the characteristic length in the same manner as shown in Subsection 2.6.3. The characteristic length of the S-ch accumulation-mode FET is

$$\frac{1}{l_{(acc)}{}^2}=G\frac{\varepsilon_{OX}}{\varepsilon_S t_S t_{OX}}+\frac{e^2 n_1}{\varepsilon_S k_B T}.$$

(2.43)

The large G gives a short characteristic length and is expected to provide high immunity against the short-channel effect.

G is 2.0 when t_S is 15 nm and W is 30 nm. Note that in Equation (2.41), the shape of the gate electrode below the S2 layer is not taken into consideration, and G is estimated to be larger than 2.0 and smaller than 3.0 ($2.0 < G < 3.0$). Figure 2.79 shows the results of the comparison between the dependences of the EBH on channel length in the cases where $G = 1.0$ (planar type), $G = 2.0$, and $G = 3.0$. The higher G offers a higher EBH in short channels, which means

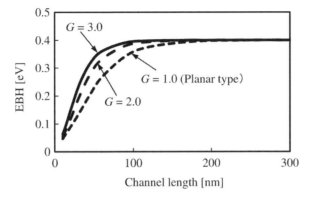

Figure 2.79   Dependence of EBH on channel length (impact of S-ch structure)

that the short-channel effect is prevented by the S-ch structure. G is saturated at 1.0 when the channel width is large, and G increases as the channel width narrows. This behavior of G supports that shown in Figure 2.63 – i.e., that the short-channel effect (increase in SS) is prevented by narrowing the channel width.

2.6.5.5 Impact of Thinning the Gate Insulator

Lastly, thinning of the gate insulator (GI) is discussed. As mentioned above, the thickness of the gate insulator of the CAAC-IGZO FET is fixed at 10 nm, and the channel length is scaled down to realize the extremely low off-state current. Making the gate insulator thinner is an effective method to prevent the short-channel effect, although the thinning causes an increase in the tunnel current. In the case where an off-state current much lower than that in a Si FET is sufficient, and the off-state current at yA/um order may not be needed, thinning of the gate-insulating film can be an option to prevent the short-channel effect.

The dependences of the EBH on channel length in planar FETs are shown in Figure 2.80, when the thicknesses of the gate insulators are 1, 2, 6, and 10 nm. The thinning of the gate insulator can avoid a decrease of the EBH. If the thickness of the gate insulator is 1 nm, the decrease rate of the EBH will only be 10% at $L = 32$ nm, even in the planar FET, which will provide favorable FET characteristics.

Aiming for further scaling down, it is necessary to make the gate insulator thinner while ensuring that the off-state current remains low enough to satisfy the specifications needed for applications.

2.6.6 Summary

The miniaturized S-ch CAAC-IGZO FET has a channel length of about 30 nm, even with a relatively thick EOT of 11 nm [62]. Several reasons why the FET has immunity against the short-channel effect have been discussed using device simulation. The manner in which the difference in the width W of the CAAC-IGZO active layer influences the controllability of the active layer in a gate has been considered. Moreover, in the above subsections, the

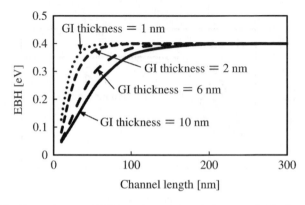

Figure 2.80   Dependence of EBH on channel length (impact of thinning of GI film)

following factors contribute to the suppression of the short-channel effect: the fact that the CAAC-IGZO FET is an accumulation-mode device, the fact that the CAAC-IGZO film has dielectric anisotropy, the thickness of the active layer, the 3D gate structure called S-ch.

The notable point is that the extremely low off-state current is unchanged, even when the S-ch CAAC-IGZO FET is miniaturized. This feature makes it possible to realize super-low-power LSIs, which has been difficult for Si FETs [13–15,65].

2.7   20-nm-Node CAAC-IGZO FET

This section introduces the latest miniaturized CAAC-IGZO FET [61]. In the TGTC structure, a gate electrode overlaps the source/drain electrodes, and thus there is a limitation on scaling down because of the alignment margin of the gate [62]. The miniaturized CAAC-IGZO FET introduced in this section is a TGSA CAAC-IGZO FET with a channel length of less than 30 nm. Source/drain electrodes are formed in a self-aligned manner at the formation of a trench, and a gate electrode is formed inside the trench. Therefore, the gate electrode does not overlap with the source/drain electrodes. The TGSA CAAC-IGZO FET has a channel length of less than 30 nm and an EOT of 11 nm. It exhibits good short-channel immunity (*DIBL* of 0.12 V/V and *SS* of 97 mV/decade), high frequency (cut-off frequency f_T > 30 GHz), and an extremely low off-state current ($I_{off} \ll 0.1$ pA).

2.7.1   TGSA CAAC-IGZO FET

The TGSA CAAC-IGZO FET was fabricated by the process shown as a flow diagram in Figure 2.81. Figure 2.82(a) and (b) shows perspective and planar views, where the gate and source/drain (S/D) electrode patterns are formed in a self-aligned manner using a trench. The FET has no gate-to-S/D overlaps, and is substantially free from disadvantages such as increases in gate parasitic capacitance, extra masks, and difficulty in miniaturization due to

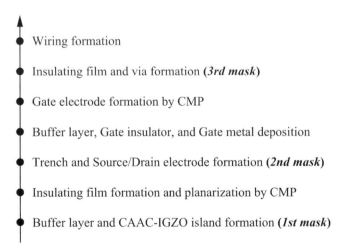

Wiring formation

Insulating film and via formation (*3rd mask*)

Gate electrode formation by CMP

Buffer layer, Gate insulator, and Gate metal deposition

Trench and Source/Drain electrode formation (*2nd mask*)

Insulating film formation and planarization by CMP

Buffer layer and CAAC-IGZO island formation (*1st mask*)

Figure 2.81   Fabrication process of TGSA CAAC-IGZO FET. *Source*: Adapted from [61]

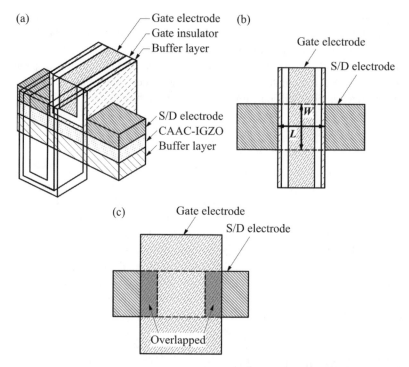

Figure 2.82   (a) Perspective and (b) planar view of the TGSA CAAC-IGZO FET; (c) planar view of the TGTC structure. *Source*: Reprinted from [61], with permission of IEEE, © 2015. *(For detail, please see color plate section)*

Table 2.14   Comparison between TGTC and TGSA. *Source*: Reprinted from [61], with permission of IEEE, © 2015

FET structure	TGTC [62]	TGSA [61]
Gate – S/D electrodes	Overlapped	Self-aligned
Parasitic gate capacitance	Large	Small
Number of masks with via	4	3
Scalability	Not good	Better

the gate-alignment margin, unlike the TGTC structure where the gate electrode partly overlaps the source/drain electrodes [see Figure 2.82(c)] [61]. Compared with the TGTC structure, the gate parasitic capacitance and the number of masks are reduced in the TGSA structure. The number of masks used for the TGSA CAAC-IGZO FET is three, including the mask for forming a via. In addition, there is no need to consider the gate-alignment margin; thus, scaling down is possible by simply shortening the width of the trench (see Table 2.14).

Figure 2.83(a) and (b) shows cross-sectional STEM images in the channel-length direction and the channel-width direction, respectively. Figure 2.83(a) elucidates how the FET is successfully scaled down to less than 30 nm in its channel length. The 15-nm-thick CAAC-IGZO film serving as an active layer is deposited by means of DC sputtering with a polycrystalline IGZO target with an In-rich composition in an Ar/O_2 atmosphere at a substrate temperature of

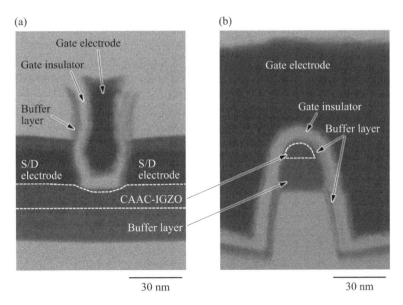

Figure 2.83   Cross-sectional STEM images in the channel-length direction (a) and the channel-width direction (b) of the TGSA CAAC-IGZO FET. *Source*: Reprinted from [61], with permission of IEEE, © 2015. *(For detail, please see color plate section)*

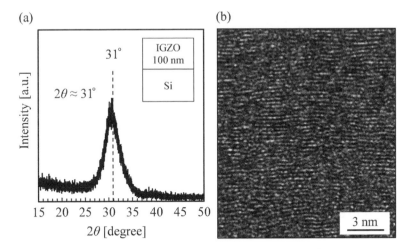

Figure 2.84   (a) Out-of-plane XRD spectrum of the CAAC-IGZO film and (b) cross-sectional TEM image of CAAC-IGZO film. *Source*: Reprinted from [61], with permission of IEEE, © 2015

200°C. Figure 2.84(a) and (b) shows an out-of-plane XRD and a cross-sectional STEM image of a 100-nm CAAC-IGZO film deposited on a Si substrate under the same conditions as those of the active layer of the TGSA CAAC-IGZO FET. A diffraction peak of (009), which derives from the CAAC, is observed at around $2\theta = 31°$ in Figure 2.84(a), and a layered structure of the IGZO crystal perpendicular to the silicon substrate is observed in Figure 2.84(b).

2.7.2 Device Characteristics

In this subsection, the device characteristics of the miniaturized TGSA CAAC-IGZO FET are explained. Figure 2.85(a) shows the I_d–V_d characteristics of the TGSA CAAC-IGZO FET with $W/L = 32\,nm\,/27\,nm$, where data are obtained for V_g in the range from 0.0 to 2.0 V in 0.2-V steps. Figure 2.85(b) shows the I_d–V_g characteristics of the same FET where the drain voltage V_d is 0.1 and 1.2 V. Good saturation characteristics and an off-state current of less than 0.1 pA are obtained. The obtained off-state current is less than 0.1 pA, which is below the lower detection limit of a semiconductor parameter analyzer. $DIBL$ is 0.12 V/V and SS at $V_d = 1.2\,V$ is 97 mV/decade, which are good values despite the short channel length of 27 nm and $EOT = 11$ nm. To examine whether the CAAC-IGZO FET with $W/L = 32\,nm\,/27\,nm$ can

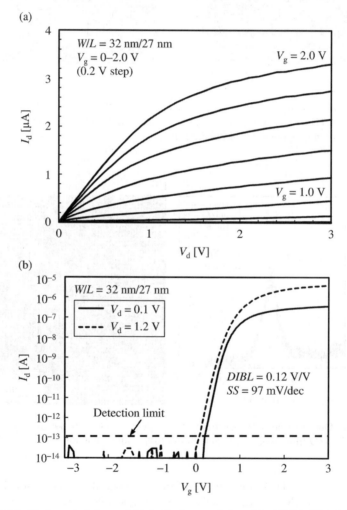

Figure 2.85 (a) I_d–V_d characteristics and (b) I_d–V_g characteristics of the TGSA CAAC-IGZO FET with $W/L = 32\,nm\,/27\,nm$. *Source*: Reprinted from [61], with permission of IEEE, © 2015

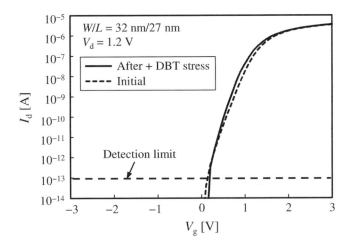

Figure 2.86 Result of the +DBT stress test (V_d = 1.8 V, V_g = 0.0 V, 150°C, 1 h) of a TGSA CAAC-IGZO FET with W/L = 32 nm /27 nm. *Source*: Reprinted from [61], with permission of IEEE, © 2015

be applied to memories, a positive drain bias temperature (+DBT) stress test (V_d = 1.8 V, V_g = 0.0 V, 150°C, during 1 hour) was conducted (see Figure 2.86). As a result, almost no degradation was seen, and the off-state current was less than 0.1 pA, even after stress. Figures 2.87(a), (b) and 2.88 show the dependences of V_{th}, V_{sh}, and SS on the channel length L of the TGSA CAAC-IGZO FETs with W = 32 nm ($\triangle$) and W = 62 nm ($\bigcirc$), respectively. Five samples were measured for each L. The filled symbol is a median. When W = 32 nm, the roll-offs of V_{th} and V_{sh} were small, and no SS degradation was observed, even with smaller L. Consequently, V_{sh} is higher than 0 V, which means that the FET is normally off even when L = 27 nm. The TGSA CAAC-IGZO FET with W = 32 nm is found to exhibit strong short-channel effect immunity. In contrast, there is a remarkable SS degradation by decreasing L when W = 62 nm [see Figure 2.88(b)], resulting in a large V_{sh} roll-off [see Figure 2.87(b)]. This is because the effect of the side gate becomes weaker with wider W and leakage current through a back channel increases as described in Section 2.6 (refer to the device simulation result for the current density distribution at V_g = −0.5 V, as shown in Figure 2.89(a) and (b) [62]). However, in light of the on-current I_{on}, the FET with W = 62 nm is superior, as shown in Figure 2.90. One reason why I_{on} per channel width is larger in the FET with W = 62 nm than in the FET with W = 32 nm is assumed to be an increase in parasitic capacitance due to W reduction. To examine the frequency characteristics, TGSA CAAC-IGZO FETs were arranged in parallel and the cut-off frequency f_T was measured. As mentioned, f_T is defined as the frequency when the current gain is 1 and serves as an indicator of the switching speed of the FETs. Figure 2.91 shows the transconductance g_m of 300 parallelized FETs with W/L = 62 nm /27 nm. At V_g corresponding to the maximal g_m, f_T was estimated as follows: 21.4 GHz at V_d = 1.2 V and 34.4 GHz at V_d = 2.0 V. Figure 2.92 indicates that f_T improves as the channel length decreases. This means that TGSA FETs obtain an additional benefit from scaling. Furthermore, when there are 600 parallelized FETs with W/L = 32 nm /27 nm, f_T is larger than 10 GHz, which is sufficiently high for application to LSIs.

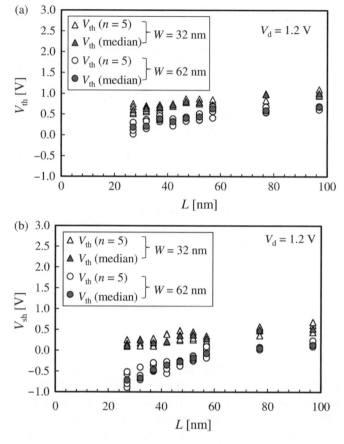

Figure 2.87 Dependences on channel length L of (a) threshold voltage V_{th} and (b) shift voltage V_{sh} (comparison between $W = 32$ nm and 62 nm). *Source*: Reprinted from [61], with permission of IEEE, © 2015

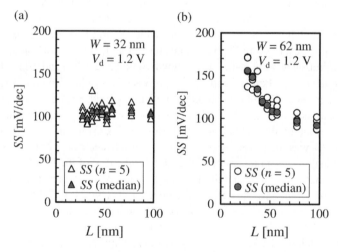

Figure 2.88 Dependences of SS upon channel length L: (a) $W = 32$ nm and (b) $W = 62$ nm. *Source*: Reprinted from [61], with permission of IEEE, © 2015

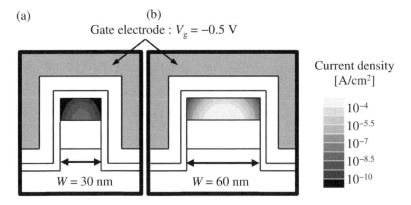

Figure 2.89 Current density distribution in cross-section in the channel-width direction of the TGSA CAAC-IGZO FET: (a) $W = 30$ nm and (b) $W = 60$ nm. *Source*: Reprinted from [61], with permission of IEEE, © 2015

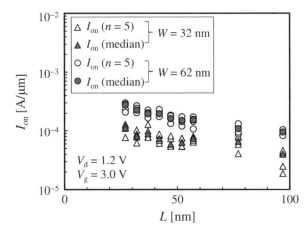

Figure 2.90 Dependence of I_{on} on channel length L at $V_d = 1.2$ V and $V_g = 3.0$ V (comparison between $W = 32$ nm and 62 nm). *Source*: Reprinted from [61], with permission of IEEE, © 2015

2.7.3 Memory-Retention Characteristics

The retention characteristics of a memory using the TGSA CAAC-IGZO FET illustrated in the circuit diagram of Figure 2.93(a) was measured by monitoring the voltage drop of a floating node V_F. The size of the FET is 51 nm in W and 55 nm in L, and the retention capacitance is 20 fF. To utilize an extremely low I_{off}, a voltage (V_g) of −1.5 V is applied. As illustrated in Figure 2.93(b), V_F decreased by only 0.3 V, even after 24 h (86,400 s) at 125°C. Based on this result, the average I_{off} is estimated to be 6.9×10^{-20} A for 24 h. This is lower than one electron per second.

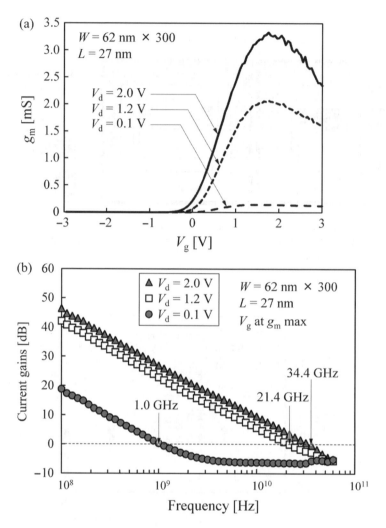

Figure 2.91 (a) Transconductance g_m and (b) current gain of 300 parallelized TGSA CAAC-IGZO FETs with $W/L = 62\,nm\,/27\,nm$ channels connected in parallel. *Source:* Reprinted from [61], with permission of IEEE, © 2015

The time dependence of the V_F drop was approximated by the stretched exponential function, and the relaxation time τ, initial voltage V_0, and power-law coefficient β were extracted. An Arrhenius plot of τ for several temperatures is shown in Figure 2.94, from which the activation energy (E_a) of τ was found to be 1.15 eV. When τ is used as an indicator for the memory-retention time, a retention time of 1×10^8 s (about 3 years) is expected, even at 85°C.

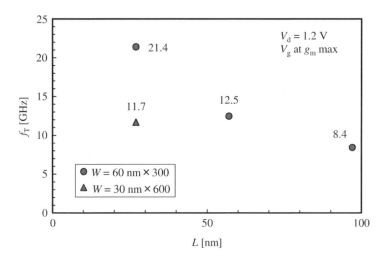

Figure 2.92 Channel-length dependence of cut-off frequency for $V_d = 1.2\,V$. *Source*: Reprinted from [61], with permission of IEEE, © 2015

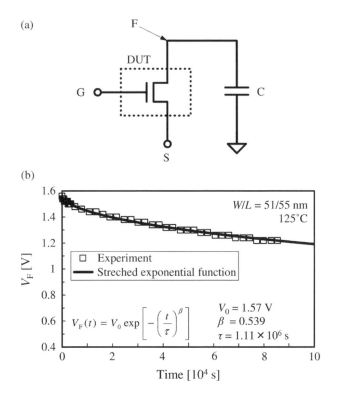

Figure 2.93 (a) Circuit diagram of memory using a TGSA CAAC-IGZO FET with $W/L = 51\,nm/55\,nm$ and (b) memory retention characteristics at 125°C and $V_g = -1.5\,V$. *Source*: Reprinted from [61], with permission of IEEE, © 2015

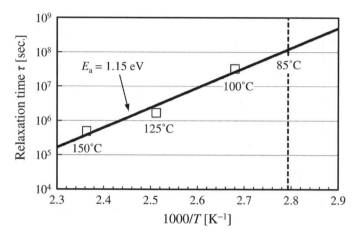

Figure 2.94 Arrhenius plot of the relaxation time τ of the memory using the TGSA CAAC-IGZO FET. *Source*: Reprinted from [61], with permission of IEEE, © 2015

2.7.4 Summary

A TGSA CAAC-IGZO FET that can be fabricated with only three masks and is suitable for scaling down has been proposed in this section. The 20-nm-node CAAC-IGZO FET with an EOT of 11 nm exhibited good short-channel characteristics, high-frequency characteristics of 10 GHz or higher, and extremely low off-state current. It was confirmed that a memory circuit could be formed with the TGSA CAAC-IGZO FETs, and the memory-retention time was estimated to be three years at 85°C with an extrapolated value. Therefore, the TGSA CAAC-IGZO FET may be a key device for realizing super-low-power LSIs.

2.8 Hybrid Structure

As described above, CAAC-IGZO FETs exhibit extremely low off-state current and short-channel-effect immunity. Examples of applications of the CAAC-IGZO FET are introduced specifically in Chapters 3 to 8. Such devices can be fabricated by replacing a part of the circuitry consisting of Si FETs with CAAC-IGZO FETs or by adding a CAAC-IGZO FET to a circuit with Si FETs [15,22–26]. In this book, the combined structure of a CAAC-IGZO FET and a Si FET is referred to as a hybrid structure.

Combining Si and CAAC-IGZO FETs is not necessarily a complicated process, because CAAC-IGZO films can be deposited even on an amorphous insulating film (please refer to *Fundamentals* [8]). Thus, a CAAC-IGZO FET fabrication process can be performed after a Si FET fabrication process is finished. In addition, a second and even a third layer of CAAC-IGZO FETs can be stacked on the Si FET layer. Figure 2.95 is a cross-sectional image of a prototype of two CAAC-IGZO FET layers stacked on the Si FET layer. Such a stack structure can lead to a reduction in chip area.

Below, the processing technology for fabricating LSI devices is described. First, the fabrication processes of a TGTC FET and a TGSA FET are explained and then an example of a hybrid structure in which the CAAC-IGZO FET is stacked on the Si FET is introduced.

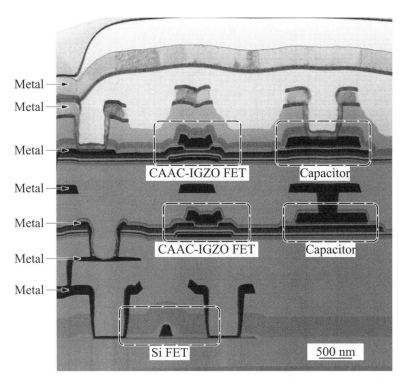

Metal
Metal
Metal
Metal
Metal
Metal
Metal

CAAC-IGZO FET   Capacitor

CAAC-IGZO FET   Capacitor

Si FET   500 nm

Figure 2.95 Cross-sectional STEM image of a prototype with two CAAC-IGZO FET layers stacked on Si FET. Channel lengths of the Si and CAAC-IGZO FETs are 0.2 μm and 0.4 μm, respectively. The image is retouched by filling voids in the interlayer film. *(For detail, please see color plate section)*

2.8.1   TGTC Structure

Figure 2.96 illustrates the process flow of a TGTC FET with a top gate and source/drain electrodes that are both in contact with the top surface of a CAAC-IGZO layer. The left and right parts of Figure 2.96 show cross-sectional views in the channel-length (L) and channel-width (W) directions, respectively. The process flow is from bottom to top.

A CAAC-IGZO film and a metal film serving as source/drain electrodes are formed on a substrate. Then, the CAAC-IGZO film is etched using the metal film serving as source/drain electrodes as a mask [Figure 2.96(a-1), (a-2), CAAC-IGZO island formation, the first mask]. Next, the metal film is etched into a slit to form the source/drain electrodes [Figure 2.96(b-1), (b-2), channel formation, the second mask]. A gate insulator and a metal film serving as a gate electrode are formed. The metal film is etched to form the gate electrode [Figure 2.96(c-1), (c-2), the third mask]. Then, an insulating film is formed and planarized. A contact hole is formed in the insulating film [Figure 2.96(d-1), (d-2), the fourth mask]. Finally, a metal material is used to fill the contact hole to form a via [Figure 2.96(e-1), (e-2)].

In this manner, the TGTC structure can be formed with only four masks. If a back gate is desirable, an additional mask step is necessary. As seen in Figure 2.96(e-2), the channel is

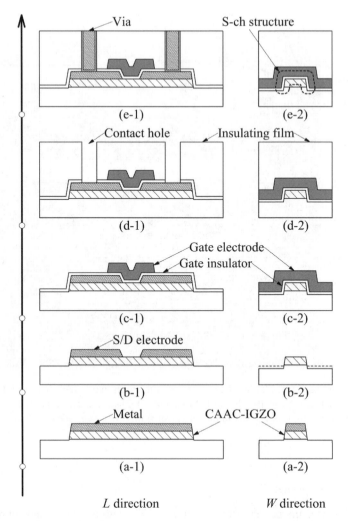

Figure 2.96 Process flow of the TGTC structure (left: channel-length L direction, right: channel-width W direction): (a) CAAC-IGZO patterning, (b) source/drain (S/D) electrode formation, (c) gate insulator and gate electrode formation, (d) insulating film and contact hole formation, (e) via formation

surrounded by the gate electrode in the channel-width direction – i.e., a so-called surrounded-channel (S-ch) structure.

2.8.2   TGSA Structure

Next, the process flow of the TGSA structure is described with reference to Figure 2.97. In this structure, a gate electrode and source/drain electrodes can be formed with use of one mask, which means that the number of masks is one fewer compared with the TGTC structure. The process flow and cross-sections are shown in Figure 2.97, similar to Figure 2.96.

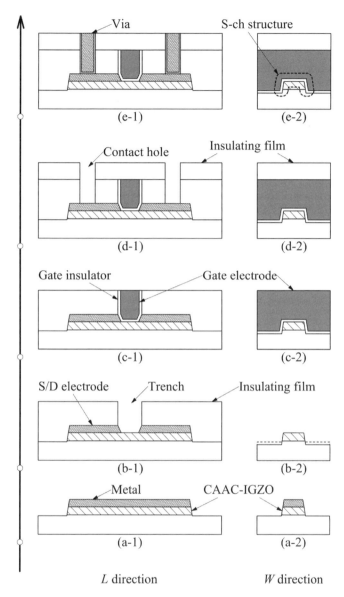

Figure 2.97 Process flow of the TGSA structure: (a) CAAC-IGZO patterning, (b) trench and source/drain (S/D) electrode formation, (c) gate insulator and gate electrode formation, (d) insulating film and contact hole formation, (e) via formation

The simple process flow of the TGSA structure of the CAAC-IGZO FET is as follows. The first step is identical to the TGTC case [Figure 2.97(a-1), (a-2), the first mask]. After, an insulating film is deposited and planarized. This film is then etched to form a trench and the metal is successively etched to form the source/drain electrodes [Figure 2.97(b-1), (b-2), the second mask]. Next, the trench is filled with the gate insulator and the metal material serving as a gate electrode, and the gate electrode is formed by chemical mechanical polishing (CMP) [see Figure 2.97(c-1), (c-2)]. Then, an insulating film is deposited and a contact hole is formed through the insulating film [Figure 2.97(d-1), (d-2), the third mask]. Finally, the contact holes are filled with a metal to form a via [Figure 2.97(e-1), (e-2)].

The TGSA structure can be formed with only three masks, but an additional mask would be needed to form a back gate.

In the TGSA process, the metal film serving as the source/drain electrodes is separated by etching in the trench formation, and the gate insulator and the gate electrode are formed by filling the trench, thereby forming the source/drain electrodes and the gate electrode in a self-alignment manner. The overlap between the gate electrode and the source/drain electrodes can be reduced or completely eliminated, and thus the parasitic capacitance between them can be reduced. As described in Section 2.7, the TGSA is particularly suitable for miniaturization, as illustrated by the electrical characteristics of the 20-nm-node CAAC-IGZO FET. As shown in the cross-sectional view in the W direction in Figure 2.97(e-2), the TGSA has an S-ch structure, which effectively reduces the short-channel effect (as previously discussed).

2.8.3 Hybrid Structure

In accordance with the process flow of the CAAC-IGZO FET described above, a hybrid structure device can be fabricated by formation of the CAAC-IGZO FET following completion of the Si FET process. Figure 2.98 illustrates an example of a cross-sectional view of such a hybrid

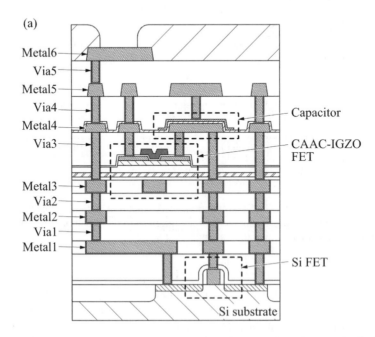

Figure 2.98   (a) Hybrid structure of the Si FET and CAAC-IGZO FET, (b) CAAC-IGZO FET

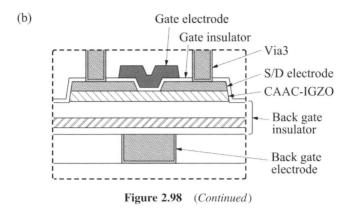

(b)

Figure 2.98   (*Continued*)

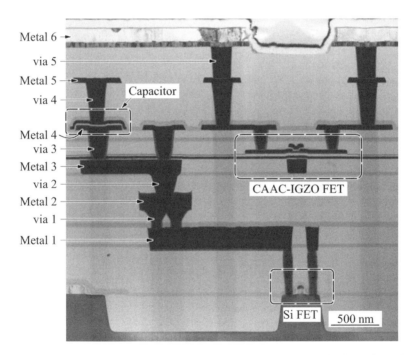

Figure 2.99   Cross-sectional STEM image of hybrid structure of Si FET with channel length of 65 nm and CAAC-IGZO FET with channel length of 60 nm (cooperative work with UMC). (*For detail, please see color plate section*)

structure. Figure 2.99 is a cross-sectional STEM image of a stacked structure in which a CAAC-IGZO FET with a channel length of 60 nm is stacked on a Si FET with a channel length of 65 nm. Components including the Si FET and Metal 2 were formed by United Microelectronics Corporation (UMC), and components including the CAAC-IGZO FET and other wirings were formed by Semiconductor Energy Laboratory Co., Ltd.

The following chapters introduce LSI devices to which the hybrid structure is applied.

Appendix: Comparison between CAAC-IGZO and Si

Table 2.15 Comparison between CAAC-IGZO and crystalline Si. *Source*: Adapted from [18,28]

	CAAC-IGZO	c-Si
1. Bandgap [eV]	2.8–3.2	1.12
2. Intrinsic carrier density [cm^{-3}]	10^{-9}	10^{11}
3. Debye length	km order	μm order
4. Conductivity type	i-type, n-type	n-type, i-type, p-type
5. Effective mass of electrons (m_e^*/m_e)	0.23–0.25	0.19 (transverse)
		0.98 (longitudinal)
6. Effective mass of holes (m_h^*/m_e)	11–40	0.16 (light)
		0.49 (heavy)

Table 2.16 Comparison between a CAAC-IGZO FET and a Si FET – device physics. *Source*: Adapted from [28]

	CAAC-IGZO FET	c-Si FET
1. Terminal	Four-terminal circuit	Four-terminal circuit
2. Short-channel effect	Small	Large
3. V_{th} control by impurity doping	Impossible	Possible
4. I_{on}/I_{off} ratio	10^{20}	10^{10}
5. CMOS fabrication	Possible when combined with Pch-Si FET	Possible
6. Gate structure in scaled FET	S-ch (Fin) structure	Fin structure

Table 2.17 Comparison between a CAAC-IGZO FET and an Nch-Si FET. *Source*: Adapted from [28]

		CAAC-IGZO FET	Nch-Si FET
1. Short-channel effect		Small	Large
2. Scaling		Possible down to 5 nm	Down to 5 nm
3. Off-state current		yA/μm order	fA/μm order
4. Mobility	Long channel	Low	High
	Short channel	Decreases slightly	Decreases significantly
5. Temperature dependence	On-state current	Increasing with temperature	Decreasing with temperature
	Off-state current	Low enough at high temperature	Drastically increases at high temperature
6. Drain breakdown voltage		High	Low
7. Impact ionization		Not observed	Observed
8. Hot carrier degradation		Not observed	Observed
9. Punch through		Not observed	Observed
10. Cut-off frequency		>30 GHz	>300 GHz
11. Drift velocity of electron		Close to Si below channel length of 10 nm	High, but saturates below channel length of 100 nm
12. Subthreshold swing (SS)		Thick gate insulator is acceptable	Very thin gate insulator is essential
13. Dielectric anisotropy		Observed	Not observed
14. S-ch (Fin) structure		Effective	Effective

References

[1] Kimizuka, N. and Mohri, T. (1985) "Spinel, $YbFe_2O_4$, and $Yb_2Fe_3O_7$ types of structures for compounds in the In_2O_3 and $Sc_2O_3–A_2O_3–BO$ systems [A: Fe, Ga, or Al; B: Mg, Mn, Fe, Ni, Cu, or Zn] at temperatures over 1000°C," *J. Solid State Chem.*, **60**, 382.

[2] Nakamura, M., Kimizuka, N., and Mohri, T. (1991) "The phase relations in the $In_2O_3–Ga_2ZnO_4–ZnO$ system at 1350°C," *J. Solid State Chem.*, **93**, 298.

[3] Orita, M., Takeuchi, M., Sakai, H., and Tanji, H. (1995) "New transparent conductive oxides with $YbFe_2O_4$ structure," *Jpn. J. Appl. Phys.*, **34**, L1550.

[4] Nomura, K., Ohta, H., Ueda, K., Kamiya, T., Hirano, M., and Hosono, H. (2003) "Thin-film transistor fabricated in single-crystalline transparent oxide semiconductor," *Science*, **300**, 1269.

[5] Nomura, K., Ohta, H., Ueda, K., Kamiya, T., Hirano, M., and Hosono, H. (2003) "Electron transport in $InGaO_3(ZnO)_m$ (m = integer) studied using single-crystalline thin films and transparent MISFETs," *Thin Solid Films*, **445**, 322.

[6] Nomura, K., Kamiya, T., Ohta, H., Ueda, K., Hirano, M., and Hosono, H. (2004) "Carrier transport in transparent oxide semiconductor with intrinsic structural randomness probed using single-crystalline $InGaO_3(ZnO)_5$ films," *Appl. Phys. Lett.*, **85**, 1993.

[7] Yamazaki, S., Hirohashi, T., Takahashi, M., Adachi, S., Tsubuku, M., Koezuka, J., *et al.* (2014) "Back-channel-etched thin-film transistor using c-axis-aligned crystal In–Ga–Zn oxide," *J. Soc. Inf. Disp.*, **22**, 55.

[8] Yamazaki, S. and Kimizuka, N. (in press) *Physics and Technology of Crystalline Oxide Semiconductor CAAC-IGZO: Fundamentals.* New York: John Wiley.

[9] Ishihara, N., Tsubuku, M., Nonaka, Y., Watanabe, R., Inoue, K., Shishido, H., *et al.* (2012) "Optical properties and evaluation of localized level in gap of In–Ga–Zn–O thin film," *Proc. AM-FPD'12 Dig.*, 143.

[10] Kato, K., Shionoiri, Y., Sekine, Y., Furutani, K., Hatano, T., Aoki, T., *et al.* (2012) "Evaluation of off-state current characteristics of transistor using oxide semiconductor material, indium–gallium–zinc oxide," *Jpn. J. Appl. Phys.*, **51**, 021201.

[11] Kobayashi, Y., Matsubayashi, D., Hondo, S., Yamamoto, T., Okazaki, Y., Nagai, M., *et al.* (2015) "Effect of surrounded-channel structure on electrical characteristics of c-axis aligned crystalline In–Ga–Zn–O field-effect transistor," *IEEE Electron Device Lett.*, **36**, 309.

[12] Matsuo, T., Mori, S., Ban, A., and Imaya, A. (2014) "Advantages of IGZO oxide semiconductor," *SID Symp. Dig. Tech. Pap.*, **45**, 83.

[13] Aoki, T., Ikeda, M., Kozuma, M., Tamura, H., Kurokawa, Y., Ikeda, T., *et al.* (2011) "Electronic global shutter CMOS image sensor using oxide semiconductor FET with extremely low off-state current," *IEEE Symp. VLSI Technol. Dig. Tech. Pap.*, 174.

[14] Inoue, H., Matsuzaki, T., Nagatsuka, S., Okazaki, Y., Sasaki, T., Noda, K., *et al.* (2012) "Nonvolatile memory with extremely low-leakage indium–gallium–zinc-oxide thin-film transistor," *IEEE J. Solid-State Circuits*, **47**, 2258.

[15] Ohmaru, T., Yoneda, S., Nishijima, T., Endo, M., Dembo, H., Fujita, M., *et al.* (2012) "Eight-bit CPU with non-volatile registers capable of holding data for 40 days at 85°C using crystalline In–Ga–Zn oxide thin film transistors," *Ext. Abstr. Solid State Dev. Mater.*, 1144.

[16] Kobayashi, Y., Matsubayashi, D., Nagatsuka, S., Yakubo, Y., Atsumi, T., Shionoiri, Y., *et al.* (2014) "Scaling to 50-nm c-axis aligned crystalline In–Ga–Zn oxide FET with surrounded channel structure and its application for less-than-5-nsec writing speed memory," *IEEE Symp. VLSI Technol. Dig. Tech. Pap.*, 170.

[17] Yamazaki, S., Suzawa, H., Inoue, K., Kato, K., Hirohashi, T., Okazaki, K., *et al.* (2014) "Properties of crystalline In–Ga–Zn-oxide semiconductor and its transistor characteristics," *Jpn. J. Appl. Phys.*, **53**, 04ED18.

[18] Murakami, M., Kato, K., Inada, K., Matsuzaki, T., Takahashi, Y., and Yamazaki, S. (2012) "Theoretical examination on a significantly low off-state current of a transistor using crystalline In–Ga–Zn-oxide," *Proc. AM-FPD'12 Dig.*, 171.

[19] Sugisawa, N., Sasaki, T., Ushikubo, T., Ohsawa, N., Seo, S., Hatano, K., *et al.* (2011) "High-definition top-emitting AMOLED display with highly reliable oxide semiconductor field effect transistors," *SID Symp. Dig. Tech. Pap.*, **42**, 722.

[20] Shishido, H., Amano, S., Toyotaka, K., Miyake, H., Murakawa, T., Nishi, T., *et al.* (2011) "Color sequential LC display using high reliable oxide semiconductors with monochrome electronic paper function," *SID Symp. Dig. Tech. Pap.*, **42**, 369.

[21] Amano, S., Harada, H., Akimoto, K., Sakata, J., Nishi, T., Moriya, K., *et al.* (2010) "Low power LC display using In–Ga–Zn-oxide TFTs based on variable frame frequency," *SID Symp. Dig. Tech. Pap.*, **41**, 626.

[22] Matsuzaki, T., Inoue, H., Nagatsuka, S., Okazaki, Y., Sasaki, T., Noda, K., *et al.* (2011) "1Mb Non-volatile random access memory using oxide semiconductor," *Proc. IEEE Int. Memory Workshop*, 185.

[23] Nishijima, T., Yoneda, S., Ohmaru, T., Endo, M., Denbo, H., Fujita, M., *et al.* (2012) "Low-power display system driven by utilizing technique using crystalline IGZO transistor," *SID Symp. Dig. Tech. Pap.*, **43**, 583.

[24] Atsumi, T., Nagatsuka, S., Inoue, H., Onuki, T., Saito, T., Ieda, Y., *et al.* (2012) "DRAM using crystalline oxide semiconductor for access transistors and not requiring refresh for more than ten days," *Proc. IEEE Int. Memory Workshop*, 99.

[25] Aoki, T., Okamoto, Y., Nakagawa, T., Kozuma, M., Kurokawa, Y., Ikeda, T., *et al.* (2015) "Normally-off computing for crystalline oxide semiconductor-based multicontext FPGA capable of fine-grained power gating on programmable logic element with nonvolatile shadow register," *IEEE J. Solid-State Circuits*, **50**, 2199.

[26] Tamura, H., Hamada, T., Nakagawa, T., Aoki, T., Ikeda, M., Kozuma, M., *et al.* (2011) "High reliable In–Ga–Zn-oxide FET based electronic global shutter sensors for in-cell optical touch screens and image sensors," *SID Symp. Dig. Tech. Pap.*, **42**, 729.

[27] Sekine, Y., Furutani, K., Shionoiri, Y., Kato, K., Koyama, J., and Yamazaki, S. (2011) "Success in measurement the lowest off-state current of transistor in the world," *ECS Trans.*, **37**, 77.

[28] Yamazaki, S. (2015) "Unique technology from Japan to the world – super low power LSI using CAAC-OS." Available at: www.umc.com/2015_japan_forum/pdf/20150527_shunpei_yamazaki_eng.pdf [accessed February 11, 2016].

[29] Murakami, M., Kato, K., Matsuzaki, T., Takahashi, Y., and Yamazaki, S. (2012) "Theoretical examination on significantly low off-state current of a transistor using crystalline In–Ga–Zn-oxide," *Ext. Abstr. Solid State Dev. Mater.*, 320.

[30] Kobayashi, Y., Matsuda, S., Matsubayashi, D., Suzawa, H., Sakakura, M., Hanaoka, K., *et al.* (2014) "Electrical characteristics and short-channel effect of *c*-axis aligned crystal indium gallium zinc oxide transistor with short channel length," *Jpn. J. Appl. Phys.*, **53**, 04EF03.

[31] Lenzlinger, M. and Snow, E. H. (1969) "Fowler–Nordheim tunneling into thermally grown SiO$_2$," *J. Appl. Phys.*, **40**, 278.

[32] Ozaki, T. and Kino, H. (2004) "Numerical atomic basis orbitals from H to Kr," *Phys. Rev. B*, **69**, 195113.

[33] Synopsys, Inc. (2010) *Sentaurus Device User Guide*. Mountain View, CA: Synopsys, Inc.

[34] Flandre, D., Terao, A., Francis, P., Gentinne, B., and Colinge, J.-P. (1993) "Demonstration of the potential of accumulation-mode MOS transistors on SOI substrates for high-temperature operation (150–300°C)," *IEEE Electron Device Lett.*, **14**, 10.

[35] Matsubayashi, D., Tsubuku, M., Takeuchi, T., Matsuda, S., Ohshima, K., Tanaka, T., *et al.* (2015) "Ideal deep-subthreshold characteristics in c-axis aligned crystalline oxide semiconductor FET," *Ext. Abstr. Solid State Dev. Mater.*, 1082.

[36] Shur, M. and Hack, M. (1984) "Physics of amorphous silicon based alloy field-effect transistors," *J. Appl. Phys.*, **55**, 3831.

[37] Hsieh, H.-H., Kamiya, T., Nomura, K., Hosono, H., and Wu, C.-C. (2008) "Modeling of amorphous InGaZnO4 thin film transistors and their subgap density of states," *Appl. Phys. Lett.*, **92**, 133503.

[38] Fung, T.-C., Chuang, C.-S., Chen, C., Abe, K., Cottle, R., and Townsend, M., *et al.* (2009) "Two-dimensional numerical simulation of radio frequency sputter amorphous In–Ga–Zn–O thin-film transistors," *J. Appl. Phys.*, **106**, 084511.

[39] Lee, S., Park, S., Kim, S., Jeon, Y., Jeon, K., Park, J.-H., *et al.* (2010) "Extraction of subgap density of states in amorphous InGaZnO thin-film transistors by using multifrequency capacitance–voltage characteristics," *IEEE Electron Device Lett.*, **31**, 231.

[40] Jeon, Y. W., Kim, S., Lee, S., Kim, D. M., Kim, D. H., Park, J., *et al.* (2010) "Subgap density-of-states-based amorphous oxide thin film transistor simulator (DeAOTS)," *IEEE Trans. Electron Devices*, **57**, 2988.

[41] Lee, S., Ghaffarzadeh, K., Nathan, A., Robertson, J., Jeon, S., Kim, C., *et al.* (2011) "Trap-limited and percolation conduction mechanisms in amorphous oxide semiconductor thin film transistors," *Appl. Phys. Lett.*, **98**, 203508.

[42] Lee, S., Ahnood, A., Sambandan, S., Madan, A., and Nathan, A. (2012) "Analytical field-effect method for extraction of subgap states in thin-film transistors," *IEEE Electron Device Lett.*, **33**, 1006.

[43] Lee, S. and Nathan, A. (2012) "Localized tail state distribution in amorphous oxide transistors deduced from low temperature measurements," *Appl. Phys. Lett.*, **101**, 113502.

[44] Germs, W. C., Adriaans, W. H., Tripathi, A. K., Roelofs, W. S. C., Cobb, B., Janssen, R. A. J., *et al.* (2012) "Charge transport in amorphous InGaZnO thin-film transistors," *Phys. Rev. B*, **86**, 155319.

[45] Silvaco, Inc. (2012) *Atlas User's Manual Device Mimulatrion Software*. Santa Clara, CA: Silvaco Inc.

[46] Togo, M., Fukai, T., Nakahara, Y., Koyama, S., Makabe, M., Hasegawa, E., *et al.* (2004) "Power-aware 65 nm node CMOS technology using variable V$_{DD}$ and back-bias control with reliability consideration for back-bias mode," *IEEE Symp. VLSI Technol. Dig. Tech. Pap.*, 88.

[47] Matsuda, S., Kikuchi, E., Yamane, Y., Okazaki, Y., and Yamazaki, S. (2015) "Channel length dependence of field-effect mobility of *c*-axis-aligned crystalline In–Ga–Zn–O field-effect transistors," *Jpn. J. Appl. Phys.*, **54**, 041103.

[48] Ohnuma, H., Hamada, T., Shimomura, A., Nagai, M., Sekiguchi, K., Kozuma, M., *et al.* (2012) "High definition 458 ppi OLED with logic circuit using low temperature single-crystal-silicon (LTSS) TFT backplane driven by 2.5 V single power supply," *SID Symp. Dig. Tech. Pap.*, **43**, 359.

[49] Sze, S. M. (1981) *Physics of Semiconductor Devices*, 2nd edn. New York: John Wiley.

[50] Clark, S. J., Segall, M. D., Pickard, C. J., Hasnip, P. J., Probert, M. I. J., Refson, K., *et al.* (2005) "First principles methods using CASTEP," *Z. Kristallogr.*, **220**, 567.

[51] Refson, K., Tulip, P. R., and Clark, S. J. (2006) "Variational density-functional perturbation theory for dielectrics and lattice dynamics," *Phys. Rev. B*, **73**, 155114.

[52] Natori, K. (1994) "Ballistic metal-oxide-semiconductor field effect transistor," *J. Appl. Phys.*, **76**, 4879.

[53] Ruch, J. G. (1972) "Electron dynamics in short channel field-effect transistors," *IEEE Trans. Electron Devices*, **19**, 652.

[54] Chou, S. Y., Antoniadis, D. A., and Smith, H. I. (1985) "Observation of electron velocity overshoot in sub-100-nm-channel MOSFETs in silicon," *IEEE Electron Device Lett.*, **6**, 665.

[55] Silvaco, Inc. (2016) "MC Device 2D Monte Carlo Device Simulator." Available at: www.silvaco.com/products/vwf/atlas/2D/mc_device/mc_device_br.html [accessed February 17, 2016].

[56] Dennard, R. H., Gaensslen, F. H., Yu, H.-N., Rideout, V. L., Bassous, E., and Reblanc, A. R. (1974) "Design of ion-implanted MOSFETs with very small physical dimensions," *IEEE J. Solid-State Circuits*, **9**, 256.

[57] Yamaguchi, K. and Tomizawa, K. (2011) *The Theory of Electron Transportation in Non-equivalent State – Semiconductor Device Simulation*. Tokyo: AdovanceSoft Corporation [in Japanese].

[58] Baccarani, G., Wordeman, M. R., and Dennard, R. H. (1984) "Generalized scaling theory and its application to a $^1/_4$ micrometer MOSFET design," *IEEE Trans. Electron Devices*, **31**, 452.

[59] Baker, R. J. (2005) *CMOS Circuit Design, Layout, and Simulation*, 2nd edn. Hoboken, NJ: John Wiley.

[60] Yakubo, Y., Nagatsuka, S., Matsuda, S., Honda, S., Hata, Y., Okazaki, Y., *et al.* (2014) "High-speed and low-leakage characteristics of 60-nm CAAC-OS FET with GHz-ordered cutoff frequency," *Ext. Abstr. Solid State Dev. Mater.*, 648.

[61] Matsubayashi, D., Asami, Y., Okazaki, Y., Kurata, M., Sasagawa, S., Okamoto, S., *et al.* (2015) "20-nm-Node trench-gate-self-aligned crystalline In–Ga–Zn-oxide FET with high frequency and low off-state current," *IEEE IEDM Tech. Dig.*, 141.

[62] Matsuda, S., Hiramatsu, T., Honda, R., Matsubayashi, D., Tomisu, H., Kobayashi, Y., *et al.* (2015) "30-nm-Channel-length *c*-axis aligned crystalline In–Ga–Zn–O transistors with low off-state leakage current and steep subthreshold characteristics," *VLSI Technol. Dig. Tech. Pap.*, 216.

[63] Lee, C.-W., Yun, S.-R.-N., Yu, C.-G., Park, J.-T., and Colinge, J.-P. (2007) "Device design guidelines for nano-scale MuGFETs," *Solid State Electron.*, **51**, 505.

[64] Banna, S. R., Chan, M., Ko, P. K., Nguyen, C. T., and Chan, M. (1995) "Threshold voltage model for deep-submicrometer fully depleted SOI MOSFET's," *IEEE Trans. Electron Devices*, **42**, 1949.

[65] Onuki, T., Kato, K., Nomura, M., Yakubo, Y., Nagatsuka, S., Matsuzaki, T., *et al.* (2014) "DRAM with storage capacitance of 3.9 fF using CAAC-OS transistor with L of 60 nm and having more than 1-h retention characteristics," *Ext. Abstr. Solid State Dev. Mater.*, 430.

3

NOSRAM

3.1 Introduction

Memories are classified as either volatile memories or non-volatile memories. Typical examples of volatile memories are static random access memory (SRAM) and dynamic random access memory (DRAM), whose features are high-speed operation and an unlimited number of write cycles. In contrast, flash memories, which are typical non-volatile memories, can store data even when no power is supplied.

A strong demand exists for novel non-volatile devices with greater memory capacities and higher operating speeds as computer memories. To satisfy this demand, new types of memory such as resistance random access memory (ReRAM) [1,2], magnetoresistive random access memory (MRAM) [3,4], and phase-change random access memory (PCRAM) [5,6] have been developed in recent years.

A non-volatile oxide semiconductor random access memory (NOSRAM) is another next-generation type of memory. NOSRAM uses a c-axis-aligned crystalline indium–gallium–zinc oxide (CAAC-IGZO) field-effect transistor (FET) in the memory cell, so it is non-volatile, operates at high speed, and offers an unlimited number of write cycles.

The use of CAAC-IGZO FETs with an extremely low off-state current allows electric charges accumulated in a cell capacitor connected to the FET to be stored for a long time, which gives it its non-volatile character. In addition, the amount of accumulated charge can be reduced and thus the cell capacitor can be made smaller, thereby reducing the area per cell.

The NOSRAM memory cell comprises a Si CMOS FET and a CAAC-IGZO FET. The Si CMOS FET is fabricated by using a standard silicon process, and the CAAC-IGZO layer is stacked on top of the Si CMOS FET to form the CAAC-IGZO FET.

NOSRAMs have a cell size of 12.32 μm^2 or smaller and can operate at 4.5 V or lower, have a write time of 10 ns or less, and have more than 10^{12} write cycles (i.e., essentially an

Physics and Technology of Crystalline Oxide Semiconductor CAAC-IGZO: Application to LSI, First Edition.
Edited by Shunpei Yamazaki and Masahiro Fujita.
© 2017 John Wiley & Sons, Ltd. Published 2017 by John Wiley & Sons, Ltd.

unlimited number of write cycles and a concomitant high reliability). NOSRAM is a non-volatile memory also capable of storing multilevel data (currently up to 4 bits/cell has been demonstrated).

This chapter explains the NOSRAM technology based on the CAAC-IGZO FETs, including its operations and characteristics.

3.2 Memory Characteristics

The most important characteristics of memories are durability (write/erase endurance), write energy, and non-volatility. Figure 3.1 shows the characteristics of various types of memories, where the hatched area at the bottom right corner represents the most desirable characteristics.

The durability and write energy for SRAM are 10^{15} writes or more and 1 fJ/bit or less, respectively, but SRAM is a volatile memory so the data are lost when power is switched off. A flash memory is a non-volatile memory with a writing energy of 1 fJ/bit or less and can store data even when no power is supplied. However, it has the drawback of low durability, approximately 10^5 writes, as shown in Figure 3.1.

The FETs used in a flash memory have a control gate and a floating gate. Depending on the voltage applied to the control gate, charge is injected or released via the floating gate by a tunnel current through a gate insulator (see Figure 3.2). The threshold voltage V_t of the FET varies depending on the presence of charge, thereby causing a characteristic shift in the current (see Figure 3.3). The current characteristics at high and low V_t are represented as data 1 and 0, respectively, and stored. However, repeated injection and release of charge from the floating

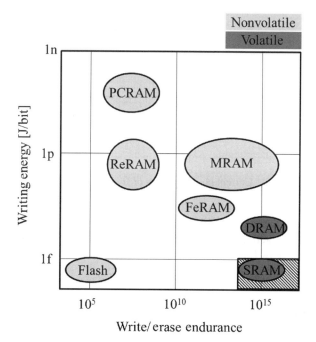

Figure 3.1 Relationship between write endurance and write energy

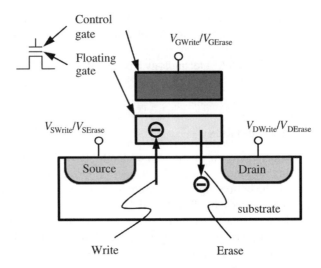

Figure 3.2   Structure of floating-gate flash memory

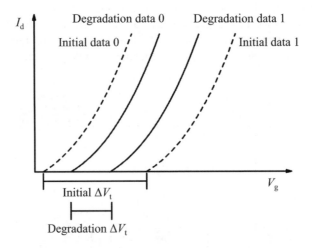

Figure 3.3   Current characteristics of floating-gate flash memory

gate degrades the gate insulator, and the difference between data 0 and 1 becomes excessively small after 10^5 writes, which makes it difficult to discriminate the stored data. For this reason, the number of write cycles for flash memory is limited.

3.3   Application of CAAC-IGZO FETs to Memory and their Operation

As already described in Chapter 2, in this volume, the off-state current of the CAAC-IGZO FET is extremely low – on the order of yoctoamps per micrometer (yA/μm). A non-volatile memory

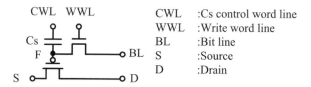

Figure 3.4   Memory cell using CAAC-IGZO FET. *Source*: Adapted from [7]

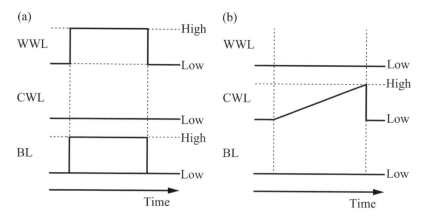

Figure 3.5   Timing diagram of memory cell: (a) write operation; (b) read operation

cell can be constructed with a CAAC-IGZO FET to which a cell capacitor (Cs) and a PMOS FET are connected (see Figure 3.4). The charge accumulated in node F is held, so its electric potential is maintained for a long time, and thus the capacitance of the capacitor Cs and hence the cell area can be small. Moreover, when the PMOS FET is used as the readout FET, a negative voltage is unnecessary.

The operation of the memory cell is described below, with reference to Figure 3.5.

Write Operation
The write word line (WWL) is set high and a Cs control word line (CWL) is set low to turn on the CAAC-IGZO FET. A write voltage is then applied to a cell capacitor Cs and node F from the bit line (BL). The write voltage is either high or low. Next, the WWL is set low to turn off the CAAC-IGZO FET. Thus, charge is accumulated in Cs and node F, and stored.

Read Operation
The WWL and BL are set low and the CWL is varied. By applying a voltage to the CWL, the gate voltage of the PMOS FET (which serves as a readout FET) varies due to the coupling capacitance between Cs and node F. The drain current I_d of the readout FET varies between data 1 and data 0 (see Figure 3.6). The difference in the drain current I_d is used to read out data.

Neither write nor read operations use a tunnel current, so the gate insulators of the CAAC-IGZO FET and the PMOS FET are not degraded; as a result, the number of writes is unlimited. Because of the low off-state current of the CAAC-IGZO FET, the capacitance of Cs connected

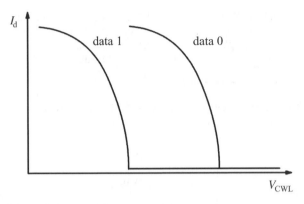

Figure 3.6 Relationship between the voltage of CWL (V_{CWL}) and I_d

to node F and hence also the write energy can be small. A memory cell based on the CAAC-IGZO FET thus offers non-volatility, unlimited writes, and low writing energy.

3.4 Configuration and Operation of NOSRAM Module

3.4.1 NOSRAM Module

Next, the configuration example of a NOSRAM module [7,8] with a memory capacity of 1 Mbit (megabits) is described. Figure 3.7 shows a block diagram of the NOSRAM module, which was prototyped and evaluated. The evaluation result is explained in Section 3.6. The module consists of a 1-Mbit cell array, row drivers, page buffers, and a source line (SL) driver. The 1-Mbit cell array consists of 1024 pages, each of which is 1024 bits (1024 bit × 1024 pages). The 1024 row drivers output 1024 WWLs and 1024 CWLs. The 1024 page buffers are connected to 1024 bit lines (BLs). The SL extending from the SL driver is connected to the source of the PMOS FET in each memory cell.

The read and write operations are executed by the page buffers. Figure 3.8 shows a circuit diagram of a page buffer, which consists of a latch circuit that stores 1 bit of data. Since the write energy is low, data can be read and written simultaneously in multiple memory cells. The 1-Mbit NOSRAM module writes data at the same time to 1024 memory cells connected to a WWL and reads data at the same time from 1024 cells connected to a CWL. The shift of the page buffers between write and read operations is performed by switching the BL connection. In each write operation, 1 bit of data is stored, and then the BL is set to the voltage of the data. In readout, 1 bit of readout data is stored. In both the write and read operations, 1024 bits of data are processed at the same time and stored temporarily in the 1024 page buffers.

3.4.2 Setting Operational Voltage of NOSRAM Module

The operational voltage of the NOSRAM module is explained below. The supply voltage is V_{DD}. The write voltage corresponding to data 0 written from the BL is V_{data0}, and the write voltage corresponding to data 1 written from the BL is V_{data1}. The high voltage of the

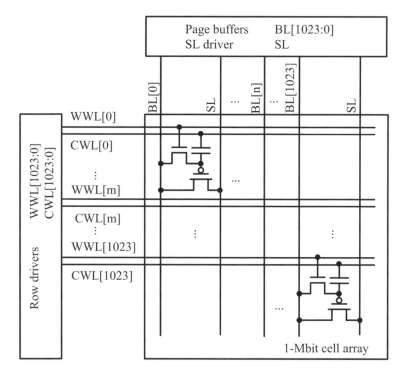

Figure 3.7 Block diagram of 1-Mbit NOSRAM module. *Source*: Adapted from [8]

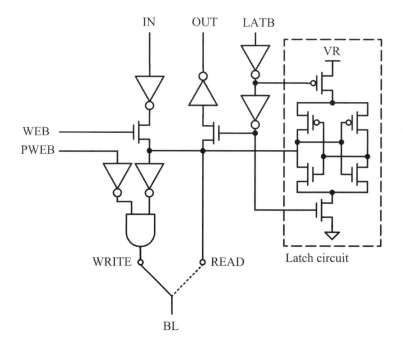

Figure 3.8 Circuit diagram of page buffer. *Source*: Adapted from [8]

WWL is V_H, which is higher than V_{DD} because the write voltage itself is written to node F so as to eliminate a drop in the threshold voltage of the CAAC-IGZO FET. The low voltage of WWL is expressed as V_L.

The high and low voltages of the CWL are V_H and V_{SS}, respectively. When the CWL is at V_H, the voltage of node F is boosted by the capacitive coupling between the CWL and node F, so it ranges from $V_{data0} + V_H - V_{SS}$ to $V_{data1} + V_H - V_{SS}$. To turn off the PMOS FET of the memory cell, the voltages of the BL and SL are set lower than the voltage of node F, which in turn is lower than V_{DD}. In other words, setting the amplitude value of CWL to $V_H - V_{SS}$ enables the memory cell in NOSRAM using PMOS FET to be non-selected.

The high voltage of the read operation for the SL is V_R, which is preferably lower than V_{DD} to extend the retention of data 1. The low voltage in standby mode and write mode of SL is V_{SS}.

3.4.3   Operation of NOSRAM Module

The operation of the NOSRAM module is explained below, with reference to the timing diagram in Figure 3.9.

Write Operation

First, data are temporarily stored in a page buffer. A latch operation control signal (LATB) of the page buffer is set low to stop the latch circuit. The write enable signal (WEB) is set high to input data written from IN to the latch circuit. LATB is set high to drive the latch circuit to store the written data.

Next, the data are written from the page buffer to the memory cell. The page buffer write-enable signal (PWEB) is set low and the BL is set to the voltage of the write data. After the BL voltage is set to V_{data1} or V_{data0}, the selected CWL is set to V_{SS} and the selected WWL is set to V_H. The CAAC-IGZO FET is thereby turned on so that V_{data1} or V_{data0} is stored in the cell capacitor Cs and data are written.

Read Operation

Data are read out from the memory cell and temporarily stored in the page buffer. The selected CWL is set to V_{SS}, SL is set to V_R, and LATB is set low. After the BL is precharged to V_{SS}, it is set to be floating. With data 0, the gate-source voltage (V_{gs}) of the PMOS FET in the memory cell is $V_{data0} - V_R$, and $V_{data0} - V_R$ is set lower than the threshold voltage V_{tp} of the PMOS FET in the memory cell. V_{tp} is a negative voltage. The PMOS FET is turned on, thereby feeding a current to BL from SL. The BL voltage is increased to V_R from V_{SS}. Conversely, with data 1, V_{gs} is $V_{data1} - V_R$, and $V_{data1} - V_R$ is set higher than V_{tp}. The PMOS FET remains off, thereby keeping the BL voltage at V_{SS}. After the BL voltage varies in accordance with data 0 and 1, LATB is set high to drive the page buffer. The latch circuit stores the readout data. A reference voltage of the readout data is an inverter turnover voltage ($V_R/2$). Data 1 or 0 is distinguished by whether the voltage of BL is higher or lower than the reference voltage.

3.5   Multilevel NOSRAM

The NOSRAM described in Section 3.4 is a non-volatile memory that stores 1 bit, data 0 or data 1. Currently, a demand exists for techniques to increase memory capacity and density.

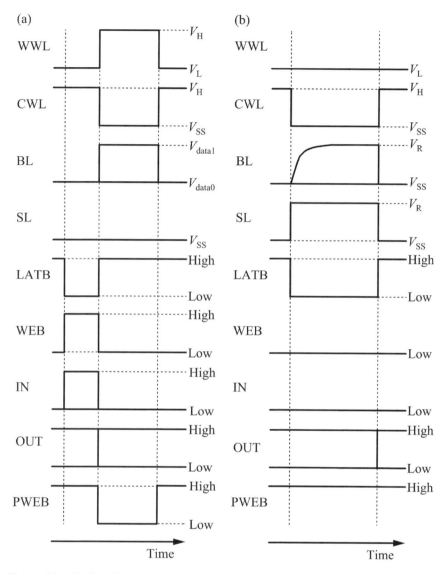

Figure 3.9 Timing diagram of NOSRAM module: (a) write operation; (b) read operation

Typical techniques to meet that demand include a multilevel-cell in which multiple bits of data are stored in one memory cell and a multilayer technique of stacking memory cells combined with the multilevel-cell technique. NOSRAM is suitable for both the multilevel-cell and multilayer techniques. For the multilevel-cell technique, 4-level (2 bits/cell), 8-level (3 bits/cell), and 16-level (4 bits/cell) are explained in Subsections 3.5.1, 3.5.2, and 3.5.3, respectively, and for the multilayer technique, the stacked multilevel NOSRAM is explained in Subsection 3.5.4.

3.5.1   4-Level (2 Bits/Cell) NOSRAM Module

To realize a multilevel memory, the distributions of readout voltages for different data in the memory cells should be small and separated without overlap. The NOSRAM module described in Section 3.4 is a 1-bit/cell NOSRAM where one memory cell contains the 2-level data. In this section, a 2-bit/cell NOSRAM that writes and reads 4-level data is explained, with special focus on the variations in readout voltages [10].

3.5.1.1   Configuration of 4-Level NOSRAM Module

The 4-level NOSRAM module has an array of 8192 cells giving a total memory capacity of 16 kbit. The number of cells, 8192, is an example. This module was prototyped and evaluated. The evaluation result is explained in Section 3.6. The module consists of a 512 × 16 cell array, row drivers, BL drivers, and an SL driver. The 512 row drivers output 512 WWLs and 512 CWLs. The 16 BL drivers are connected to 16 BLs. The SL extended from the SL driver is connected to the source of the PMOS FETs in all memory cells (Figure 3.10). Inverters in column drivers are used as a read circuit.

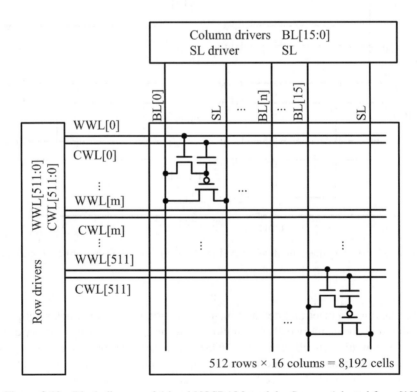

Figure 3.10   Block diagram of 4-level NOSRAM module. *Source*: Adapted from [10]

3.5.1.2 Voltage Settings of 4-Level NOSRAM Module

The operation voltages of the 4-level NOSRAM module are now described. The power supply voltage is V_{DD}. The write voltage is one of V_{00}, V_{01}, V_{10}, and V_{11}. The range of the four write voltages is from V_{DD} to V_{SS}. The high and low voltages for WWL are V_H and V_L, respectively. The high and low voltages of CWL are V_H and V_{SS}, respectively.

3.5.1.3 Operation of 4-Level NOSRAM Module

The operation of the 4-level NOSRAM module is now described, with reference to the timing diagram of Figure 3.11.

Write Operation
The write voltage from the BL driver is used to set the BL. The selected CWL is set to V_{SS}, and a selected WWL is set to V_H to turn on the CAAC-IGZO FET and charge the cell capacitor Cs with the write voltage, thereby storing the data. The selected WWL is set to V_L, and the selected CWL is set to V_H to turn off the CAAC-IGZO FET, thereby finishing the write operation.

Read Operation
The selected CWL is set to V_{SS}. After the BL is precharged to V_{SS}, it is set to be floating. The SL voltage and the voltage of an inverter in the column driver connected to the BL are set to V_{DD}. The gate voltage of the PMOS FET in the selected memory cell is the write voltage. When the PMOS FET is turned on, the BL voltage is boosted from V_{SS}. When the BL voltage is increased from the turnover voltage of the inverter, the inverter outputs a low signal. On the contrary, when the PMOS FET is turned off, the BL remains V_{SS} so the inverter outputs a high signal. The voltage of the SL and the voltage of the inverter connected to the BL are decreased stepwise from V_{DD} to obtain a switching voltage from the high to the low signal as the output of the inverter. The switching voltage from the high to the low signal is the readout voltage V_{read}. The distribution of variations in V_{read} corresponds to the distribution of the data readout.

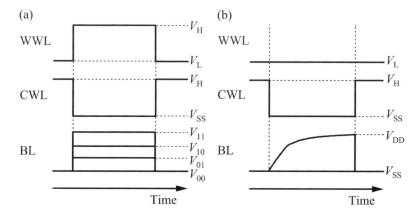

Figure 3.11 Timing diagram of 4-level NOSRAM module: (a) write operation; (b) read operation

3.5.2   8-Level (3 Bits/Cell) NOSRAM Module

Over the same range of voltage as that for the above 4-level NOSRAM, the distributions of eight readout voltages are separated without overlap, which leads to an 8-level NOSRAM [11]. While the write mechanism realizing an 8-level NOSRAM remains essentially the same as that of 2- or 4-level NOSRAMs, it requires variations in the inverter turnover voltages of the readout circuit to be removed, and the distributions of readout voltages have to be narrowed. A different readout method wherein the BL voltage is discharged to read data (here referred to as the discharge method) can narrow the distributions of the readout voltages, shorten the time for acquiring data, and improve the accuracy.

3.5.2.1   Configuration of 8-Level NOSRAM Module

The 8-level NOSRAM module, which was prototyped and evaluated, is configured with 6144 cells and consequently has a memory capacity of 18 kbit. The number of cells, 6144, is an example. The evaluation result is explained in Section 3.6. The configuration of the 8-level NOSRAM module is a 512×12 array of 6144 cells, 512 row drivers, 12 write switches, and 12 voltage followers. The 512 row drivers output 512 WWLs and 512 CWLs. The 12 writing switches are connected to the 12 BLs. The voltage followers are used as a read circuit. The voltage followers are connected to BLs and output the same voltage as the voltage input from the BLs. The SLs are connected to the drains of the PMOS FETs in all memory cells. Figure 3.12 shows a block diagram of the 8-level NOSRAM. Alternatively, a write voltage is output from a 3-bit digital-to-analog (D/A) converter and read data are output from the 3-bit analog-to-digital (A/D) converter instead of using the voltage followers.

3.5.2.2   Voltage Settings of 8-Level NOSRAM Module

The operation voltages for the 8-level NOSRAM module are now described. The power supply voltage is V_{DD}. Write voltages are V_{000}, V_{001}, V_{010}, V_{011}, V_{100}, V_{101}, V_{110}, and V_{111}. The range of the eight write voltages is from V_{DD} to V_{SS}. This range is the same as that of the four write voltages, and the number of write voltages increases from four to eight. The high and low voltages for the WWL are V_H and V_L, respectively. The high and low voltages for the CWL are V_H and V_{SS}, respectively.

3.5.2.3   Operation of 8-Level NOSRAM Module

The operation of an 8-level NOSRAM module is now described, with reference to the timing diagram of Figure 3.13.

Write Operation
The write voltage is set on the BL. The selected CWL is set to V_{SS}, and the selected WWL is set to V_H to turn on the CAAC-IGZO FET and charge the cell capacitor Cs with the write voltage, thereby storing the data. The selected WWL is set to V_L, and the selected CWL is set to V_H to turn off the CAAC-IGZO FET, thereby finishing the write operation.

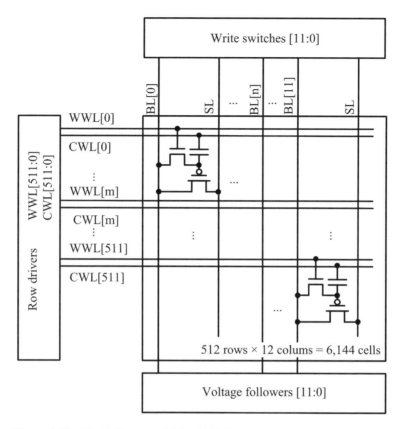

Figure 3.12 Block diagram of 8-level NOSRAM. *Source*: Adapted from [11]

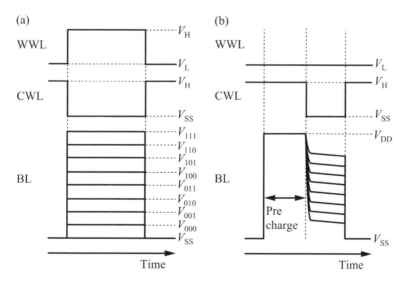

Figure 3.13 Timing diagram of 8-level NOSRAM module: (a) write operation; (b) read operation. *Source*: Adapted from [11]

Read Operation

The discharge method used for readout is as follows: the voltage of the BL is discharged to a voltage that is the difference between the write voltage and the threshold voltage V_{tp} of the PMOS FET. V_{tp} is a negative voltage. The BL is precharged to V_{DD} and then is set to be floating. The selected CWL is set to V_{SS} and the SL voltage is set to V_{SS}. The gate voltage of the PMOS FET in the selected memory cell is the write voltage. The PMOS FET is turned on to discharge the BL voltage. When the BL voltage decreases, the gate-source voltage V_{gs} of the PMOS FET in the memory cell increases. When V_{gs} is higher than V_{tp}, the PMOS FET is turned off and the BL voltage is saturated. The saturated BL voltage serves as the readout voltage V_{read}. The distribution of V_{read} corresponds to the distribution of data readout. The selected CWL is set to V_H to turn off the PMOS FET, thereby finishing the read operation.

The readout voltage of the 4-level NOSRAM is a turnover voltage from the high to the low signal, which is an inverted output of the inverter. The readout voltage of the 8-level NOSRAM is a voltage obtained by discharge of the BL. The 8-level NOSRAM does not vary the supply voltage of the inverter. Thus, the time for reading can be shortened. The readout voltage of the 8-level NOSRAM is not affected by variation in the turnover voltage; therefore, the variation of the readout voltage of the 8-level NOSRAM is smaller than that of the 4-level NOSRAM.

3.5.3   16-Level (4 Bits/Cell) NOSRAM Module

To realize a 16-level NOSRAM, 16 readout voltages should be distributed and separated without overlap. Also, with the same range of voltage as that of the above 8-level NOSRAM, the distribution width of 16 readout voltages should be narrower than that of the 8-level NOSRAM. The distributions of readout voltages explained in Subsection 3.5.2 are influenced by variations in the threshold V_{tp} of the PMOS FET in the memory cell. To cancel the V_{tp} variations, a V_t cancel-write method was proposed [12,13]. With the V_t cancel-write method, the distributions of readout in the NOSRAM are sharpened to narrow the voltage distribution width. In addition, the V_t cancel-write method eliminates any threshold voltage shift by temperature [14]. By this method, a 16-level (4 bit/cell) NOSRAM can be realized.

Table 3.1 summarizes write and read operations of multilevel NOSRAMs.

3.5.3.1 V_t Cancel-Write Method

Figure 3.14 depicts write methods to a multilevel NOSRAM. The 8-level NOSRAM described in Subsection 3.5.2 adopts the BL write method as follows: the CAAC-IGZO FET is turned on and data are written to node F from the BL [see Figure 3.14(a)]. The read operation consists of

Table 3.1   Write/read operation of multilevel cell NOSRAM

	4-Level NOSRAM	8-Level NOSRAM	16-Level NOSRAM
Write operation	BL write method	BL write method	V_t cancel-write method
Read operation	Inverter voltage	BL discharge	SL discharge
Read circuit	Inverter	Voltage follower or A/D converter	Voltage follower or A/D converter

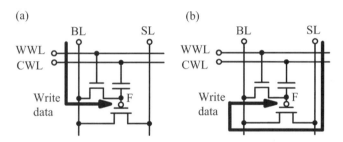

Figure 3.14 Write method for multilevel NOSRAM: (a) BL write method; (b) V_t cancel-write method. *Source*: Adapted from [12]

the discharge method explained in Subsection 3.5.2. The SL charge precharged through the PMOS FET in the memory cell is discharged to the BL to decrease the SL voltage. When the voltage of the SL decreases, the gate-source voltage (V_{gs}) of the PMOS FET in the memory cell increases. V_{gs} is close to the threshold V_{tp} of the PMOS FET in the memory cell, thereby stopping the voltage drop of the SL. The voltage of the SL serves as the readout voltage V_{read}. V_{tp} is a negative voltage. In the BL write method, V_{read} depends on V_{tp} and V_{read} satisfies

$$V_{read} = V_F - V_{tp} \tag{3.1}$$

where V_F is the voltage of node F. V_{tp} fluctuates more as the PMOS FET is scaled down; thus, the multilevel NOSRAM has a broad V_{read} distribution.

In the V_t cancel-write method, the voltage of node F at the write operation includes V_{tp} in the write data. In the read operation, V_{tp} is canceled from V_{read}. The write operation is as follows [see Figure 3.14(b)]: V_{SS} is input to the BL, the CAAC-IGZO FET is turned on, and V_F is set to V_{SS}. Then, the BL is floated and a write voltage V_{write} is input to the SL. In the write operation, SL serves as the source of the PMOS FET in the memory cell. Thereby, the PMOS FET in the memory cell is turned on, and current flows to the BL from the SL to increase the voltage of the BL. This increase in the voltage of the BL results in an increase in V_F. When V_F increases up to $V_{write} + V_{tp}$, V_{gs} equals V_{tp}. The PMOS FET is turned off to stop the increase in the voltage of the BL. When the increase in the voltage of the BL stops, the increase of V_F also stops. The CAAC-IGZO FET is turned off, thereby finishing the write operation. The relationship between V_F and V_{tp} after the write operation is

$$V_F = V_{write} + V_{tp} + V_c \tag{3.2}$$

where V_c is the sum of the voltage variations of the parasitic capacitance Cp 1 between the WWL and node F, the parasitic capacitance Cp 2 between the BL and node F, and the parasitic capacitance Cp 3 between the SL and node F shown in Figure 3.15. Thus, V_{read} satisfies

$$V_{read} = V_F - V_{tp} = V_{write} + V_{tp} + V_c - V_{tp} = V_{write} + V_c \tag{3.3}$$

where V_{tp} is canceled. In this way, V_{tp} has less influence on V_{read}, and the distribution width of V_{read} is narrowed.

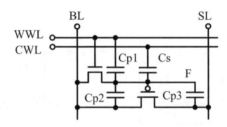

Figure 3.15 Capacitor and parasitic capacitances (Cp) in NOSRAM memory cell. *Source*: Adapted from [13]

Next, we consider the temperature shift in the threshold voltage (ΔV_{tp}). With the V_t cancel-write method, the relationship between V_F and V_{tp} is

$$V_F = V_{\text{write}} + \left(V_{\text{tp}} + \Delta V_{\text{tp}}\right) + V_c. \tag{3.4}$$

Similarly, V_{read} satisfies

$$V_{\text{read}} = V_F - \left(V_{\text{tp}} + \Delta V_{\text{tp}}\right). \tag{3.5}$$

From these equations, the following equation is obtained:

$$V_{\text{read}} = V_{\text{write}} + \left(V_{\text{tp}} + \Delta V_{\text{tp}}\right) + V_c - \left(V_{\text{tp}} + \Delta V_{\text{tp}}\right) = V_{\text{write}} + V_c. \tag{3.6}$$

In this manner, $V_{\text{thp}} + \Delta V_{\text{thp}}$ can be canceled, and thus it is possible to avoid widening the read-out voltage distribution.

3.5.3.2 Configuration of 16-Level NOSRAM Module

The 16-level NOSRAM is configured with 32,768 4 bit/cell NOSRAM cells and consequently has a memory capacity of 128 kbit. The number of cells, 32,768, is an example. This module was prototyped and evaluated. The evaluation result is explained in Section 3.6. The 16-level NOSRAM module is configured with 32,768 cell arrays, 128 row drivers, eight 4-bit D/A converters, eight SL comparators, and an output selector. The 32,768 cell arrays have 128 rows and 256 columns. The 128 row drivers output 128 WWL and 128 CWL. The input selector and the output selector select eight SLs. A group of eight memory cells conducts the write and read operations, and thus eight 4-bit D/A converters, eight voltage followers, and eight SL comparators are provided. The voltage followers function as a read circuit. The voltage follower outputs the same voltage as that of the selected SL. The SL comparator switches the voltage of the selected BL from V_{BL} to V_{SS} when the voltage of the selected SL drops below a reference voltage V_{REF}. Therefore, the drain-source voltage V_{ds} of the PMOS FET can constantly be half or less of V_{DD}. Consequently, the reliability of the PMOS FET is improved. Figure 3.16 shows a block diagram of the 16-level NOSRAM. Alternatively, data can be read out by a 4-bit A/D converter instead of using the voltage followers [12].

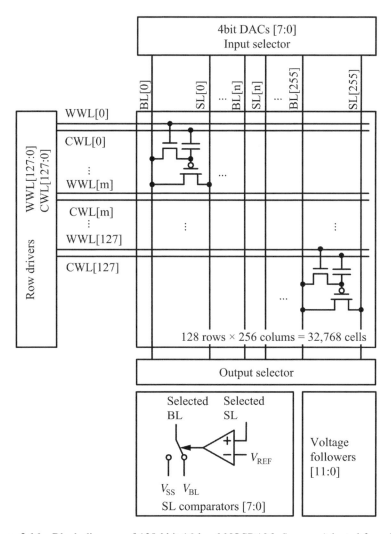

Figure 3.16 Block diagram of 128-kbit 16-level NOSRAM. *Source*: Adapted from [13]

3.5.3.3 Voltage Settings of 16-Level NOSRAM Module

The operation voltages for a 16-level NOSRAM module are now described. The power supply voltage is V_{DD}. The BL provides write voltages: V_{0000}, V_{0001}, V_{0010}, V_{0011}, V_{0100}, V_{0101}, V_{0110}, V_{0111}, V_{1000}, V_{1001}, V_{1010}, V_{1011}, V_{1100}, V_{1101}, V_{1110}, and V_{1111}. The range of the 16 write voltages is from V_{DD} to V_{SS}. This range is the same as that of the eight write voltages, but the number of write voltages increases from eight to sixteen. The high and low voltages of the WWL are V_H and V_L, respectively. The high and low voltages of the CWL are V_H and V_{SS}, respectively.

3.5.3.4   Operation of 16-Level NOSRAM Module

The operation of the 16-level NOSRAM module is now described, with reference to the timing diagram of Figure 3.17.

Write Operation

V_{write} is output from the 4-bit D/A converter to the selected SL by an input selector. Data are written to node F from the selected SL.

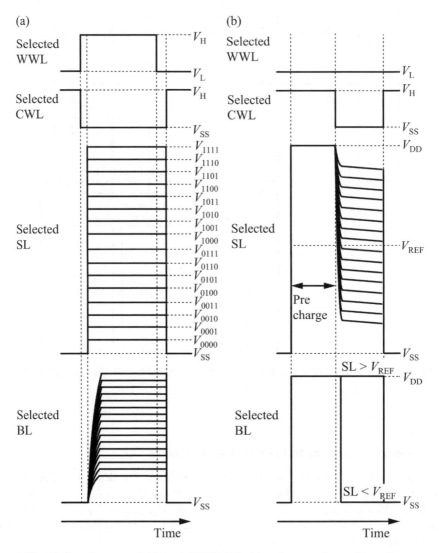

Figure 3.17   Timing diagram of 16-level NOSRAM: (a) write waveform; (b) readout waveform. *Source*: Adapted from [13]

Read Operation

The SL selected by the output selector is precharged to V_{DD} and floated, then the CWL is set to V_{SS}. The PMOS FET of the memory cell is turned on to decrease the voltage of the SL down to V_{read}. The voltage follower outputs the SL voltage from the input terminal to read out the data.

3.5.4   Stacked Multilevel NOSRAM

As described above, a NOSRAM cell consists of a CAAC-IGZO FET and a Si FET. If the Si FET is replaced with a CAAC-IGZO FET, the NOSRAM cell can be made with only oxide semiconductor layers, and thus allow stacking of memory cell layers. This leads to high integration and is one of the approaches to increase the memory capacity and density, as mentioned at the beginning of this section.

While circuits other than memory cells (such as the control circuit, a word-line driver circuit, and a bit-line driver circuit) are formed within a silicon layer, multiple oxide semiconductor layers are stacked to constitute the memory cell. With four layers each with 16 levels (4 bits), the memory cell corresponds to a 16-bit memory cell, which is often used in computers. Figure 3.18 shows a conceptual diagram of a stacked multilevel NOSRAM.

The stacked multilevel NOSRAM has four oxide layers stacked on the silicon layer and uses a 4 bit/cell NOSRAM, which is a memory suitable for 16-bit processing. This memory cell configuration is called "word operation."

In a 15-nm technology node, a stacked multilayer NOSRAM with four oxide layers corresponds to the cell size per bit of a 256 gigabyte solid state drive (SSD) commercialized in 2015. In a 10-nm technology node, a stacked multilayer NOSRAM with six oxide semiconductor layers corresponds to the cell size per bit of a 1 terabyte SSD (see Figure 3.19).

The stacked multilayer NOSRAM has the potential to implement a memory suitable for computers and increase the memory capacity thereof.

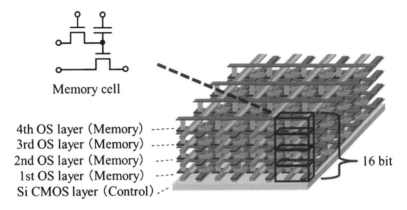

Figure 3.18   Conceptual diagram of word-processing NOSRAM

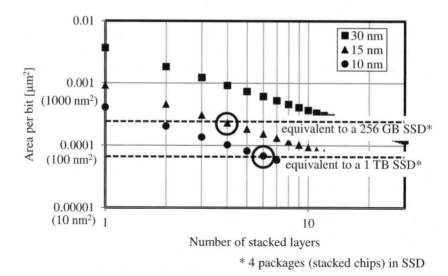

Figure 3.19   Scaling of $4 \times N$-bit NOSRAM, where N is the number of stacked layers

3.6   Prototype and Characterization

This section describes the prototyping and characterization of NOSRAMs.

The fabricated prototypes include 2-, 4-, and 8-level NOSRAMs (i.e., 1, 2, and 3 bits per cell), described in [7–11], respectively. The higher number of levels was achieved by scaling down the technology node, narrowing of the readout voltage distribution, and contriving driving methods. Furthermore, a prototype of the 16-level (4 bit/cell) NOSRAM was also fabricated [12–14].

Prototypes of 2-, 4-, 8-, and 16-level NOSRAMs are described in Subsections 3.6.1, 3.6.2, 3.6.3, and 3.6.4, respectively.

3.6.1   2-Level NOSRAM

The basic characteristics of the 2-level NOSRAM cell were evaluated to confirm that the NOSRAM works as a memory. Next, a prototype of a 1-Mbit, 2-level NOSRAM module (1-Mbit NOSRAM) was fabricated and examined [7,8]. The retention characteristics of the 2-level NOSRAM were also evaluated [9].

3.6.1.1 ΔV_t of Memory Cell

With the memory cell of the NOSRAM illustrated in Figure 3.20, a threshold voltage V_t is determined by the electrical characteristics of the PMOS FET and its difference ΔV_t corresponds to the difference between data 1 and 0 in terms of V_{CWL} variation. Note that V_t of the PMOS FET is calculated by square-root extrapolation.

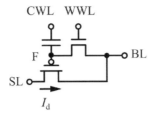

Figure 3.20   NOSRAM memory cell. *Source*: Adapted from [8]

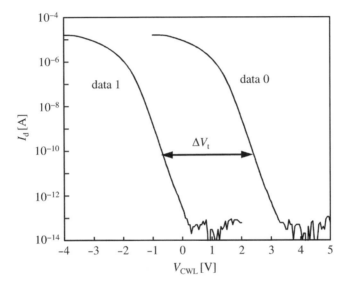

Figure 3.21 I_d–V_{CWL} characteristics when the NOSRAM memory cell receives data 1 and data 0. *Source*: Reprinted from [8], with permission of IEEE, © 2012

In the write operation of data, 0 V is applied to the CWL and 4.5 V is applied to the WWL to turn on the CAAC-IGZO FET and input data from the BL. Then, −1 V is applied to the WWL to turn off the CAAC-IGZO FET. 3 V corresponds to data 1, and 0 V corresponds to data 0. The drain current of the PMOS FET is measured by sweeping CWL. Figure 3.21 depicts the measurement results of the I_d–V_{CWL} characteristics. As can be seen from Figure 3.21, ΔV_t is 3 V or higher. The current ratio in the drain current between data 1 and 0 is 10^7 when the CWL is 0 V, which indicates a large tolerance to noise-related reading errors.

3.6.1.2   Write Time of Memory Cell

A voltage pulse of 4.5 V was applied to the gate electrode of the CAAC-IGZO FET, the pulse width was swept, and the write time was measured. Figure 3.22 shows measurement results for

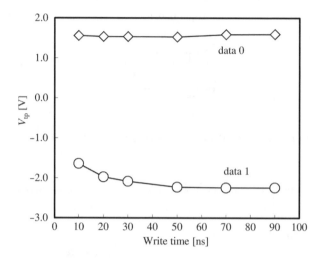

Figure 3.22 Write time of NOSRAM memory cell. *Source*: Reprinted from [8], with permission of IEEE, © 2012

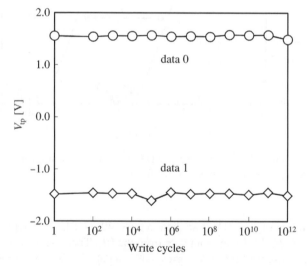

Figure 3.23 Write endurance of NOSRAM memory cell. *Source*: Reprinted from [8], with permission of IEEE, © 2012

the write time. As is apparent from Figure 3.22, the write time is as short as 10 ns and ΔV_t is 3 V or higher. This means that the data are correctly written in at least 10 ns.

3.6.1.3 Write Endurance of Memory Cell

Using a write time of 10 ns, the endurance was measured up to 10^{12} cycles. As shown in Figure 3.23, the write cycle endurance of the NOSRAM is at least seven orders of magnitude better than that of flash memory (e.g., 10^5 cycles in general).

3.6.1.4 Write and Readout Times of 1-Mbit NOSRAM Module

Next, the write and readout times of the fabricated 1-Mbit NOSRAM module were measured. V_{DD} is 3 V, V_{data0} is 0 V, V_{data1} is 3 V, V_H is 4.5 V, V_L is −1V, V_{SS} is 0 V, and V_R is 1.5 V.

The 1-Mbit NOSRAM module was fabricated with Si CMOS and CAAC-IGZO processes, both of which have a 0.8-μm technology node. Peripheral circuits (such as a row decoder, a page buffer, and the PMOS FETs in the memory cells) were fabricated with the Si CMOS process, whereas the writing FET and cell capacitors Cs in the memory cells were fabricated with the CAAC-IGZO process. The size of the memory cell is 12.32 μm^2.

Figure 3.24(a) shows the measured waveforms of the write operation. The BL voltage increases upon writing data 1, while it remains unchanged upon writing data 0. The write time is the period between the rise and fall of the WWL select signal (i.e., 150 ns/page).

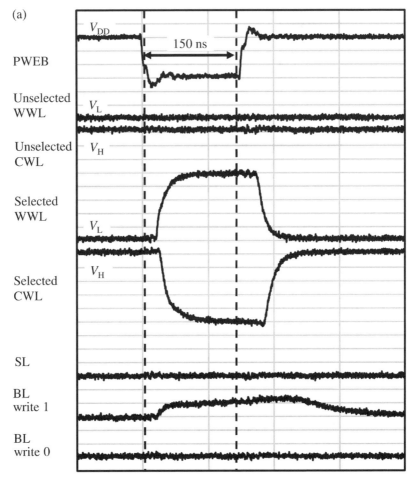

Figure 3.24 Operation waveform of 1-Mbit NOSRAM module: (a) write operation; (b) read operation. *Source*: Reprinted from [8], with permission of IEEE, © 2012

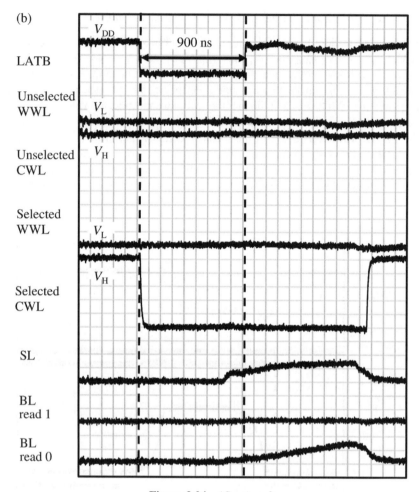

Figure 3.24 (*Continued*)

Figure 3.24(b) shows the measured waveforms of the read operation. The BL voltage remains unchanged upon readout of data 1, whereas it increases upon readout of data 0. The readout time is the period between the rise and fall of the LATB, or 900 ns.

3.6.1.5 Shmoo Plot of 1-Mbit NOSRAM Module

Figure 3.25 shows a Shmoo plot for the write time. The horizontal axis is the write time, and the vertical axis is V_{DD}/V_H. The write time of 150 ns/page (1024 bits) was obtained when V_{DD} is 3 V and V_H is 4.5 V, and the corresponding current was 3 mA. The write time is shorter than the write time (millisecond order) of a flash memory.

3.6.1.6 Readout Voltage Distribution of 1-Mbit NOSRAM Module

Figure 3.26 shows the voltage distribution of read data 1. NOSRAM can provide a narrower read distribution than that of flash memory, because the write voltage is given to the cell

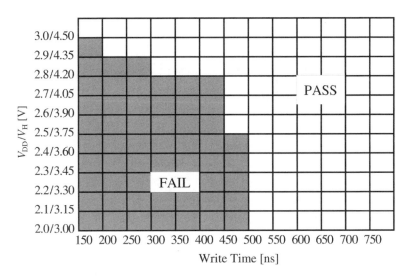

Figure 3.25 Shmoo plot for write time of 1-Mbit NOSRAM module. *Source*: Reprinted from [8], with permission of IEEE, © 2012

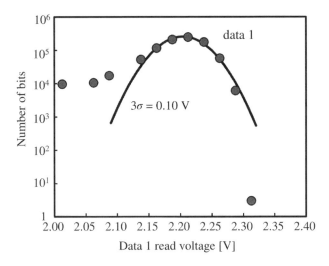

Figure 3.26 Distribution in readout of data 1 in 1-Mbit NOSRAM module. *Source*: Reprinted from [8], with permission of IEEE, © 2012

capacitor Cs from the BL through the CAAC-IGZO FET. The distribution was fit to the normal distribution to calculate the voltage standard deviation σ. As seen from Figure 3.26, a readout voltage of 2.10 V or higher yielding a width of $3\sigma = 0.10$ V is confirmed.

A photograph and the specifications of the fabricated 1-Mbit NOSRAM module are shown in Figure 3.27 and Table 3.2, respectively.

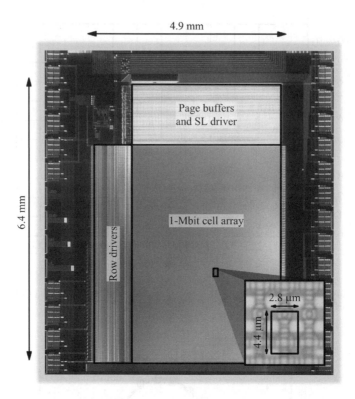

Figure 3.27   Photograph of fabricated 1-Mbit NOSRAM. *(For color detail, please see color plate section)*

Table 3.2   Specifications of 1-Mbit NOSRAM. *Source*: Adapted from [8]

Process	0.8-μm CMOS, 0.8-μm CAAC-IGZO
Die size	6.4 mm × 4.9 mm
Memory capacity	1 Mbit
Array organization	1024 bit/page × 1024 pages
Cell size	12.32 μm^2 (4.4 μm × 2.8 μm)
Write time	150 ns/page
Read time	900 ns/page
Supply voltage	$V_{DD}/V_H/V_L$ = 3 V/4.5 V/−1 V

3.6.1.7   Retention Characteristics of 2-Level NOSRAM [9]

A 1-kbit NOSRAM module was used to measure the retention characteristics. Figure 3.28 shows a circuit of a memory cell and its timing diagram. The memory cell operates as follows. The voltage of the WWL is set to 3.3 V and 1.8 V or 0 V from the BL is applied to node F. After that, the voltage of the WWL is set to 0 V to hold the charge in the cell capacitor Cs. Because the voltage of node F serves as the gate voltage of the readout FET, data are read out from the current of the readout FET.

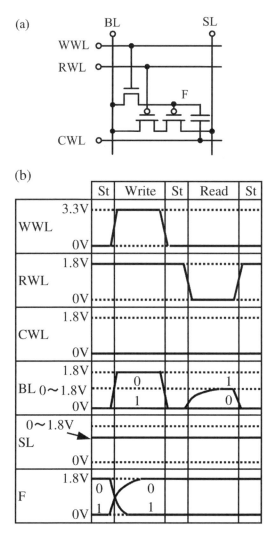

Figure 3.28 (a) Memory cell circuit of a 1-kbit NOSRAM and (b) its timing diagram. *Source*: Adapted from [9]. Copyright 2015 The Japan Society of Applied Physics

To analyze the data, the SL voltage is swept to measure the voltage V_{RM} to turn off the readout FET connected to node F. The distribution of charge accumulated in the cell capacitor Cs is measured indirectly. Figure 3.29 shows the retention characteristics at the operating temperature of 85°C. The write voltages for data 1 and 0 are 0 V and 1.8 V, respectively. For data 1, charge leakage of node F does not occur because there is no difference between the source and drain voltages of the CAAC-IGZO FET; this does not cause the change of readout voltage. In contrast, for data 0, the readout voltage decreases because there is a difference between the source and drain voltages of the CAAC-IGZO FET causing charge leakage. Figure 3.29 shows that the data are separated even after 2000 h.

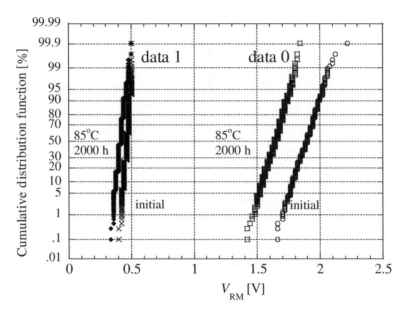

Figure 3.29 Distributions of voltages for retention in 2-level NOSRAM. *Source*: Adapted from [9]. Copyright 2015 The Japan Society of Applied Physics

3.6.2 4-Level NOSRAM

The measurement results of the fabricated 4-level (16-kbit) NOSRAM are now explained [10]. A 1.2-μm CAAC-IGZO FET process and a 0.8-μm Si CMOS FET are used. The technology node was the latest one at the time when the prototype was fabricated. V_{DD} is 3 V, V_{00} is 0 V, V_{01} is 1.6 V, V_{10} is 2.1 V, V_{11} is 3 V, V_H is 4.5 V, V_L is −1 V, and V_{SS} is 0 V.

Figure 3.30 shows the distribution of the readout voltage V_{read} in 8192 cells. Table 3.3 summarizes the peak values and calculated 3σ assuming the standard distribution. V_{read} has sharp peaks, with a maximum 3σ of 142 mV. In addition, the distribution of V_{read} after 10^8 write cycles is plotted in the figure. As is seen from Figure 3.30 and Table 3.3, the distribution of V_{read} does not change even after 10^8 write cycles. This confirms that the distributions of V_{read} for four values in the 8192 cells are sharp and do not overlap with each other.

3.6.3 8-Level NOSRAM

Here, the measurement results of the fabricated 8-level NOSRAM [11] are explained. A 0.45-μm CAAC-IGZO FET and a 0.45-μm Si CMOS FET were used. The technology node was the latest one at the time when the prototype was fabricated. V_{DD} is 3 V, V_{000} is 0.6 V, V_{001} is 0.9 V, V_{010} is 1.2 V, V_{011} is 1.5 V, V_{100} is 1.8 V, V_{101} is 2.1 V, V_{110} is 2.4 V, V_{111} is 2.7 V, V_H is 4.5 V, V_L is −1 V, and V_{SS} is 0 V.

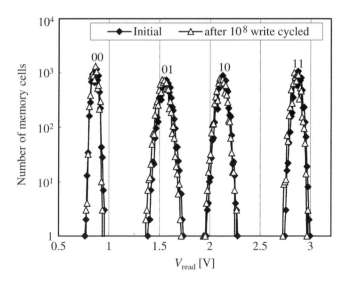

Figure 3.30 Distributions of V_{read} of 4-level NOSRAM. *Source*: Adapted from [10]. Copyright 2012 The Japan Society of Applied Physics

Table 3.3 Peak of V_{read} and calculated values of 3σ in V_{read} distribution (Figure 3.30). *Source*: Adapted from [10]

	Initial		After 10^8 cycled	
	Peak [V]	3σ [mV]	Peak [V]	3σ [mV]
00	0.86	86	0.86	83
01	1.56	142	1.54	144
10	2.13	126	2.11	131
11	2.87	106	2.85	109

3.6.3.1 8-Level NOSRAM Module

The distribution of the readout voltage V_{read} of the 8-level NOSRAM in 6144 cells (18 kbit) is confirmed (see Figure 3.31). The V_{read} voltage distributions are separated without overlap. Table 3.4 summarizes the peak values and the calculated values of 3σ.

A photograph and the specifications of the 8-level NOSRAM are shown in Figure 3.32 and Table 3.5.

3.6.4 16-Level NOSRAM

A prototype of the 16-level NOSRAM formed with a 0.18-μm Si CMOS process and a 0.35-μm CAAC-IGZO process is now explained. The technology node was the latest one at the time when the prototype was fabricated. V_{DD} is 3.5 V, V_{0000} is 0.95 V, V_{0001} is 1.12 V, V_{0010} is

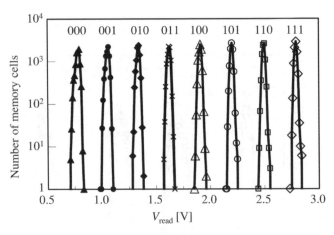

Figure 3.31 Distributions of V_{read} of 8-level NOSRAM. *Source*: Reprinted from [11], with permission of IEEE, © 2013

Table 3.4 Distribution peak values of V_{read} and 3σ of 8-level NOSRAM. *Source*: Adapted from [11]

Data	000	001	010	011	100	101	110	111
Write voltage [V]	0.6	0.9	1.2	1.5	1.8	2.1	2.4	2.7
Peak of V_{read} [V]	0.79	1.06	1.34	1.62	1.91	2.20	2.50	2.79
3σ [mV]	55	49	43	42	40	37	37	30

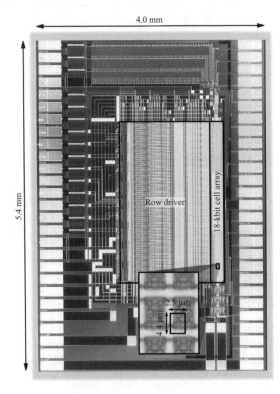

Figure 3.32 Photograph of prototype of 8-level NOSRAM cell. *(For color detail, please see color plate section)*

Table 3.5 Specifications of 8-level NOSRAM cell. *Source*: Adapted from [11]

Process	0.45-μm CMOS 0.45-μm CAAC-IGZO
Die size	5.4 mm × 4 mm
Memory capacity	18 kbit
Array organization	512 rows × 12 columns
Cell size	12.63 μm² (5.05 μm × 2.5 μm)
Number of levels	8
Write time	100 ns
Read time	8 μs
Supply voltage	$V_{DD}/V_H/V_L$ = 3 V/4.5 V/−1 V

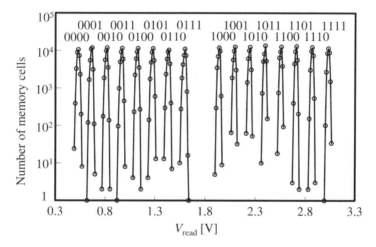

Figure 3.33 Distributions of V_{read} of 16-level NOSRAM. *Source*: Reprinted from [13], with permission of IEEE, © 2015

1.29 V, V_{0011} is 1.46 V, V_{0100} is 1.63 V, V_{0101} is 1.80 V, V_{0110} is 1.97 V, V_{0111} is 2.14 V, V_{1000} is 2.31 V, V_{1001} is 2.48 V, V_{1010} is 2.65 V, V_{1011} is 2.82 V, V_{1100} is 2.99 V, V_{1101} is 3.16 V, V_{1110} is 3.33 V, V_{1111} is 3.50 V, V_H is 4.6 V, V_L is −1 V, and V_{SS} is 0 V.

Figure 3.33 shows the distributions of readout voltages V_{read} of the 16-level NOSRAM [12,13]. The number of NOSRAM cells is 32,768 (128 kbit). With the V_t cancel-write method using the CAAC-IGZO FET, the distributions of V_{read} are narrow and have sharp peaks which are completely separated. The distribution width 3σ is 37 mV (maximum; see Table 3.6), i.e., 33% narrower than $3\sigma = 55$ mV (max) of the 8-level NOSRAM.

In Figure 3.33, the interval between peaks of distributions of data 1000 and 0111 is larger than those between other peaks. The BL voltage is switched from V_{BL} (1.7 V) to V_{SS} (0 V) when V_{read} becomes lower than V_{REF} (1.8 V) by the SL comparator. Thus, V_F is shifted because of the parasitic capacitance Cp 2 (Figure 3.15) between the BL and node F, thereby decreasing the SL voltage and widening the interval between peaks of distributions of data 1000 and 0111.

Table 3.6 Peak of V_{read} and 3σ of 16-level NOSRAM. *Source*: Reprinted from [13], with permission of IEEE, © 2015

Data	V_{write} [V]	Peak of V_{read} [V]	3σ [mV]
1111	3.50	3.03	31
1110	3.33	2.87	29
1101	3.16	2.71	30
1100	2.99	2.56	30
1011	2.82	2.40	31
1010	2.65	2.25	31
1001	2.48	2.10	31
1000	2.31	1.94	31
0111	2.14	1.60	37
0110	1.97	1.44	36
0101	1.80	1.28	35
0100	1.63	1.13	34
0011	1.46	0.97	33
0010	1.29	0.82	33
0001	1.12	0.67	31
0000	0.95	0.53	35

Table 3.7 Capacitances in NOSRAM (Cp are parasitic capacitances). *Source*: Reprinted from [13], with permission of IEEE, © 2015

Parameter	Capacitance [fF]
Cp 1	0.67
Cp 2	0.34
Cp 3	0.49
PMOS gate	0.72
Cell capacitor	7.10
C_{total}	9.31

Table 3.7 shows the calculated capacitances of the capacitors depicted in Figure 3.15.

Parasitic capacitances are the parameters that determine the shift in peak values of the V_{read} distribution. Cp 1 is the parasitic capacitance between the WWL and node F. The difference between the high and low voltages of the WWL is 5.6 V, and the voltage variation of node F by Cp 1 is 0.40 V. V_c is the sum of the voltage differences of Cp 1, Cp 2, and Cp 3. V_c ranges from 0.40 V to 0.46 V. The voltage variation by Cp 1 ranges from 87% to 100% of V_c. Table 3.8 shows the calculated readout voltage V_{read} for the 16 voltage values V_{write}.

The calculated result of V_{read} in Table 3.8 corresponds to the peaks of the measurement result V_{read} in Table 3.6, which means the V_{read} variation due to parasitic capacitance. Cp 1 is a capacitance generated from overlapping gate and drain electrodes of the CAAC-IGZO FET. Scaling down the CAAC-IGZO FET reduces the overlap capacitance, reducing the voltage

Table 3.8 Calculation results of V_{read}. *Source*: Reprinted from [13], with permission of IEEE, © 2015

Data	V_{write} [V]	V_c [V]	V_{read} [V]
1111	3.50	0.44	3.06
1110	3.33	0.44	2.89
1101	3.16	0.43	2.73
1100	2.99	0.43	2.56
1011	2.82	0.42	2.40
1010	2.65	0.41	2.24
1001	2.48	0.41	2.07
1000	2.31	0.40	1.91
0111	2.14	0.46	1.68
0110	1.97	0.45	1.52
0101	1.80	0.44	1.36
0100	1.63	0.44	1.19
0011	1.46	0.43	1.03
0010	1.29	0.43	0.86
0001	1.12	0.42	0.70
0000	0.95	0.41	0.54

variation of Cp 1. In addition, reducing the voltage difference between high and low voltages of the WWL diminishes the voltage variation in Cp 1. Thus, the distribution width can be narrowed.

Figure 3.34 depicts the temperature dependence of the V_{read} voltage distribution for data 0011 from −40°C to 85°C [14]. With the BL write method [Figure 3.34(a)], V_{read} shifts depending on temperature and the distribution range is 0.32 V. With the V_t cancel-write method [Figure 3.34(b)], the distribution range is 0.13 V, which is 59% narrower than that of the BL write method. The step voltage of the write voltage (V_{write}) is 0.17 V, which is larger than the distribution range of 0.13 V. Therefore, the voltage distributions of V_{read} do not overlap in the indicated temperature range.

Figure 3.35 shows the temperature dependence of the V_{read} peak. The variations of temperature in the BL write method and the V_t cancel-write method are −1.12 mV/°C and −0.24 mV/°C, respectively, which indicates that the V_t cancel-write method provides a 78% smaller value than the BL write method.

Figure 3.36 shows a photograph of the prototype of the 16-level NOSRAM, and Table 3.9 summarizes its specifications.

3.6.5 Comparison of Prototypes

Table 3.10 summarizes the specifications of the prototypes. Note that the distribution width of readout can be reduced to 37 mV at 3σ by correcting the threshold voltage of the PMOS FET in the memory cell with the V_t cancel-write method, while variations resulting from miniaturization of the FET are avoided [13]. In this manner, operation of the 16-level NOSRAM was

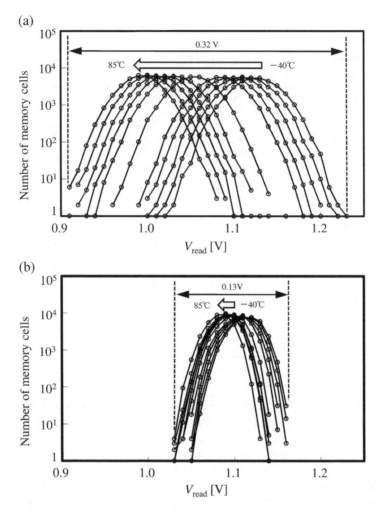

Figure 3.34 Shift depending on temperature of V_{read} distribution: (a) BL write method; (b) V_t cancel-write method. *Source*: Adapted from [14]. Copyright 2015 The Japan Society of Applied Physics

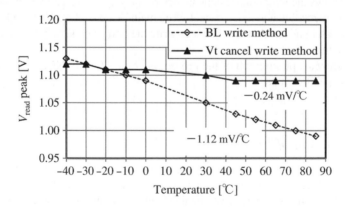

Figure 3.35 Temperature dependence of V_{read} peak. *Source*: Adapted from [14]. Copyright 2015 The Japan Society of Applied Physics

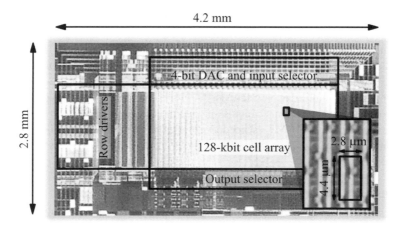

Figure 3.36   Photograph of prototype of 16-level NOSRAM. *(For color detail, please see color plate section)*

Table 3.9   Specification of 16-level NOSRAM. *Source*: Adapted from [13]

Process	0.18-μm CMOS, 0.35-μm CAAC-IGZO
Die size	2.8 mm × 4.2 mm
Memory capacity	128 kbit
Array organization	128 rows × 8 × 32 columns
Access cell number	8
Cell size	34.23 μm^2 (8.15 μm × 4.2 μm)
Number of levels	16
Write time	200 ns
Read time	10 μs
Supply voltage	$V_{DL}/V_{DD}/V_H/V_L$ = 1.8 V/3.5 V/4.6 V/−1 V

Table 3.10   Specifications of NOSRAM

		2-Level NOSRAM [7,8]	4-Level NOSRAM [10]	8-Level NOSRAM [11]	16-Level NOSRAM [12–14]
Bits per cell		1 bit/cell	2 bits/cell	3 bits/cell	4 bits/cell
Distribution of 3σ		100 mV	120 mV	55 mV	37 mV
Memory capacity		1 Mbit	8 kbit	18 kbit	128 kbit
Technology	CMOS	0.8 μm	0.8 μm	0.45 μm	0.18 μm
	CAAC-IGZO	0.8 μm	1.2 μm	0.45 μm	0.35 μm
Write time		150 ns	—	100 ns	200 ns
Read time		900 ns	—	900 ns	900 ns
Supply voltage		V_{DD} = 3 V V_H/V_L = 4.5 V/−1 V	V_{DD} = 3 V V_H/V_L = 4.5 V/−1 V	V_{DD} = 3 V V_H/V_L = 4.5 V/−1 V	V_{DL} = 1.8 V V_{DD} = 3.5 V V_H/V_L = 4.6 V/−1 V

demonstrated under almost the same operational conditions ($V_{DD} = 3$ V, $V_H = 4.6$ V, $V_L = -1$ V) as those of the 2-level NOSRAM.

With further scaling down of the CAAC-IGZO technology, the memory size will become smaller. In addition, the distance between terminals in memory cells is shortened. Thus, the capacitance between terminals is increased and a variation in capacitance leads to a variation in readout voltages of multilevel NOSRAMs. As future challenges, a technology to reduce the variations in capacitance between terminals, or a technology to reduce the variations in readout voltage in multilevel NOSRAMs, is needed. It is expected that the number of levels will increase from the 16-level NOSRAM.

References

[1] Liu, S. Q., Wu, N. J., and Ignatiev, A. (2000) "Electric-pulse-induced reversible resistance change effect in magnetoresistive films," *Appl. Phys. Lett.*, **76**, 2749.

[2] Beck, A., Bednorz, J. G., Gerber, Ch., Rossel, C., and Widmer, D. (2000) "Reproducible switching effect in thin oxide films for memory applications," *Appl. Phys. Lett.*, **77**, 139.

[3] Durlam, M., Andre, T., Brown, P., Calder, J., Chan, J., Cuppens, R., et al. (2005) "90 nm Toggle MRAM array with 0.29 μm² cells," *IEEE Symp. VLSI Technol. Dig. Tech. Pap.*, 186.

[4] Noguchi, H., Ikegami, K., Shimomura, N., Tetsufumi, T., Ito, J., and Fujita, S. (2014) "Highly reliable and low-power nonvolatile cache memory with advanced perpendicular STT-MRAM for high-performance CPU," *IEEE Symp. VLSI Circuits. Dig. Tech. Pap.*, 97.

[5] Lai, S. (2003) "Current status of the phase change memory and its future," *IEEE IEDM Tech. Dig.*, 10.1.1.

[6] Kang, D., Ahn, D., Kim, K., Webb, J., and Yi, K. (2003) "One-dimensional heat conduction model for an electrical phase change random access memory device with 8F2 memory cell (F = 0.15 μm)," *J. Appl. Phys.*, **94**, 3536.

[7] Matsuzaki, T., Inoue, H., Nagatsuka, S., Okazaki, Y., Sasaki, T., Noda, K., et al. (2011) "1Mb Non-volatile random access memory using oxide semiconductor," *Proc. IEEE Int. Memory Workshop*, 185.

[8] Inoue, H., Matsuzaki, T., Nagatsuka, S., Okazaki, Y., Sasaki, T., Noda, K., et al. (2012) "Nonvolatile memory with extremely low-leakage indium–gallium–zinc-oxide thin-film transistor," *IEEE J. Solid-State Circuits*, **47**, 2258.

[9] Tsubuku, M., Takeuchi, T., Ohshima, K., Murakawa, T., Fujita, M., Shimada, D., et al. (2015) "Analysis and experimental proof of deterioration-free memory device using CAAC-IGZO FET," *Ext. Abstr. Solid State Dev. Mater.*, 1148.

[10] Ishizu, T., Inoue, H., Matsuzaki, T., Nagatsuka, S., Okazaki, Y., Onuki, T., et al. (2012) "Multi-level cell memory with high-speed, low-voltage writing and high endurance using crystalline In–Ga–Zn oxide thin film FET," *Ext. Abstr. Solid State Dev. Mater.*, 590.

[11] Nagatsuka, S., Matsuzaki, T., Inoue, H., Ishizu, T., Onuki, T., Ando, Y., et al. (2013) "A 3bit/cell nonvolatile memory with crystalline In–Ga–Zn–O TFT," *Proc. IEEE Int. Memory Workshop*, 188.

[12] Matsuzaki, T., Onuki, T., Nagatsuka, S., Inoue, H., Ishizu, T., Ieda, Y., et al. (2015) "A 128kb 4b/cell nonvolatile memory with crystalline In–Ga–Zn oxide FET using V_t cancel write method," *IEEE Int. Solid-State Circuits Conf. Dig. Tech. Pap.*, 306.

[13] Matsuzaki, T., Onuki, T., Nagatsuka, S., Inoue, H., Ishizu, T., Ieda, Y., et al. (2015) "A 16-level-cell nonvolatile memory with crystalline In–Ga–Zn oxide FET," *Proc. IEEE Int. Memory Workshop*, 125.

[14] Matsuzaki, T., Onuki, T., Nagatsuka, S., Inoue, H., Ishizu, T., Ieda, Y., et al. (2015) "A 16-level-cell memory with 0.24 mV/°C temperature characteristics comprising crystalline In–Ga–Zn oxide FET," *Ext. Abstr. Solid State Dev. Mater.*, 122.

4

DOSRAM

4.1   Introduction

A dynamic random access memory (DRAM) in which a c-axis-aligned crystalline indium–gallium–zinc oxide (CAAC-IGZO) field-effect transistor (FET) is used is referred to as a dynamic oxide semiconductor random access memory (DOSRAM). This chapter introduces the functions, operations, and characteristics of DOSRAMs.

Large-scale integrated (LSI) memory circuits used for general-purpose computing systems have advanced noticeably over recent decades. In particular, they comprise an ever-larger memory capacity, faster operation, and lower power consumption. DRAM, which is a typical memory component in computer systems, is a high-performance component, and its simple memory cell structure is suitable for very fine fabrication; thus, the cost of DRAMs is low and they are used in many systems. However, DRAM is a volatile memory, which is a major drawback because it makes reducing power consumption difficult when it is not accessed. This situation has spurred research into LSI memory technologies of new materials.

Because CAAC-IGZO FETs have an extremely low off-state current, on the order of yoctoamps per micrometer (yA/μm; see Section 2.2), charges in a capacitor connected to such a transistor can be stored for a long time. This characteristic is used in DOSRAM memory cells, which are expected to replace the DRAM and become the next-generation memory.

The following sections first explain the problems of DRAMs and then describe the circuit configurations, properties, and characteristics of DOSRAMs.

Physics and Technology of Crystalline Oxide Semiconductor CAAC-IGZO: Application to LSI, First Edition.
Edited by Shunpei Yamazaki and Masahiro Fujita.
© 2017 John Wiley & Sons, Ltd. Published 2017 by John Wiley & Sons, Ltd.

4.2   Characteristics and Problems of DRAM

The memory cell of a DRAM comprises an n-channel (Nch) Si FET and a cell capacitor connected to the FET. The structure is simple and suitable for microfabrication, and requires only a small number of manufacturing steps. Thus, DRAMs can be produced at low cost.

However, the Nch-Si FET has a large off-state current as well as a leakage current from its source/drain to the Si substrate, so charge leaks from the capacitor relatively quickly. Meanwhile, the memory cell reading voltage, which should be sufficiently high to operate a sense amplifier to determine the data 0 or 1 in the memory, is specified by the amount of charge stored. Thus, rewriting (i.e., refreshing) is required to compensate for the loss of the stored charge. The refresh operation makes it difficult to reduce power consumption. In addition, to ensure the minimum storage capacitance required for correct operations, the size of the cell capacitor cannot be too small.

To scale down to 36-nm technology node, the capacitance of the cell capacitor is ensured by thinning the capacitor dielectric and developing materials with high dielectric permittivity. However, ensuring the capacitance in a small-technology node becomes difficult because the capacitance of the cell capacitor has to be at least ~20 fF, even for a technology node of 32 nm or smaller [1].

4.3   Operations and Characteristics of DOSRAM Memory Cell

A DOSRAM cell comprises a circuit where the Nch-Si FET of the DRAM is replaced with a CAAC-IGZO FET (see Figure 4.1). The CAAC-IGZO FET and a cell capacitor (Cs) are located in the same layer.

Data reading and writing are described below. To write data 1, a supply voltage V_{DD} is applied to a bit line (BL), and the voltage V_{WL1}, which is greater than the sum of V_{DD} and the threshold voltage V_{th} of the CAAC-IGZO FET, is applied to the word line (WL). Accordingly, the CAAC-IGZO FET is turned on, and the charge for data 1 is stored in Cs. Next, WL is set to 0 V to turn off the FET, and the writing of data 1 terminates.

Because of the very low off-state current of the CAAC-IGZO FET, the charge stored in Cs is not lost [2], but held for a long time. Therefore, the interval between refresh operations can be significantly extended, so a lower refresh frequency is possible. During the period without refresh, the data can be stored without supplying voltage to the BL and WL; thus, the memory cell practically serves as a non-volatile memory. This is the biggest difference in comparison with volatile DRAM, which requires frequent refreshing.

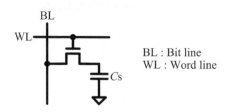

Figure 4.1   Circuit diagram of DOSRAM cell. *Source*: Adapted from [4]

A common DRAM requires a refresh operation every 64 milliseconds (ms). In contrast, in DOSRAMs, the interval between refresh operations is seven orders of magnitude longer than DRAM (in terms of time), so the power consumed by DOSRAMs is significantly lower than that by DRAMs [3].

To write data 0, 0 V is applied to the BL so that the charge in Cs is reduced to 0.

To read data (1 or 0), the BL is first precharged to $V_{DD}/2$, and then V_{WL1} is applied to WL to turn on the CAAC-IGZO FET. In response to the charge stored in Cs, the BL voltage changes from $V_{DD}/2$. This changed BL voltage is a readout voltage corresponding to data 1 or 0 being read from the memory cell.

This scheme for DOSRAM cell operation uses the on–off characteristics of the CAAC-IGZO FET. No charge is injected into the insulating film of the gate, and thus the elements do not degrade. Therefore, the memory cells may be read or written an unlimited number of times.

4.4 Configuration and Basic Operation of DOSRAM

From Section 4.4 to Section 4.6, an 8-kbit DOSRAM that was actually fabricated is explained.

4.4.1 Circuit Configuration and Operation of DOSRAM

The DOSRAM with a 8-kbit capacity comprises eight circuit blocks (see Figure 4.2). Basic reading and writing operations are as follows: signals such as a read–write bar (RWB), a column line enable (CLE), or a word line enable (WLE) are input to a controller; a 7-bit column address (CA) signal is input to a column decoder; a 6-bit row address (RA) signal is input to a row decoder. This selects a memory cell from the array of 8-kbit (8192) memory cells. Writing data that has been supplied to data input (DIN) passes through a data-in buffer and the IO gating of the sense amplifiers, following which they are written into the selected memory cell. Readout data from the selected memory cell is amplified by the sense amplifier and sent out through the data output (DOUT).

The 8-kbit memory cell array consists of an arrangement of 64 rows × 128 columns. The sense amplifier is connected to the cell array with folded BLs, which are resistant to noise generated in the memory cell array when a WL is selected.

4.4.2 Hybrid Structure of DOSRAM

Both a CAAC-IGZO FET and Cs, which together constitute a memory cell, are formed in the CAAC-IGZO FET layer. Peripheral circuits other than the cell array are formed in a Si FET layer on the substrate, below the CAAC-IGZO FET layer as shown in Figure 4.3, thereby reducing the die area by approximately 21% compared with a structure without stacking. In addition, the distance between the peripheral circuits in the lower layer and the cell array in the upper layer can be small, which accelerates operations. A schematic cross-sectional view of the hybrid structure is shown in Figure 4.4.

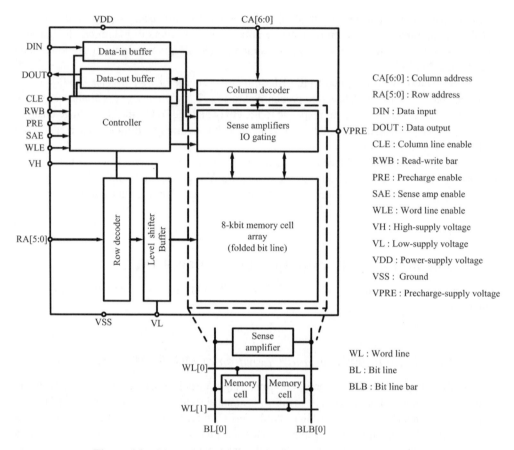

Figure 4.2   Block diagram of DOSRAM. *Source*: Adapted from [3]

4.5   Operation of Sense Amplifier

Figure 4.5 shows the basic circuit diagram of the sense amplifier connected to the cell array with folded BLs.

4.5.1   Writing Operation

To write data, V_H is first applied to a precharge enable (PCE) to turn FETs on in the equalizer. The BL and BLB enter the conducting state and have precharge voltage V_{PRE} (here, $V_{DD}/2$). A voltage of 0 V is applied to the PCE to turn off the equalizer transistors. Next, the global bit line (GBL) and the global bit line bar (GBLB) are each supplied with the voltage to be written to memory. V_H is applied to a column line select (YSW) to turn on the YSW transistors, whereby BL and BLB have the same voltages as GBL and GBLB, respectively. Subsequently, V_H is applied to the nth word line (WL[n]) to turn on the CAAC-IGZO FET in memory cell [n]. To write data 1, V_{DD} is applied to the BL so that the charge corresponding to data 1 is stored in Cs. To write data 0, 0 V is applied to the BL so that zero charge is stored in Cs. After that, the CAAC-IGZO FET in memory cell [n] is turned off by applying V_L to WL[n]. The CAAC-IGZO

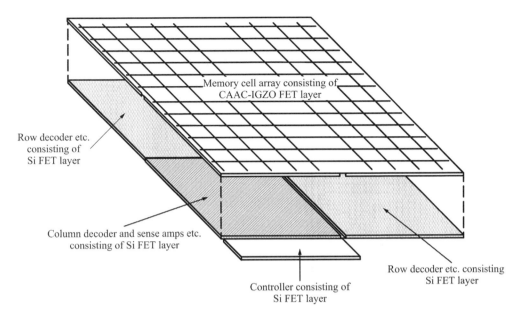

Figure 4.3 Hybrid structure of Si FET layer and CAAC-IGZO FET layer

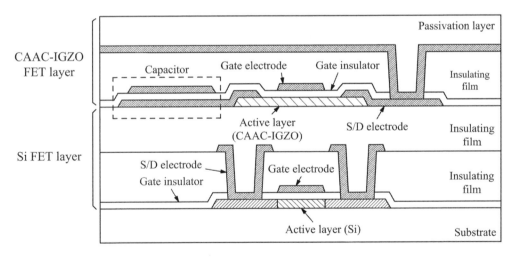

Figure 4.4 Cross-sectional view of stacked layers of DOSRAM. *Source*: Adapted from [3]

FET in memory cell [$n-1$] is kept off by applying V_L to WL[$n-1$]. Finally, V_L is applied to YSW to turn off the YSW transistor, and the writing operation terminates.

4.5.2 Reading Operation

To read data 1 or 0, the BL and BLB are first precharged to $V_{DD}/2$, and then V_H is applied to WL [n] to turn on the CAAC-IGZO FET in memory cell [n]. The voltage V_L is applied to WL[$n-1$]

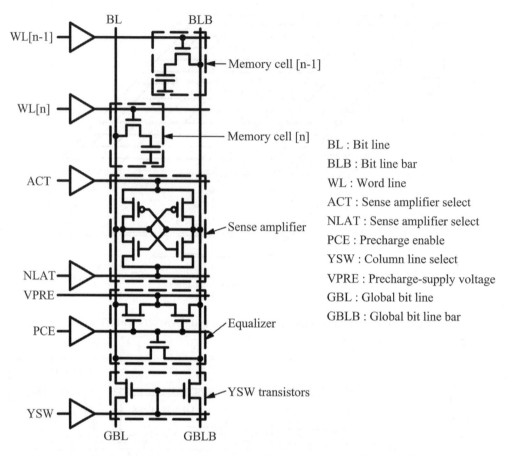

BL : Bit line
BLB : Bit line bar
WL : Word line
ACT : Sense amplifier select
NLAT : Sense amplifier select
PCE : Precharge enable
YSW : Column line select
VPRE : Precharge-supply voltage
GBL : Global bit line
GBLB : Global bit line bar

Figure 4.5   Basic circuit diagram of sense amplifier connected to cell array with folded BLs

so that the CAAC-IGZO FET in memory cell [n−1] remains in the off state. The BL voltage varies from $V_{DD}/2$ in response to the charge stored in Cs. The varied BL voltage is the readout voltage corresponding to data 1 or 0 being read from memory cell [n]. The readout voltage V_{sig} is expressed by

$$V_{sig} = (V_{write} - V_{leak} - V_{PRE}) \times \left(\frac{C_s}{C_s + C_{bl}}\right) \qquad (4.1)$$

where V_{write} is a voltage written into the memory cell, V_{PRE} is the precharge BL voltage, V_{leak} is the voltage drop due to the loss of charge stored in Cs, C_s is the capacitance of Cs, and C_{bl} is the parasitic capacitance of the BL. Next, a sense amplifier select (ACT) power source supplies the voltage V_{DD} to p-channel (Pch) Si FETs and another sense amplifier select (NLAT) supplies 0 V to Nch-Si FETs; in other words, ACT and NLAT supply power and ground to the sense amplifier, respectively. The BL voltage becomes V_{DD} (0 V) in the case of reading data 1 (0).

According to Equation (4.1), V_{sig} depends on V_{leak}. Unlike for an Nch-Si FET, however, V_{leak} of the CAAC-IGZO FET is negligibly small, and thus a high V_{sig} is obtained and the sense amplifier operates stably.

4.6 Characteristic Measurement

This section describes the characteristic measurement results of the fabricated 8-kbit DOSRAM. The gate lengths L of the Si and CAAC-IGZO FETs are 0.8 μm and 1.2 μm, respectively; the capacitance C_s is 31 fF; and the die area is 1.1 mm × 1.4 mm. The supply voltages used for the characteristic measurements were V_{DD} = 1.8 V, V_H = 3.3 V, and V_L = −1 V. V_{PRE} is $V_{DD}/2$.

4.6.1 Writing Characteristics

Figure 4.6 shows the operation waveforms for writing data 1. 1.8 V is applied to DIN. A column line is selected when SAE and CLE are at 1.8 V. The memory cell is selected when WLE is at 1.8 V (boosted to 3.3 V in the circuit). Next, data 1 is written to the selected memory cell. The write time elapsed from applying voltage to the DIN to switching off the CAAC-IGZO FET (i.e., from CLE rising to WLE falling) is 75 ns. Figure 4.7 shows the Shmoo plot for the writing operation with respect to write time. In the graph, "Pass" means that the output is obtained under corresponding measurement conditions, whereas "Fail" means that no output is obtained. The writing operation for data 0 is examined in a similar manner, following which normal operation is confirmed.

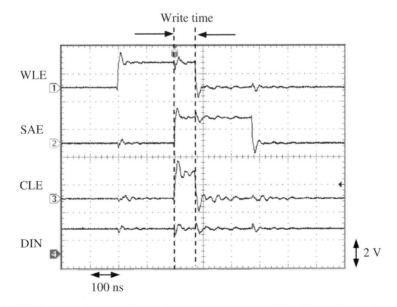

Figure 4.6 Writing-operation waveforms. *Source*: Reprinted from [3], with permission of IEEE, © 2012

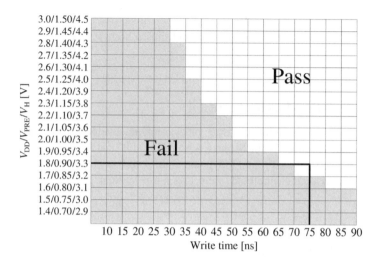

Figure 4.7   Shmoo plot of write time. *Source*: Reprinted from [3], with permission of IEEE, © 2012

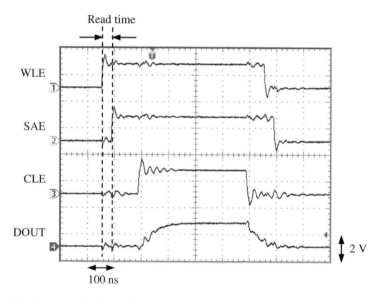

Figure 4.8   Reading-operation waveforms. *Source*: Reprinted from [3], with permission of IEEE, © 2012

4.6.2   *Reading Characteristics*

To read data 1, 1.8 V (boosted to 3.3 V in the circuit) is applied to WLE and 1.8 V is applied to SAE and CLE; then, data is read out from the memory cell, and DOUT is changed from 0 V to 3.3 V. The read time is 35 ns (Figure 4.8). This is the time elapsing between switching on the CAAC-IGZO FET and driving the sense amplifier (i.e., from WLE rising to SAE rising).

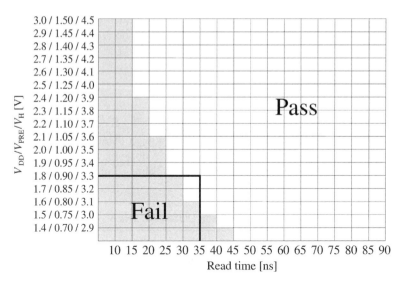

Figure 4.9 Shmoo plot of read time. *Source*: Reprinted from [3], with permission of IEEE, © 2012

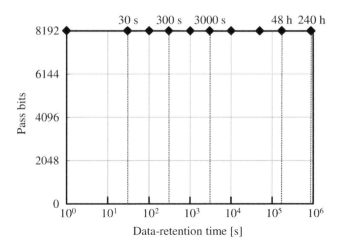

Figure 4.10 Data-retention characteristics at 85°C. *Source*: Reprinted from [3], with permission of IEEE, © 2012

Figure 4.9 shows the Shmoo plot of the reading operation. The reading operation for data 0 was also examined and normal operation was confirmed.

4.6.3 Data-Retention Characteristics

Figure 4.10 shows the data-retention characteristics for data 1 and 0 at 85°C. The vertical axis represents the number of memory cells from which stored data is normally read out, and the horizontal axis represents the data-retention time. After measuring for 240 h, no defect bit occurred. This result suggests that a refresh operation can be done once every 240 h or longer.

4.6.4 Summary of 8-kbit DOSRAM

The data reading and writing operations for DOSRAMs were confirmed, as discussed above. The data-retention time is seven orders of magnitude longer than that of DRAMs, and it has been verified that DOSRAMs can serve practically as a non-volatile memory. The specifications and a die photograph of a DOSRAM are given in Table 4.1 and Figure 4.11, respectively.

Table 4.1 Specifications of the 8-kbit DOSRAM. *Source*: Adapted from [3]

Process	1.2-μm CAAC-IGZO/0.8-μm Si	
Die size	1.1 mm × 1.4 mm	
Memory capacity	8 kbit	
Array organization	64 (rows) × 128 (columns)	
Storage capacitance (C_s)	31 fF	
Bit-line architecture	Folded	
Supply voltage	Nominal	1.8 V
	Level shifter	3.3 V/–1 V

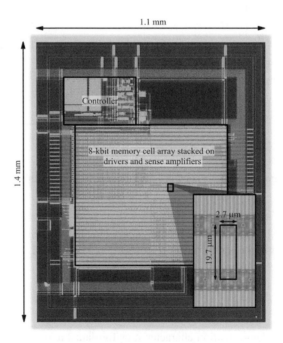

Figure 4.11 Die photograph of the 8-kbit DOSRAM. *(For color detail, please see color plate section)*

4.7   Prototype DOSRAM Using 60-nm Technology Node

In Sections 4.4 to 4.6, we observed the characteristics and operational properties of the 8-kbit DOSRAM with an oxide semiconductor in a memory cell. In this section, the suitability for fine processing of the memory cell will be discussed, as well as the test result of a DOSRAM fabricated by a 60-nm technology node.

4.7.1   Configuration of Prototype

Transfer gates supplying writing data to the BL and a source follower for outputting readout data are connected to the memory cell (see Figure 4.12). The FETs in the memory cell are all CAAC-IGZO FETs, each with a gate length $L = 60$ nm and a channel width $W = 40$ nm. The cell array has 16 bits (4 × 4), and dummy cells are arranged around the cell array to eliminate

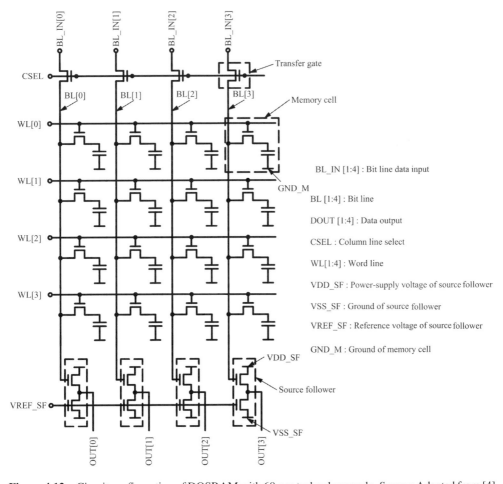

Figure 4.12   Circuit configuration of DOSRAM with 60-nm technology node. *Source*: Adapted from [4]

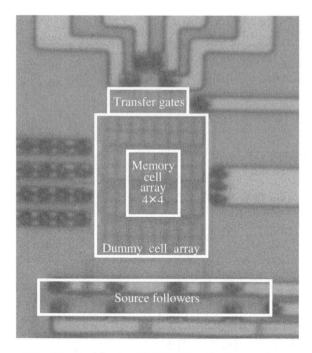

Figure 4.13 Die photograph of DOSRAM with 60-nm technology node

optical proximity effects during exposure to light (see Figure 4.13). To evaluate the effect of cell capacitance C_s, which decreases with finer processing, C_s is set to 3.9 fF, which is less than a fifth of the capacitance of a common DRAM. The bit-line capacitance C_{bl} = 8.7 fF, and the ratio of C_{bl} to C_s is 2.2.

4.7.2   Measurements of Prototype Characteristics

4.7.2.1   Writing Characteristics

To write data 1, 3 V is first applied to the column line select (CSEL) to turn on the FET of the transfer gate. Next, the bit-line data input (BL_IN) and BL are set to 1.8 V. Simultaneously, a line for the reference voltage of the source follower (VREF_SF), a line for the power-supply voltage of the source follower (VDD_SF), and a line for the ground of the source follower (VSS_SF) are set to −1, 0, and −3 V, respectively, to activate the source follower. Next, 3 V is applied to WL to select the memory cell, and charge is stored in Cs. After that, −1 V is applied to WL to turn off the memory cell FET, and the writing operation terminates. To write data 0, 0 V is applied to BL_IN, and zero charge is stored in Cs of the similarly selected memory cell.

Figure 4.14 shows the operation waveforms for writing data 1. This graph shows the result in the case of a writing time of 10 μs, as an example. The write time is defined as the time elapsing from source follower activation until setting the WL voltage to 3 V. The operation for writing data 0 was examined in a similar manner, and normal operation was confirmed.

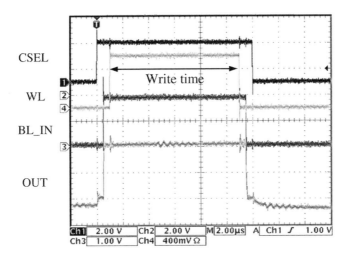

CSEL

WL

BL_IN

OUT

Figure 4.14 Writing-operation waveforms. *Source*: Reproduced from [4], with permission of *Japanese Journal of Applied Physics*

4.7.2.2 Reading Characteristics

To read data 1 or 0, BL_IN is set to a precharge voltage of 0.8 V, and 3 V is applied to CSEL to turn on the FETs of the transfer gates. BL and BL_IN begin conducting when 3 V is applied to CSEL (precharge time), so that BL is precharged to 0.8 V. Next, VREF_SF, VDD_SF, and VSS_SF are supplied with −1, 0, and −3 V, respectively, to activate the source follower. From the output voltage V_{out} of the source follower and from knowledge of the source follower characteristics, we can monitor the BL voltage. Next, 3 V (no boosting necessary in this case) is applied to WL to turn on the CAAC-IGZO FET of the memory cell. Here, the read time is defined as the time during which the CAAC-IGZO FET of the memory cell is on. During the read time, the BL voltage changes from 0.8 V because of the charge stored in Cs. The changed BL voltage is a readout voltage V_{sig} expressed by Equation (4.1). V_{sig} can be determined from the difference between V_{out} before and after the read time (ΔV).

Figure 4.15 shows the waveforms for reading data 1. This graph shows the result in case of a reading time of 4 µs, as an example. The graph shows ΔV corresponding to data 1. While reading data 0, ΔV corresponding to data 0 is similarly observed.

4.7.2.3 Measurement of Operation Speed

Figure 4.16 shows the relationship between V_{sig} and the operation speed. Because the DOSRAM uses the CAAC-IGZO FET, which has almost no leakage, a high V_{sig} (±100 mV) is obtained. In addition, the cell operates quickly (read time and write time are each 5 ns).

4.7.2.4 Data-Retention Characteristics

To measure the data-retention characteristics, data is written in a write time of 100 ns (see Subsection 4.7.2.1), following which the WL, BL_IN, and CSEL voltages are set to −1, 0, and 0 V, respectively, so that the charge stored in Cs is maintained. The read time is 100 ns.

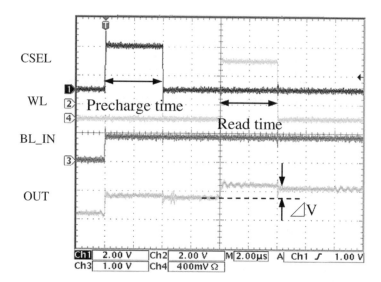

Figure 4.15 Reading-operation waveforms. *Source*: Reproduced from [4], with permission of *Japanese Journal of Applied Physics*

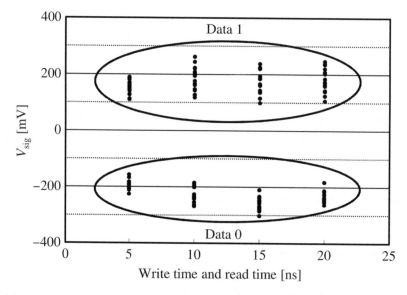

Figure 4.16 Relationship between V_{sig} and operation speed (read time = write time). *Source*: Reproduced from [4], with permission of *Japanese Journal of Applied Physics*

Figure 4.17 shows the results of retention tests at 27°C and 85°C. At each temperature, $V_{sig} = 150$ mV or higher was maintained after 1 h, and no defect bit occurred. This suggests that the interval between refresh operations can be 1 h or even longer.

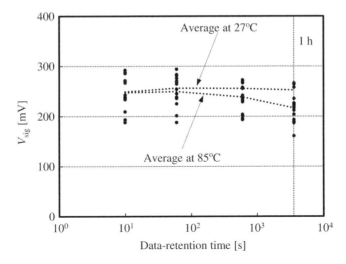

Figure 4.17   Data-retention test at 27°C and 85°C. *Source*: Reproduced from [4], with permission of *Japanese Journal of Applied Physics*

4.7.3   Summary for Prototype DOSRAM

The characteristic measurements of the prototype confirm that the DOSRAM, even with L as short as 60 nm and C_s less than a fifth that of the DRAM, has excellent read and write characteristics. In addition, the data-retention time for DOSRAM fabricated based on a 60-nm technology node is four orders of magnitude longer than for DRAM. Thus, memory cells with CAAC-IGZO FETs are suitable for fine processing.

4.8   Conclusion

DOSRAM is a memory component that uses a CAAC-IGZO FET in the memory cell of a DRAM.

The fabricated 8-kbit DOSRAM verified that the memory cell array in the CAAC-IGZO FET layer can be stacked above a Si FET layer that includes a decoder and a sense amplifier, and that the data-retention time is seven orders of magnitude longer than that of a corresponding DRAM. These results show that the refresh frequency required for DOSRAM is much lower than that of DRAM, making them highly promising for reducing power consumption.

The prototype DOSRAM fabricated with a 60-nm technology node confirmed that even a very small storage capacitance enables a retention time four orders of magnitude longer than that of DRAM, and that the read and write speed is 5 ns or less.

Correct operations without read/write errors of the 8-kbit DOSRAM and prototype have been confirmed. With its non-volatility and low power consumption, DOSRAM may replace DRAM as the next-generation mainstream memory technology.

References

[1] ITRS (2012) "Table FEP5 DRAM stacked capacitor technology requirement." Available at: www.itrs.net/ITRS%201999-2014%20Mtgs,%20Presentations%20&%20Links/2012ITRS/2012Tables/FEP_2012Tables.xlsx [accessed October 7, 2015].

[2] Kato, K., Shionoiri, Y., Sekine, Y., Furutani, K., Hatano, T., Aoki, T., *et al.* (2012) "Evaluation of off-state current characteristics of transistor using oxide semiconductor material, indium–gallium–zinc oxide," *Jpn. J. Appl. Phys.*, **51**, 021201.

[3] Atsumi, T., Nagatsuka, S., Inoue, H., Onuki, T., Saito, T., Ieda, Y., *et al.* (2012) "DRAM using crystalline oxide semiconductor for access transistors and not requiring refresh for more than ten days," *Proc. IEEE Int. Memory Workshop*, 99.

[4] Onuki, T., Kato, K., Nomura, M., Yakubo, Y., Nagatsuka, S., Matsuzaki, T., *et al.* (2015) "Fabrication of dynamic oxide semiconductor random access memory with 3.9 fF storage capacitance and greater than 1 h retention by using *c*-axis aligned crystalline oxide semiconductor transistor with L of 60 nm," *Jpn. J. Appl. Phys.*, **54**, 04DD07.

5

CPU

5.1 Introduction

Mobile devices such as smartphones and tablets are getting more functions and higher perform-ance, and they will take more essential roles in the Internet of Things (IoT). To satisfy these expectations, it is necessary to develop a new technology for enabling higher-speed operation and lower power consumption simultaneously. This is the most significant issue for battery-powered mobile devices.

This chapter begins with normally-off computing [1], which achieves sufficiently high per-formance at minimum power consumption. It then introduces normally-off central processing units (CPUs) and c-axis-aligned crystalline indium–gallium–zinc oxide (CAAC-IGZO) cache memory based on CAAC-IGZO field-effect transistors (FETs).

5.2 Normally-Off Computing

Before discussing normally-off computing, we explain typical technologies for lowering the power consumption. Two widely known examples are clock gating and power gating.

Clock gating reduces the power by stopping the clock supply to non-operating parts of the circuit [Figure 5.1(a)]. This technology can be applied to circuit blocks and reduce the power consumption for clock supply. Furthermore, the method is effective when the power consump-tion for clock supply is dominant in the total power consumption of the die. However, in the forefront of the miniaturization process, sufficiently reducing the power consumption with clock gating alone is difficult because the power consumption caused by leakage current of the FET is increasing.

Power gating stops the power supply to each circuit block that need not operate [Figure 5.1(b)]. This method is extremely effective in transistor technologies with high leakage current, because

Physics and Technology of Crystalline Oxide Semiconductor CAAC-IGZO: Application to LSI, First Edition.
Edited by Shunpei Yamazaki and Masahiro Fujita.
© 2017 John Wiley & Sons, Ltd. Published 2017 by John Wiley & Sons, Ltd.

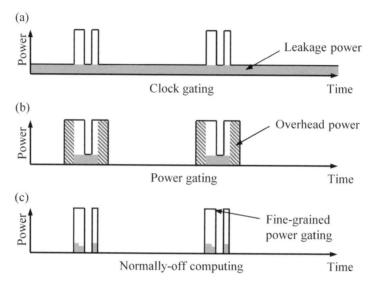

Figure 5.1   Power-reducing schemes of clock and power gating compared with normally-off computing (bottom)

it completely stops the power supply to a circuit block. However, volatile memory elements such as flip-flops, static random access memory (SRAM), and dynamic random access memory (DRAM) lose their data unless supplied with power. Therefore, power gating is of limited applicability in data-retentive circuit blocks. Although power gating may be executed in volatile blocks by saving the data of the volatile memory section into an external non-volatile memory section, the backup and restoration of the data incur considerable energy and performance overheads, because a circuit is required to backup/restore data of the circuit block into/from the external non-volatile circuit. Indeed, the overhead of power gating offsets its power reduction.

Contrary to these technologies, normally-off computing supplies power to a circuit only when needed [Figure 5.1(c)]. The data memory block is a non-volatile memory that does not require power in the normal state. Because the non-volatile memory retains data in the absence of a power supply, it is amenable to circumstantial power supply control by power gating. Therefore, this technology can effectively reduce wasted power.

To achieve normally-off computing, memory elements have been developed using CAAC-IGZO FETs [2, 3]. Unlike Si FETs, CAAC-IGZO FETs (even with short channel lengths) have significantly low off-state current. A capacitor and CAAC-IGZO FET can therefore provide a simple memory configuration, whereby the CAAC-IGZO FET can be turned off once the data are written into the capacitor. Using this configuration, we can create memory elements requiring neither constant power supply nor frequent refresh operations, thus reducing the power consumption. The reading and writing operations in the memory element are performed by charging and discharging the capacitor, and theoretically there is no device degradation due to reading and writing operations.

CAAC-IGZO FETs also have the advantage of an extremely small area overhead, which is the circuit area that increases when a CAAC-IGZO FET is added. As shown in Figure 5.2, a circuit with CAAC-IGZO FETs can be stacked on top of Si FETs. For example, a backup flip-flop circuit fabricated with CAAC-IGZO FETs can be positioned on a normal flip-flop

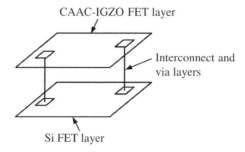

Figure 5.2   Schematic view of CAAC-IGZO FETs stacked on Si FETs

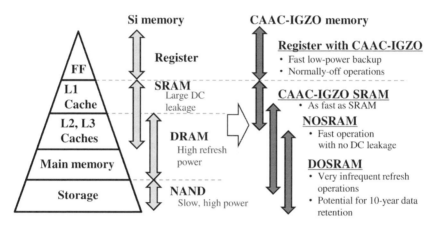

Figure 5.3   Memory hierarchy of normally-off computing

section formed with Si FETs. This configuration provides the flip-flop with a backup function without increasing its area, while maintaining the speed advantages of Si FETs. It also provides reduced interconnection delay, which enables high-speed backup.

Figure 5.3 shows the memory hierarchy of normally-off computing. Flip-flops with CAAC-IGZO FETs can be used in register files, the fastest-operating storage units in CPUs. A CAAC-IGZO SRAM (see Section 5.4) can operate at a speed comparable to that of an SRAM formed with Si FETs, and thus satisfies the required operation speed for an L1 cache. A non-volatile oxide semiconductor random access memory (NOSRAM), which was discussed in Chapter 3, can be applied to L1, L2, and L3 caches. For main memories and storages that require larger volume and higher density, we can apply DOSRAM, a high-density memory device with CAAC-IGZO FETs.

Utilization of a non-volatile memory and straightforward power switching may not reduce power consumption by themselves. In particular, if the power required for data backup into non-volatile memories exceeds the power saved by utilizing the non-volatile memory, the power consumption will instead increase. Figure 5.4 illustrates the break-even time (BET), which is defined as the time during which the energy required for executing backup (including the increased power usage by the backup mechanism during normal operation) equals the saved

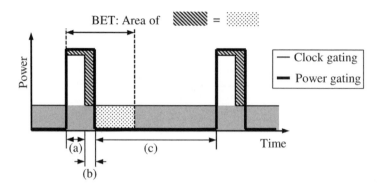

Figure 5.4 Schematic diagram of the break-even time

energy. In period (a) of this figure, the circuit is processing data. After processing, the device enters period (b) and the circuit is turned off. The data backup power is consumed at this time. Finally the circuit enters period (c), where the power consumption is minimal. Even if the circuit enters period (c) after the backup operation, the power savings will not override the total power consumption unless period (c) exceeds the BET. The memory elements in normally-off computing must satisfy two essential requirements: low power consumption for access, and a power-off period exceeding the BET. The latter depends on the former.

5.3 CPUs

Computers are incorporated in many everyday electrical appliances, such as mobile phones, television devices, and automobiles. They are often called embedded systems.

Generally, an embedded system is a circuit consisting of five basic parts: output, input, memory, data path, and control (Figure 5.5). The output is related to, for example, display units and/or speakers via interfaces that output signals for controlling such devices. The inputs are acquired through devices such as touch panels, keyboards, mice, and other devices providing data for processing. The signals transmitted from these devices are received through interfaces. The memory retains software, text information, images, sounds, and other data. DRAM is the typical memory in smartphones and personal computers. The flexibility of software increases with increasing memory capacity, spurring significant developments in high-capacitance DRAM. On-chip SRAM, a highly integrated device that realizes small size at low cost, is commonly used in computer-based electrical appliances. The data path executes the required calculations, including the four basic arithmetic operations and comparison processing. The control mediates the output, input, memory, and data path in accordance with the software program. In particular, a combination of the data path and control is often called a CPU.

As mentioned above, the CPU consists of a data path and control. Referring to Figure 5.6, we now discuss the details of these components.

The CPU is synchronized with an external clock signal output. The data path has a register file and an arithmetic logic unit (ALU). During proper operation, the CPU interprets an instruction readout from the memory, and the control synchronizes the data path with the clock.

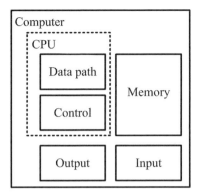

Figure 5.5   Basic configuration of a computer

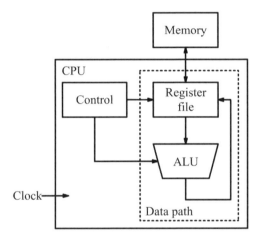

Figure 5.6   Simplified diagram of a CPU

For accurate processing, the CPU should store its own state. In general, the state of the CPU is stored in a flip-flop and SRAM in the control and the register file.

The flip-flop and SRAM provided in the CPU store temporary data necessary for continuing the processing, as mentioned above, but the stored data are generally lost when the power supply to the flip-flop and SRAM is discontinued. Accordingly, the CPU resides in an unstable or reset state during power off. In normally-off computing, however, the CPU can continue to process the data stored in the flip-flop and SRAM without any glitches, even after a power-off period. Efficiently switching the power on and off while retaining information in the CPU is therefore important for reducing power consumption.

As explained in Section 5.2, CAAC-IGZO FETs can realize a memory element that retains data in the absence of a power supply. Researchers have developed a variety of low-power-consumption CPUs based on CAAC-IGZO memory elements, and examined their operation. Some of their findings are summarized below.

- In 2012, Ohmaru *et al.* [2] reported an 8-bit normally-off CPU capable of storing data for 40 days at 85°C.
- In 2013, Sjökvist and co-workers [3, 4] reduced the overhead area of the flip-flops in a 32-bit normally-off CPU.
- In 2014, Tamura *et al.* [5] reported a normally-off CPU with an ARM® Cortex®-M0 and on-chip SRAM incorporating CAAC-IGZO FETs.
- Also in 2014, Ishizu *et al.* [6] reported a cache memory based on CAAC-IGZO FETs.

Examples of normally-off CPUs fabricated from CAAC-IGZO FETs are shown below.

5.3.1   Flip-Flop (FF)

Examples of flip-flops (FFs) with CAAC-IGZO are explained in the following subsections.

5.3.1.1   Backup FF (Type A)

Figure 5.7 shows the circuit diagram of a Type-A backup FF [2, 9], which comprises backup circuit and D-FF sections. The D-FF employs Si FETs and is generally used in digital circuit design, whereas the backup circuit is constructed from both Si and CAAC-IGZO FETs.

The backup FF operates in two modes: active mode, wherein the FF operates normally with input clock signals, and sleep mode, wherein the power supply terminates and the D-FF portion transfers its data to the backup section. Forcing the circuit into sleep mode during quiescent periods reduces the standby power, which is caused by the leakage current of the Si FETs. A circuit operation that switches from active to sleep mode is called a backup operation; the reverse operation is called a restoration operation.

In active mode, the Type-A backup FF operates as a D-FF. However, as the D-FF and backup circuit sections are functionally separated, the additional backup circuit increases the delay time of the D-FF. The time increase is caused by the one-stage multiplexer (the central part of the circuit diagram in Figure 5.7) that connects the two sections. Any speed reduction of the D-FF because of the addition of the backup function can be minimized in the backup FF. This is one feature of the backup FF.

The operation of the Type-A backup FF is detailed in the timing diagram of Figure 5.8. In this diagram, VDD is the power supply voltage supplied to the entire FF. High and low levels denote that power is supplied to and blocked from the entire FF, respectively. In active mode (T1), the input signals WE and RE are set to a low level to functionally separate the D-FF section from the backup circuit section. In this mode, the backup FF functions as a D-FF. T2 indicates the period of backup operation. The data from the D-FF section is sent to the backup circuit section, with the CLK fixed at a high level to distinguish the data. The input signal WE is then set to a high level to charge or discharge the capacitor Cs. In sleep mode (T3), the input signal WE is fixed at a low level, and the data stored as charge in the capacitor Cs is retained. Since the FET controlled by the WE is a CAAC-IGZO FET with an extremely small off-state current, the capacitor Cr retains the charge over a long period, even without the power supply. T4 and T5 denote different periods of the restoration operation. During T4, the power supply is resumed and the input signal RE is set low to charge the capacitor Cr. During T5, the input signal RE is set high to retain or release the charge in the capacitor Cr, depending on the voltage of node N1. The restoration

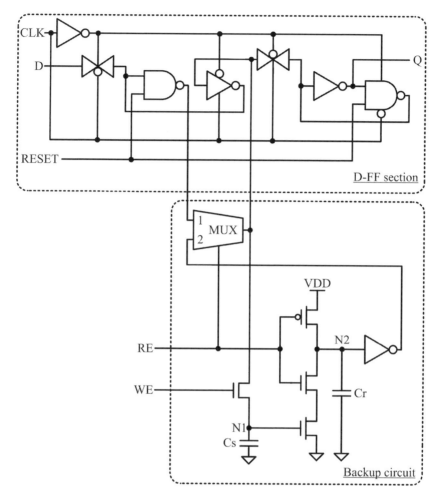

Figure 5.7 Circuit diagram of Type-A backup FF. *Source*: Adapted from [9]

operation is complete when the potential of node N2 is input to the D-FF section via the inverter and the multiplexer. During period T6, the input signal RE is returned to a low level, and the CLK signal switches the FF to active mode.

Simulation results of backup FFs with miniaturization processes are shown below.

Figure 5.9 shows a micrograph of a fabricated Type-A backup FF. Figure 5.10 shows the layout of an assumed backup FF with a 30-nm Si FET. The *W/L* of the CAAC-IGZO FET was 0.3 μm/0.3 μm, and the capacitance of the storage capacitor Cr was 2 fF. The layout area was 8.19 μm^2, 25% larger than that of a D-FF formed only using Si FETs. The increase in area is contributed by the reading circuit and the multiplexer used for connecting the CAAC-IGZO FET and the D-FF. The CAAC-IGZO FET and storage capacitor are stacked over the Si FET.

An FF with 30-nm Si and 0.3-μm CAAC-IGZO technologies was simulated (FF2). For comparison, an FF with 0.5-μm Si and 0.8-μm CAAC-IGZO technologies was also simulated

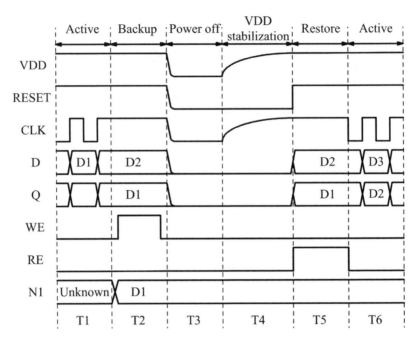

Figure 5.8   Timing diagram of Type-A backup FF. *Source*: Adapted from [9]

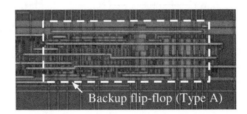

Figure 5.9   Photograph of a Type-A backup FF. *Source*: Adapted from [2]. Copyright 2012 The Japan Society of Applied Physics

(FF1). The results are listed in Table 5.1. The write time and write energy of the FF2 are 6.4 ns and 3.7 fJ, respectively; the write energy is particularly favorable, being less than the previously reported value of 30-nm magnetic tunnel junction (MTJ) [7]. These results suggest that the use of CAAC-IGZO effectively achieves normally-off computing.

5.3.1.2   Backup FF (Type B)

Figure 5.11 is a circuit diagram of a Type-B backup FF [5], which comprises a standard D-FF section and a backup-circuit section. The backup-circuit section is fabricated from one

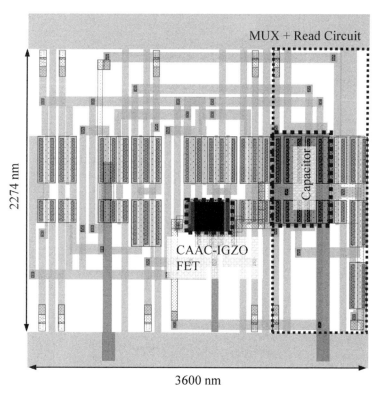

Figure 5.10 Assumed layout with 30-nm Si FET. *Source*: Reprinted from [9], with permission of IEEE, © 2013

Table 5.1 Simulation results of backup FF (FF1, FF2). *Source*: Reprinted from [9], with permission of IEEE, © 2013

	FF1	FF2
L_{IGZO}	0.8 μm	0.3 μm
Cs	1 pF	2 fF
L_{SI}	0.5 μm	30 nm
Endurance	$>10^{11}$ (1)	
Write time	500 ns (1)	6.4 ns (2)
Write energy	3.1 pJ (2)	3.7 fJ (2)
Read time	10 ns (1)	
Retention (3)	>10 years	69 days

1. Experimental results.
2. Estimated by HSPICE simulation at (V_{dd}, V_h) = (2.5 V, 3.2 V) (FF1) and (1 V, 3 V) (FF2).
3. Estimated as $t \sim Cs \times \Delta V / I_{off}$ for I_{off} = 135 yA/μm and ΔV = 1 V (FF1), 0.4 V (FF2).

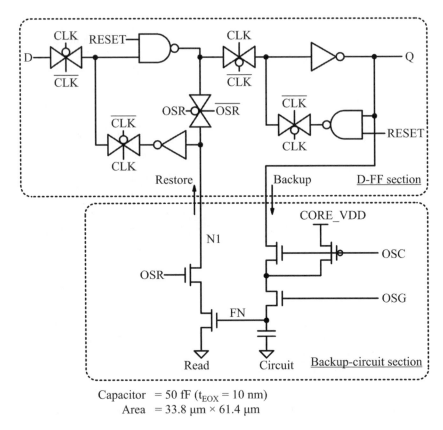

Capacitor = 50 fF (t_{EOX} = 10 nm)
Area = 33.8 μm × 61.4 μm

Figure 5.11 Circuit diagram of Type-B backup FF. *Source*: Adapted from [5]

CAAC-IGZO FET, four Si FETs, and one storage capacitor. The Type-B configuration requires fewer FETs than the Type-A configuration; therefore, the area overhead of Type B is smaller than that of Type A.

The backup time of the Type-B FF is shortened by pre-charging the storage capacitor. The charge is retained and discharged when the backup data are 1 and 0, respectively. The discharge follows the precharge, because the discharge rate of a CAAC-IGZO FET (n-type) is higher than its charge rate.

Figure 5.12 shows a timing diagram of the Type-B backup FF. T1 denotes an operation period in active mode. During this period, the precharge (OSC) and gate-control signals (OSG) of the CAAC-IGZO FET are set to low and high, respectively. Under this signaling, the retention node FN of the backup circuit is charged to a high level. During the backup operation period T2, OSC is set to a high level; consequently, the potential of the retention node FN is charged by the data of the slave part of the FF. T3 denotes a sleep-mode period, wherein all the control signals are set low to block the power supply. T4 and T5 denote different periods of the restoration operation. In period T4 of the restoration operation, the power supply is restarted to resume the power-supply voltage VDD, and the clock signal CLK is set to high. In period T5, the reset signal RESET and the restoration control signal OSR are set to high, and the data are restored to the master part of the FF.

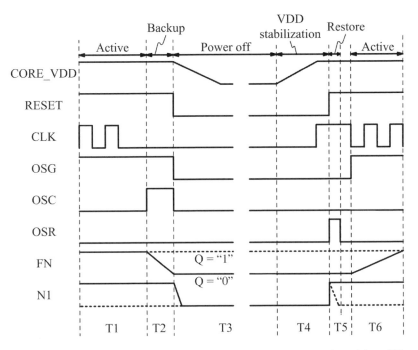

Figure 5.12 Backup sequence of Type-B backup FF. *Source*: Adapted from [5]

5.3.1.3 Serial Backup FF

Although the above-mentioned backup FF in active mode operates at an equivalent rate to a standard D-FF, the backup circuit section is constructed from multiple Si FETs. Accordingly, the circuit area is larger than for a standard D-FF. To avoid the area increase incurred by the backup function, a serial backup FF was configured [3, 4].

Figure 5.13 is a circuit diagram of a serial backup FF. In this design, a CAAC-IGZO FET and a storage capacitor (CAP) are added to the latch of the master part in a normal D-FF. This configuration has the advantage of a small number of additional FETs required for the backup function and control-signal lines. However, the delay time is considerably larger than that in the normal D-FF due to the different driving performances of the CAAC-IGZO and Si FETs.

Figure 5.14 is a timing diagram of the serial backup FF. Like the backup FF, the serial backup FF operates in active, sleep (power-off), backup, and restoration modes. The power-supply voltage VDD is supplied to the entire FF. Again, high and low levels denote that power is supplied to and blocked from the entire FF, respectively. The input signal IGZO_G is set to high in active mode, and CLK is set to low in backup mode. Consequently, to complete the backup operation, the potential FN is inverted from potential D. Next, the input signal IGZO_G is set low to turn off the CAAC-IGZO FET, thus driving the FF into sleep mode, where a power supply is not required. Power supply resumes in the restoration operation and the reset signal RESET is released. The restoration operation is completed by setting the input signal IGZO_G to high and resuming the CLK supply.

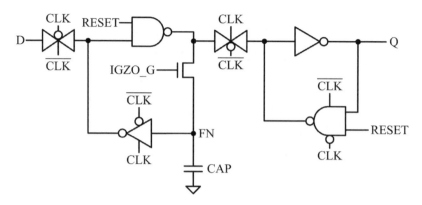

Figure 5.13 Circuit diagram of a serial backup FF. *Source*: Adapted from [4]

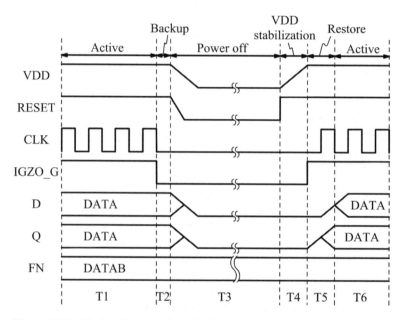

Figure 5.14 Timing diagram of a serial backup FF. *Source*: Adapted from [3]

Figures 5.13 and 5.15 show the circuit diagram and layout of the serial backup FF, respectively. The CAAC-IGZO FET and storage capacitor are stacked on the Si FET layer. Note that the backup function does not increase the area, because the serial backup FF requires no additional Si FET.

5.3.1.4 Two-Step Backup FF

The two-step backup FF [8] was developed from the design concept of the backup FF.

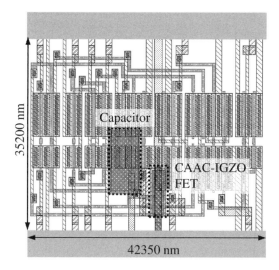

Figure 5.15 Layout of a serial backup FF. *Source*: Adapted from [3]. Copyright 2013 The Japan Society of Applied Physics

As mentioned above, the backup FF achieves backup functionality while properly performing as a D-FF. However, the time required for the backup operation depends largely on the driving performance of the CAAC-IGZO FET, through which the storage capacitor is charged or discharged. Because the driving performances of the CAAC-IGZO and Si FETs differ, the speeds of the backup operation and D-FF are difficult to achieve at the same order. A longer backup operation period limits the opportunity for transferring the FF into sleep mode. Therefore, to reduce the standby power even for short periods, we require a high-speed backup operation. The two-step backup FF adopts a two-stage backup mechanism: the first stage ensures a high-speed backup operation, whereas the second stage achieves long-term retention. This scheme reduces the power consumption over both long and short periods.

Figure 5.16 shows a circuit diagram of the two-step backup FF. Like the Type-A backup FF, this design comprises a D-FF section and a backup circuit section. The D-FF section is fabricated from Si FETs, whereas the backup circuit section contains both CAAC-IGZO and Si FETs. Distinctly different from the backup FF, the two-step FF possesses two backup circuit sections (labeled SRC1 and SRC2 in Figure 5.16). Sections SRC1 and SRC2 provide the high-speed backup operation and long-term retention, respectively.

Figure 5.17(a) and (b) shows timing diagrams of the two-step backup FF. Like the backup FF, the two-step backup FF operates in active, sleep (power-off), backup, and restoration modes. VDD is supplied to the entire FF. In active mode, the reset signal RESET is set to high, and capacitor Cs1 is continuously charged or discharged depending on the FF data [Figure 5.17(a)]. In active mode, the input signal OSG is continuously set to high, and capacitor Cs2 is charged through the diode-connected Si FET. During the backup operation, backup and power-off are simultaneously performed merely by setting the reset signal RESET to low. The backup data can be transferred from capacitor Cs1 to capacitor Cs2 without the power supply [Figure 5.17(b)]. In active mode, the FF data are placed in the first backup circuit SRC1; therefore, power-off is possible without the backup operation, and a high-speed backup with zero clock signal is achieved.

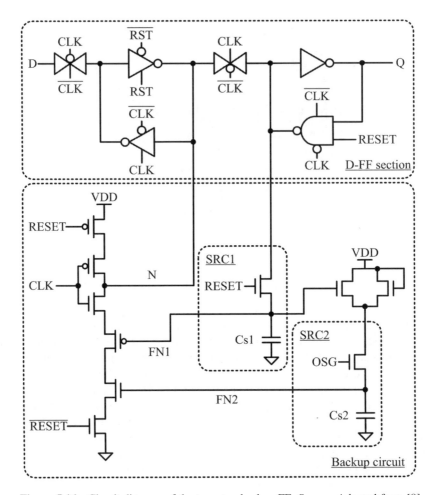

Figure 5.16   Circuit diagram of the two-step backup FF. *Source*: Adapted from [8]

The restoration operation begins when VDD is stabilized by power-on. Next, the CLK and the reset signal RESET are set to high. The node to be restored by the FF is charged or discharged by an operation corresponding to the potential of capacitor Cs1 or Cs2. The reset signal RESET is then set to high, and the data are latched in the FF. The two backup circuit sections enable both short-time data backup and restoration and long-time data retention.

5.3.2   8-Bit Normally-Off CPU

From 2012 to 2013, a prototype 8-bit normally-off CPU [2, 9] was fabricated using a backup FF. Figure 5.18 shows a photograph of the CPU. A Type-A backup FF is incorporated in a block denoted 8-bit CPU. The architecture is a Z80-like 8-bit CISC. The die was fabricated by a hybrid process of 0.5-μm Si and 0.8-μm CAAC-IGZO FETs.

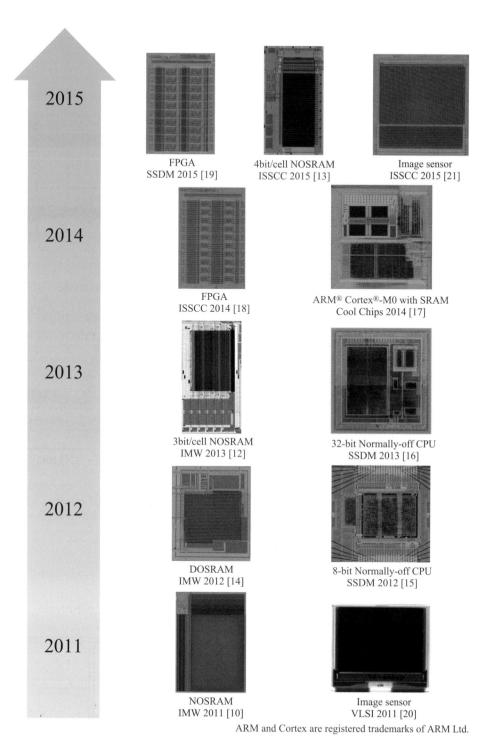

2015

FPGA
SSDM 2015 [19]

4bit/cell NOSRAM
ISSCC 2015 [13]

Image sensor
ISSCC 2015 [21]

2014

FPGA
ISSCC 2014 [18]

ARM® Cortex®-M0 with SRAM
Cool Chips 2014 [17]

2013

3bit/cell NOSRAM
IMW 2013 [12]

32-bit Normally-off CPU
SSDM 2013 [16]

2012

DOSRAM
IMW 2012 [14]

8-bit Normally-off CPU
SSDM 2012 [15]

2011

NOSRAM
IMW 2011 [10]

Image sensor
VLSI 2011 [20]

ARM and Cortex are registered trademarks of ARM Ltd.

Figure 1.6   Examples of CAAC-IGZO LSIs fabricated between 2011 and 2015

Physics and Technology of Crystalline Oxide Semiconductor CAAC-IGZO: Application to LSI, First Edition.
Edited by Shunpei Yamazaki and Masahiro Fujita.
© 2017 John Wiley & Sons, Ltd. Published 2017 by John Wiley & Sons, Ltd.

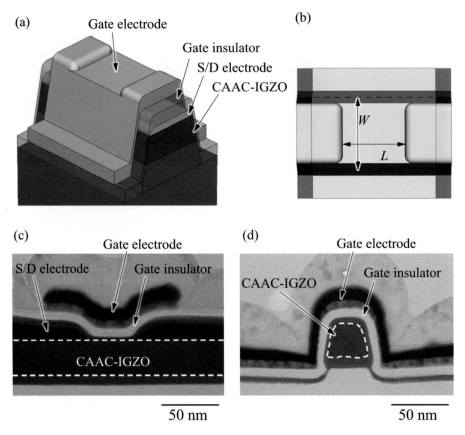

Figure 2.60 S-ch CAAC-IGZO FET using CAAC-IGZO film as active layer: (a) schematic view; (b) plane view; (c) cross-sectional STEM image in channel-length direction; and (d) cross-sectional STEM image in channel-width direction. *Source*: Reprinted from [16], with permission of IEEE, © 2014

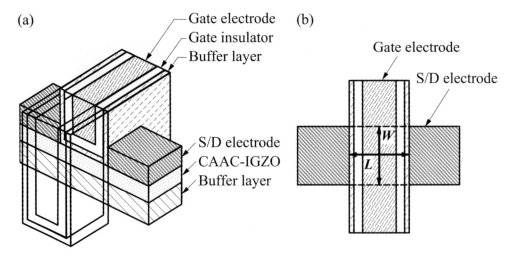

Figure 2.82 (a) Perspective and (b) planar view of the TGSA CAAC-IGZO FET. *Source*: Reprinted from [16], with permission of IEEE, © 2014

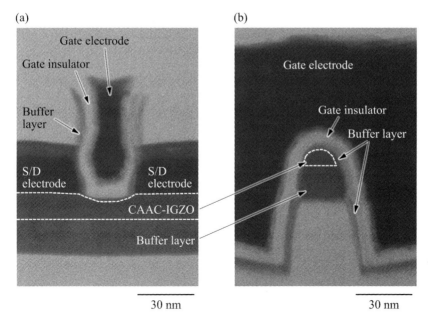

Figure 2.83 Cross-sectional STEM images in the channel-length direction (a) and the channel-width direction (b) of the TGSA CAAC-IGZO FET. *Source*: Reprinted from [61], with permission of IEEE, © 2015

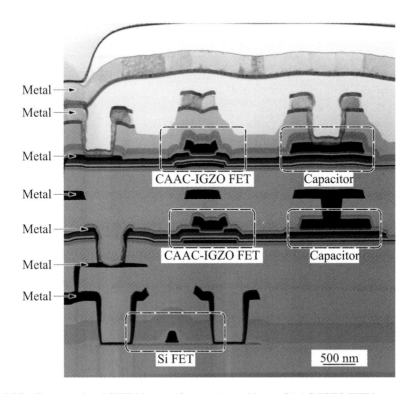

Figure 2.95 Cross-sectional STEM image of a prototype with two CAAC-IGZO FET layers stacked on Si FET. Channel lengths of the Si and CAAC-IGZO FETs are 0.2 μm and 0.4 μm, respectively. The image is retouched by filling voids in the interlayer film

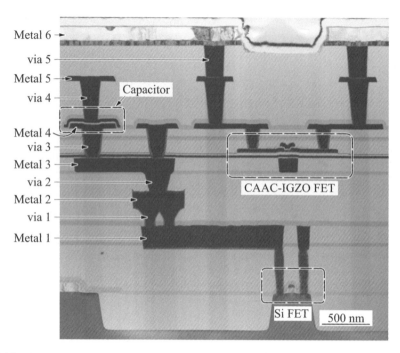

Figure 2.99 Cross-sectional STEM image of hybrid structure of Si FET with channel length of 65 nm and CAAC-IGZO FET with channel length of 60 nm (cooperative work with UMC)

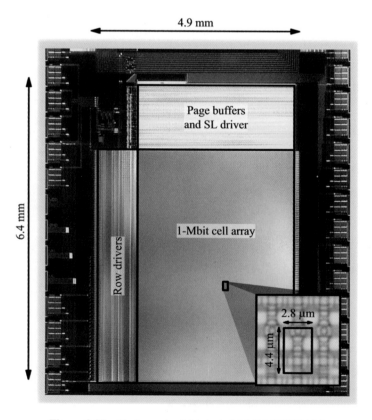

Figure 3.27 Photograph of fabricated 1-Mbit NOSRAM

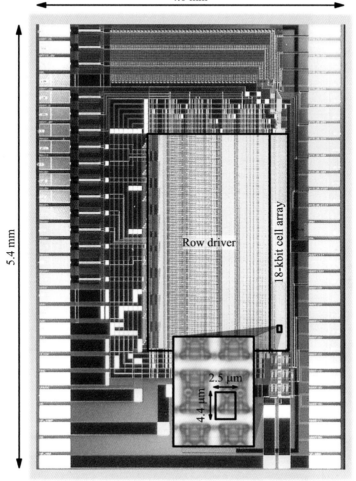

Figure 3.32 Photograph of prototype of 8-level NOSRAM cell

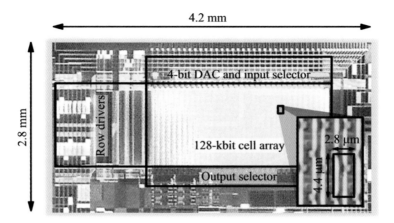

Figure 3.36 Photograph of prototype of 16-level NOSRAM

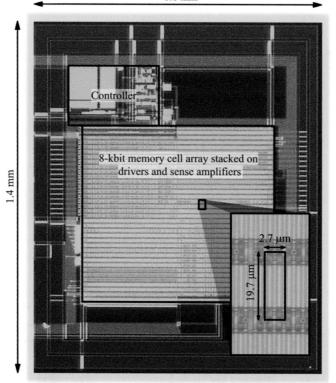

Figure 4.11 Die photograph of the 8-kbit DOSRAM

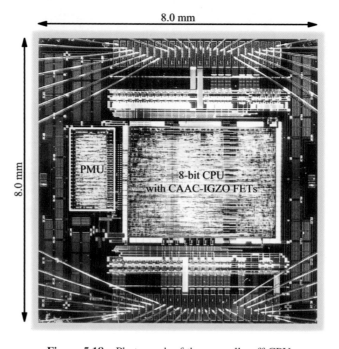

Figure 5.18 Photograph of the normally-off CPU

Figure 5.25 Photograph of the 32-bit normally-off CPU

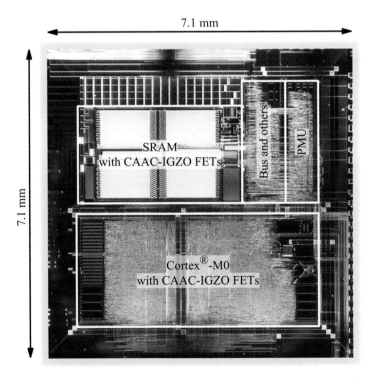

Figure 5.30 Micrograph of the 32-bit normally-off CPU (ARM® Cortex®-M0)

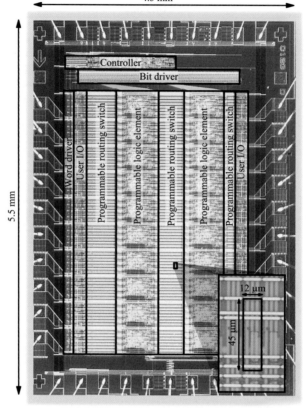

Figure 6.23 Micrograph of a prototype fabricated by a hybrid process

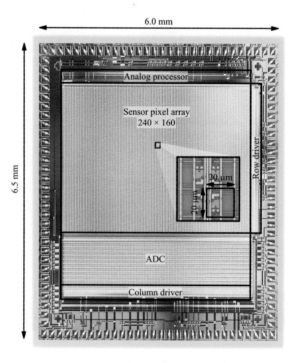

Figure 7.34 Micrograph of motion sensor

(a)

(b)

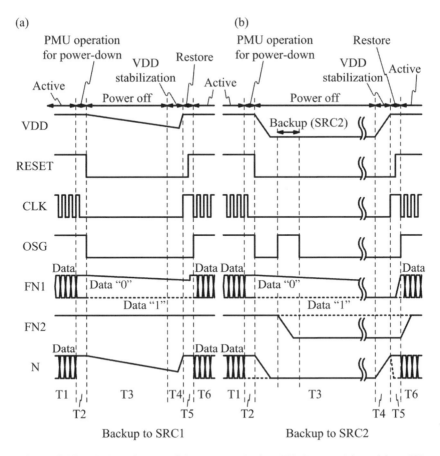

Figure 5.17 Timing diagram of the two-step backup FF. *Source*: Adapted from [8]

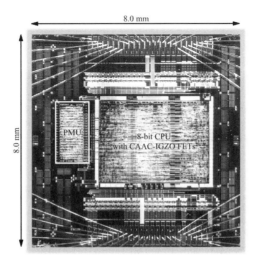

Figure 5.18 Photograph of the normally-off CPU. *(For color detail, please see color plate section)*

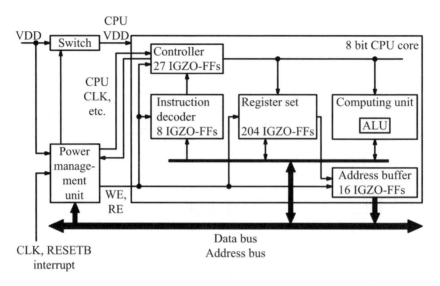

Figure 5.19   Block diagram of an 8-bit normally-off CPU. *Source*: Adapted from [9]

Table 5.2   Specifications of 8-bit normally-off CPU

Metric	
Architecture	8-bit CISC
CAAC-IGZO technology node (μm)	0.8
Si technology (μm)	0.5
Core size (mm × mm)	4.5 × 3.3
Clock frequency (MHz)	25
Si power supply (V)	2.5
CAAC-IGZO power supply (V)	3.2

Figure 5.19 shows a block diagram of the die. The 8-bit normally-off CPU consists of an 8-bit CPU core incorporating a backup FF, a power management unit (PMU), and a power switch. The number of incorporated Type-A backup FFs (CAAC-IGZO-FFs) is written in each function block. The PMU controls the timing of the backup and restoration operations and the on–off operation of the power switch.

Its specifications are listed in Table 5.2. Considering the threshold voltage drop caused by the CAAC-IGZO FET when used as a pass FET, the gate-driving voltage for the CAAC-IGZO FET is set higher than for the Si FET.

The normally-off CPU operates under the PMU state transitions shown in Figure 5.20. The backup operation is initiated by a program running on the CPU. The PMU controls each signal line in the backup FF. The restoration operation is initiated by an interrupt signal, and executed under PMU control of each signal line of the FF.

Operational waveforms in the power gating for a 1.8-μs backup sequence and a 2.2-μs restoration sequence at 25 MHz are shown in Figure 5.21. According to the operation test, the data of the accessible 88 FFs in the CPU are identical before and after power-off, verifying

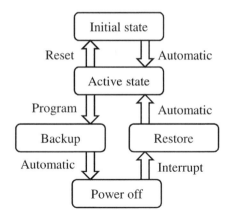

Figure 5.20   State transition diagram of normally-off CPU. *Source*: Adapted from [9]

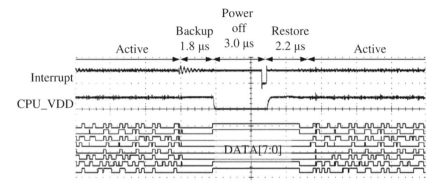

Figure 5.21   Measured waveforms of the 8-bit normally-off CPU. *Source*: Reprinted from [9], with permission of IEEE, © 2013

the correct function of the power gating. Furthermore, the CPU gave the correct results in a test program involving LOAD, ADD, and STORE operations, even with a power gating inserted between two consecutive instructions.

The data-retention characteristics of the 8-bit CPU were estimated. During the estimation procedure, the Type-A backup FF in the 8-bit CPU executed a backup operation, and the CPU was powered off. After a certain period, the CPU was powered on and the data in the backup circuit section were read out to the D-FF. Next, the read data were compared with the data written to the backup circuit section. Finally, the CPU was powered off without backup operation and the processing steps were repeated. Because the data are not rewritten to the backup circuit section in this test, we can constantly verify the long-term data retention of the CAAC-IGZO FET.

The data-retention characteristics were measured at 85°C. Figure 5.22 illustrates the retention time, number of errors, and other performance data. The retention time is the product of the off-time and the number of repetitions (labeled "A" and "B" in the figure, respectively). Thus, the retention time is calculated as 300 s × 11,334 times = 3,400,200 s. C denotes the number of

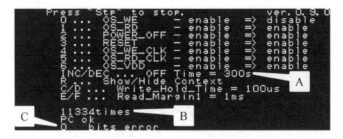

Figure 5.22 Screenshot of retention characteristics after approximately 40 days. *Source*: Adapted from [2]

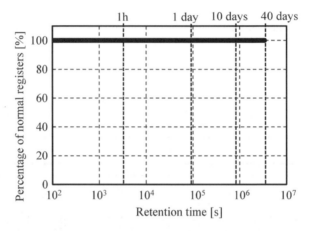

Figure 5.23 Data-retention characteristics of Type-A backup FF at 85°C. *Source*: Adapted from [2]. Copyright 2012 The Japan Society of Applied Physics

bit errors in the FF of the backup circuit section. Figure 5.23 plots the percentage of the registers that hold data correctly versus the data-retention time. These figures verify long-term data retention by the Type-A backup FF (3,400,000 s, or 40 days) at 85°C after power-off. This indicates that the CPU can operate correctly for 40 days with only at least one refresh (backup and restore) operation.

5.3.3 32-Bit Normally-Off CPU (MIPS-Like CPU)

Figure 5.24 is a block diagram of the 32-bit normally-off CPU fabricated in 2013 [8]. The CPU specifications are listed in Table 5.3. The 32-bit normally-off CPU comprises a MIPS I-compatible core, cache, cache controller, bus interface, PMU, and power switch.

The MIPS I-compatible core has a 32-bit RISC architecture and includes 31 32-bit general-purpose registers. The pipeline has three stages. All of the 32-bit general-purpose and pipeline registers are fabricated with two-step backup FFs. The cache memory has a capacity of 2 kB and adopts two-way set association and a write-through mode as the writing mode. The memory array of the cache has a SRAM-based circuit configuration, which retains data even during

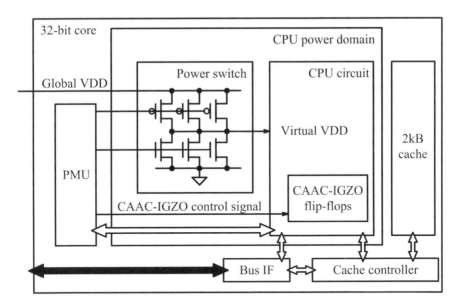

Figure 5.24 Block diagram of the 32-bit normally-off CPU. *Source*: Reprinted from [8], with permission of IEEE, © 2014

Table 5.3 Specifications of the 32-bit normally-off CPU

Metric	
Architecture	MIPS I-compatible (32-bit RISC)
Pipeline stage	3
Clock frequency (MHz)	15
Cache memory	2-way set associative, 2 kB
Si technology node (nm)	350
CAAC-IGZO technology node (nm)	180
Si power supply (V)	2.5
CAAC-IGZO power supply (V)	3.2/−1

power-off by power gating. Such an SRAM configuration, which is called backup SRAM (CAAC-IGZO SRAM) herein, will be detailed in Section 5.4. Here, the operation of the CPU using backup FFs is explained. The cache controller is connected to the core, the bus interface, and the cache. When the core sends a request, the cache controller controls access to the cache. Furthermore, when processing cache misses or write-through operations, it accesses an external memory via the 32-bit bus interface. Triggered by an instruction from the core or an interrupt signal and acting through the power switch, the PMU supplies or blocks the power supply to each block.

Figure 5.25 shows a photograph of the 32-bit normally-off CPU. The die was fabricated by a hybrid process of 350-nm Si and 180-nm CAAC-IGZO FETs.

Figure 5.26(a) presents waveforms of the on-stage backup circuit measured during power gating. During these measurements, the test program switched the state between active and

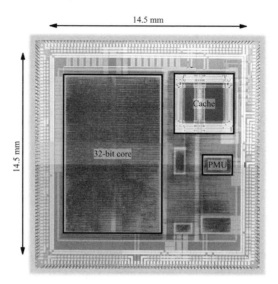

Figure 5.25 Photograph of the 32-bit normally-off CPU. *(For color detail, please see color plate section)*

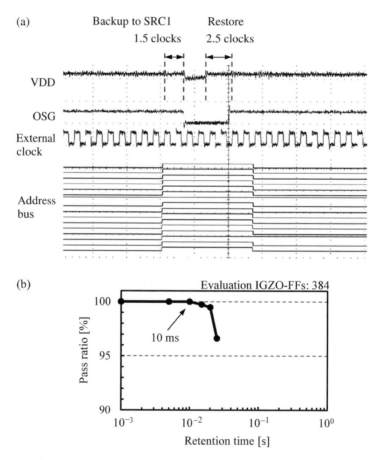

Figure 5.26 (a) Measured waveforms and (b) measured retention time of the first-stage backup circuit in power gating. *Source*: Reprinted from [8], with permission of IEEE, © 2014

sleep. The voltages of the Si and CAAC-IGZO FETs, measured at room temperature, were 2.5 V (the standard condition) and 3.3 V, respectively. The backup and restoration operations were performed in 1.5 and 2.5 clocks, respectively. Note that the 1.5 clocks in the backup operation is the delay time necessary for accurate operation of the PMU, not the time of writing data to the backup circuit. The data stored in the first-stage backup circuit are being updated every clock; therefore, no additional clock cycle is required to back up data to it.

The retention time of the general-purpose registers in the CPU was also measured with a test program. To check data retention, the data values in 12 general-purpose registers were compared before and after power gating. The values of the 12 general-purpose registers can be freely set up. Figure 5.26(b) shows that the first-stage backup circuit can hold data for up to 10 ms without an error.

Figure 5.27(a) shows the measured waveforms of the second-stage backup circuit during power gating. The retention time was estimated similarly to that of the first-stage backup circuit. The results [Figure 5.27(b)] verified that the second-stage backup circuit retains data for at least one day.

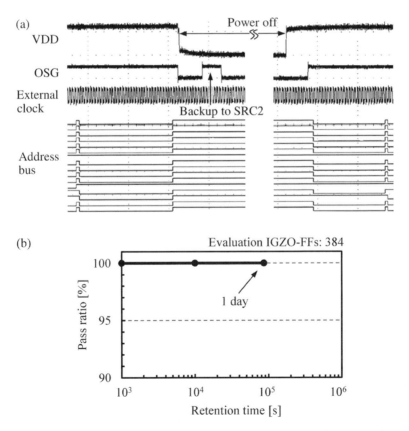

Figure 5.27 (a) Measured waveforms and (b) measured retention time of the second-stage backup circuit in power gating. *Source*: Reprinted from [8], with permission of IEEE, © 2014

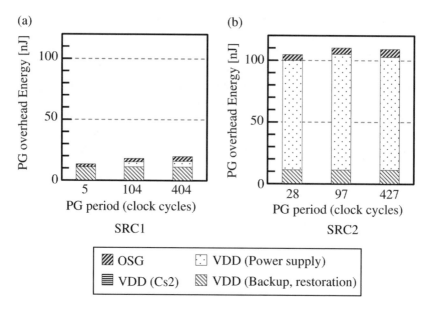

Figure 5.28 Measured overhead energies of SRC1 and SRC2. *Source*: Reprinted from [8], with permission of IEEE, © 2014

As shown in Figure 5.28, the overhead energies demanded by the backup and restoration operations were 1.77 and 11.64 nJ, respectively. The execution speeds of backup and restoration are higher in this die than in conventional dies with equivalent energy consumption (see Table 5.4).

To investigate whether miniaturizing the CAAC-IGZO FET affects the operation performance, the circuit characteristics of an FF (see Figure 5.29), fabricated by a hybrid process of 45-nm Si and 180-nm CAAC-IGZO FETs, were simulated. The simulation results are listed in Table 5.5.

5.3.4   32-Bit Normally-Off CPU (ARM® Cortex®-M0)

We now explain a 32-bit normally-off CPU with an ARM® architecture using Type-B backup FFs [5]. Figure 5.30 and Table 5.6 present a micrograph and an overview of the die specifications, respectively. The fabricated die contains an ARM® Cortex®-M0, an on-chip 4-kbit SRAM, PMU, and other functional circuits (see block diagram in Figure 5.32). The 32-bit normally-off CPU was targeted for sensor networks, and the basic specifications were determined accordingly. All FFs in the CPU, including the general-purpose registers, are Type-B backup FFs. Therefore, the CPU retains its internal data even during power-off by power gating, and immediately resumes processing after power-on. The SRAM (detailed in Section 5.4) incorporates a CAAC-IGZO FET and hence retains data even in the power-off state. The CPU was fabricated by a hybrid process of 180-nm Si and 60-nm CAAC-IGZO FETs, where the latter is stacked on the former as shown in Figure 5.31.

As shown in the block diagram of the 32-bit normally-off CPU (Figure 5.32), the main blocks of the 32-bit normally-off CPU are a Cortex®-M0, SRAM, and a PMU. These blocks are connected by a bus operating at 30 MHz in accordance with the 32-bit AHB® Lite standard.

Table 5.4 Comparisons of 32-bit normally-off CPU and an MCU with ferroelectric RAM (FeRAM) reported at a conference. *Source*: Adapted from [8]

	32-bit Noff-CPU	Bartling [10]
Si implementation	Yes	Yes
Architecture	32-bit MIPS I-like	ARM® Cortex®-M0
Technology node	Si: 350 nm	Si: 130 nm
	CAAC-IGZO: 180 nm	FeRAM: 130 nm
Supply voltage	Si: 2.5 V	1.5 V
	CAAC-IGZO: 3.2 V	
Area	289 mm^2	4.4 mm^2
Clock frequency	15 MHz	8 MHz/125 MHz
Memory circuit implementation	2-step backup flip-flop with CAAC-IGZO	FeRAM
Core area overhead	4.8%*	12%
Power gating energy overhead	13.42 nJ	7.25 nJ
	Power shutdown	Power shutdown
	1.11 nJ: CPU w/o CAAC-IGZO-FFs**	0.86 nJ: w/o NVL
	0.66 nJ: CAAC-IGZO-FFs**	4.72 nJ: NVL
	Wake-up	Wake-up
	5.82 nJ: CPU w/o CAAC-IGZO-FFs**	0.33 nJ: w/o NVL
	5.82 nJ: CAAC-IGZO-FFs**	1.34 nJ: NVL
Sleep operation delay	1.5 clocks	40 clocks@125 MHz*
Wake-up operation delay	2.5 clocks	48 clocks@125 MHz*
Comment	*Including routing overhead	*Needs two types of clock
	**Ratio by simulation	

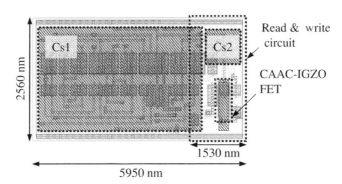

Figure 5.29 Layout of simulated FF with a hybrid process of 45-nm Si and 180-nm CAAC-IGZO FETs. *Source*: Reprinted from [8], with permission of IEEE, © 2014

The bus is provided with input and output pins for connecting the CPU with peripheral circuits. The Cortex®-M0 is a 32-bit processor with an ARM® architecture having a three-stage pipeline that operates at the bus frequency (30 MHz). The processor incorporates 841 Type-B backup FFs and an SRAM as a working memory and program and data storage.

Table 5.5 Comparison of CPU simulated using a hybrid process of 45-nm Si and 180-nm CAAC-IGZO FETs with CPU using a hybrid process of 350-nm Si and 180-nm CAAC-IGZO FETs. *Source*: Reprinted from [8], with permission of IEEE, © 2014

Metric		45-nm Si 180-nm CAAC-IGZO	350-nm Si 180-nm CAAC-IGZO
Restore time		1.1 ns	2.3 ns
SRC1	Backup time	0.4 ns	1.4 ns
	Retention time	489 ns	2.6 ms
	Backup & restore energy	31 fJ and 2 fJ: IGZO and Cs2	1390 fJ and 7 fJ: IGZO and Cs2
SRC2	Backup time	61 ns	135 ns
	Retention time	>8.1 h	>17.6 days
	Backup & restore energy	86 fJ 40 fJ: IGZO and Cs2	3649 fJ 973 fJ: IGZO and Cs2
Simulation condition	Cs1	2 fF	13 fF
	Cs2	27.5 fF	133 fF
	VDD	Si 1.1 V CAAC-IGZO 1.8 V	Si 2.5 V CAAC-IGZO 3.2 V
Impact on conventional technique		8% Performance 3% Power 35% Area	16% Performance 11% Power 35% Area

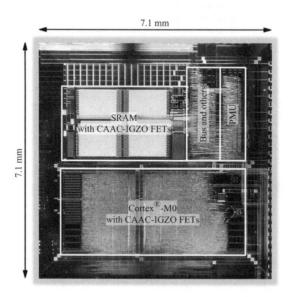

Figure 5.30 Micrograph of the 32-bit normally-off CPU (ARM® Cortex®-M0). *(For color detail, please see color plate section)*

Table 5.6 Specifications of the 32-bit normally-off CPU (ARM® Cortex®-M0)

Metric	
Technology	Si 180 nm
	CAAC-IGZO 60 nm
Supply voltage	Si 1.8 V
	CAAC-IGZO 2.5 V
Architecture	ARM® Cortex®-M0 Design Start Edition
Clock frequency	30 MHz
Backup circuit implementation	Embedded SRAM and flip-flop

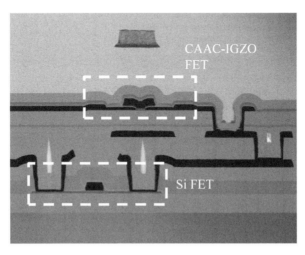

Figure 5.31 Cross-sectional view of SRAM formed by a hybrid process of 180-nm Si and 60-nm CAAC-IGZO FETs. *Source*: Reprinted from [5], with permission of IEEE, © 2014

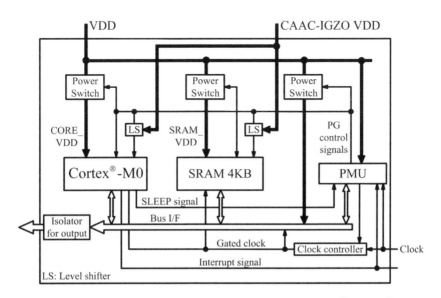

Figure 5.32 Block diagram of the 32-bit normally-off CPU (ARM® Cortex®-M0). *Source*: Adapted from [5]

The PMU manages the power supply in several power domains: Cortex®-M0, the SRAM, and the bus. Each power domain is controlled through an individual power switch, whose on–off state responds to a control signal from the PMU. The Cortex®-M0 provides a SLEEP signal line, which informs the PMU of a pending sleep state. When the SLEEP signal goes high, the PMU begins to manage the backup FF and controls the power supply of each power domain. Subsequently, the PMU enters the power-off state. Upon an interrupt signal input to the PMU, the power supply resumes. The interrupt signal is conveyed through a peripheral circuit of the CPU, such as a timer or sensor.

The high- and low-level gate voltages of the CAAC-IGZO FETs are controlled at 2.5 and −1 V, respectively [i.e., different from the Si FET voltage (1.8 V)]. Therefore, prior to the CAAC-IGZO FET, the control signal from the PMU is passed through a level shifter provided for the corresponding block.

The operation of the backup FF was examined by measuring the shortest possible backup and restoration times (see Figure 5.33) at room temperature and all voltages at their default values. The results indicate that the backup proceeds in two clocks (approximately 66 ns), and that it is limited by on-state characteristics of the CAAC-IGZO FET and the storage capacitance. By contrast, the data-restoration time, which is limited by the characteristics of the Si FET, was executed in 1 clock. Since the power supply needs to be stabilized after power-on, however, the CPU returns to active mode only after 6 clocks. This stabilization time is expended in raising the potential of the power supply line, which is connected to a decoupling capacitor. The control voltage of the CAAC-IGZO FET was then reduced from 2.5 to 2.25 V (−10%), and further measurements were conducted at a CPU voltage of 1.8 V and a die temperature of 85°C. While the backup time was unchanged, the restoration time was increased by 3 clocks.

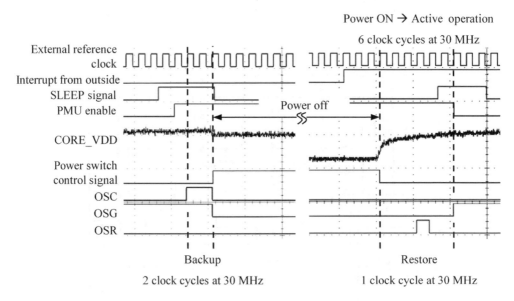

Figure 5.33 Measured waveforms of the Type-B backup flip-flop during operation. *Source*: Reprinted from [5], with permission of IEEE, © 2014

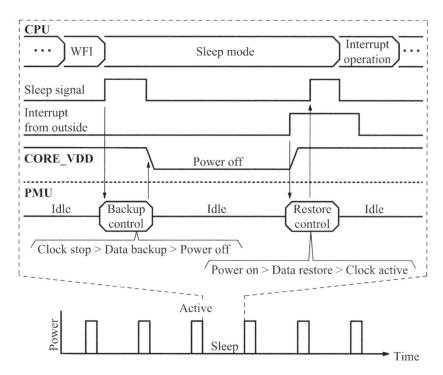

Figure 5.34 Evaluation program of alternating active and sleep modes. *Source*: Adapted from [5]

The energy required from power-off to power-on (150 fJ/bit) was estimated from the energy requirements of control backup and restoration. This calculation includes the power consumed by the 2.5-V power supply line controlling the CAAC-IGZO FET and the energy of charging the storage capacitor.

The power-reduction effects were evaluated under repeated alterations of active and sleep (power-off) modes, as shown in Figure 5.34. A wait-for-interrupt (WFI) instruction triggers the SLEEP signal from the Cortex®-M0, indicating that the Cortex®-M0 will enter sleep mode and the PMU thereby begins the data backup. Power-on is implemented by an external interrupt signal. The temporal ratio of active to sleep modes was decided from the CPU's intended use in sensor network applications. The power consumption was measured for three cases. In Case 1, an interrupt signal was generated from an acceleration sensor at 1-ms intervals; in Case 2, temperature sensor data were obtained at 1-s intervals; in Case 3, the sleep mode was maintained over a long period (100 s). During the active mode (which lasts for 1 ms), an instruction to access the memory and the external interface is executed. Figure 5.35 shows the measured temporally averaged power consumption of the power gating (PG) and clock gating (CG). In all three cases, the PG greatly reduced the power consumption compared with the CG, and it was found that the standby power of the 32-bit normally-off CPU can be reduced by 99% or more for some use cases. Table 5.7 shows a comparison between a CAAC-IGZO-based normally-off CPU and an MCU with ferroelectric RAM.

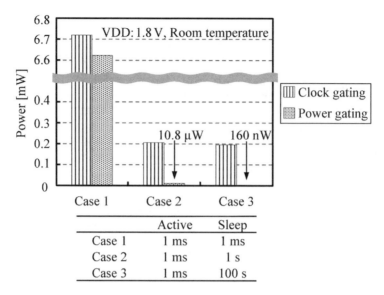

Figure 5.35 Evaluation results of power reduction. *Source*: Reprinted from [5], with permission of IEEE, © 2014

Table 5.7 Comparison between a CAAC-IGZO-based normally-off CPU and an MCU with ferroelectric RAM (FeRAM). *Source*: Adapted from [5]

	CAAC-IGZO (embedded SRAM)	MRAM (Sakimura [11])	FeRAM (Bartling [10])
Data-retention principle	Charge on a capacitor and CAAC-IGZO FET	Magnetization direction	Ferroelectric
Non-volatility	Yes	Yes	Yes
Endurance	$>10^{12}$	$>10^{15}$	$>10^{12}$
Fabrication technology	180 nm Si, 60 nm CAAC-IGZO	90 nm, MVT	130 nm, HVT
Memory technology	SRAM memory cell + 2Tr(CAAC-IGZO) + 2C	3T-SpinRAM	FeRAM (mini arrays and FeCaps)
Supply voltage	1.8 V (Si), 2.5 V/−1 V (CAAC-IGZO)	1.8–3.3 V (Dvcc), 1.0 V (VCORE)	1.5 V (single supply)
Clock frequency	30 MHz	20 MHz	8 MHz/125 MHz
Backup time	66 ns	4 ns	320 ns
Backup energy/bit	142.5 fJ	6 pJ	2.2 pJ
Restoration time	132 ns	120 ns	384 ns
Restoration energy/bit	61.5 fJ	0.3 pJ	0.66 pJ

5.4   CAAC-IGZO Cache Memory

Cache memory refers to a high-speed memory for temporary data storage. As shown in the memory hierarchy in Figure 5.3, some architectures have plural cache memories, which are numbered according to their closeness to the CPU. For instance, caches L1 and L2 are the closest and second-closest caches to the CPU, respectively. Most cache memories comprise small-capacitance static RAM (SRAM), but NOSRAM and DOSRAM (mentioned in Chapters 3 and 4) are also applicable. This section discusses cache memories with SRAM based on CAAC-IGZO FETs.

As mentioned in Section 5.1, the standby power of recent LSIs increases significantly by the high leak current caused by FET downsizing and high integration. To reduce that power, power gating must be implemented either by employing fast non-volatile memories (NVM) or by backing up registers into an external NVM. However, replacing SRAM (a high-speed memory) with a conventional NVM would compromise the operation speed, so a hybrid SRAM that combines a volatile circuit with a non-volatile element has been proposed [12–14]. Some challenges of the hybrid SRAM design include the cell area overheads, operation speed, operation power, and other factors. Large overheads may, in the worst case, cancel the power savings achieved by power gating.

The hybrid SRAM introduced in this section was fabricated from CAAC-IGZO FETs with extremely low off-state current [6]. The process design was aided by the process design kit (PDK) called FreePDK45^{TM} [15]. Simulation Program with Integrated Circuit Emphasis (SPICE) is usually implemented by a high-performance (HP) model called VTL, assuming a register file and an L1 cache memory. The VTL model is a transistor parameter in which the threshold voltage is set to a small value and is suitable for high-speed operation.

Figure 5.36(a) and (b) shows a circuit diagram and a power gating sequence of the CAAC-IGZO SRAM cell, respectively. The CAAC-IGZO SRAM cell is composed of a six-FET standard SRAM, two CAAC-IGZO FETs, and two capacitors. The standby power is reduced by backing up the data stored at the bistable nodes Q and $\bar{Q}$ into storage nodes SN1 and SN2. The virtual V_{DM} ($V - V_{DM}$) is then interrupted. For restoration, the data stored at SN1 and SN2 are returned to Q and $\bar{Q}$, and normal operation resumes from the state before the power gating.

Figure 5.37 shows the layout of the CAAC-IGZO SRAM cell. The backup components (i.e., two CAAC-IGZO FETs and two capacitors) are stacked on the layer containing the standard SRAM based on Si FETs.

Using SPICE simulations, the backup and restoration times in the power gating sequence of the CAAC-IGZO SRAM cell were estimated to 3.9 ns and 2.0 ns, respectively. The energies required for each period are listed below (see also Figure 5.38).

- Backup period: When backing up data stored at the bistable nodes, the charge and discharge of the gate capacitances of the CAAC-IGZO FETs and capacitors require the energy E_{backup}.
- Power-off period: The power-off reduces the standby power consumption to almost zero (although, strictly speaking, the power supply is interrupted by a small leakage current flowing through the power switch). Because this leakage current can be reduced to negligibly small by an appropriate design, it was set to 0 ($I_{off} = 0$) in the simulation.
- Restoration period: The return of data from the backup section to the bistable nodes requires the $E_{restore}$, which includes charging the gate capacitance of the CAAC-IGZO FETs, the parasitic capacitance of the $V - V_{DM}$ wiring, and the Joule losses by flow-through current at the start of bistable-node operations.

(a)

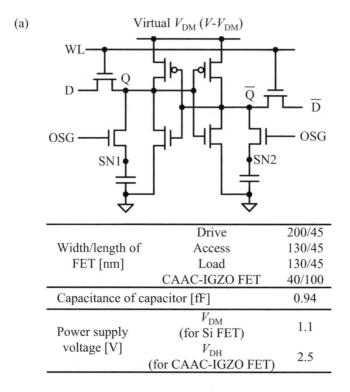

		Drive	200/45
Width/length of	Access	130/45
FET [nm]	Load	130/45
	CAAC-IGZO FET	40/100
Capacitance of capacitor [fF]		0.94
Power supply voltage [V]	V_{DM} (for Si FET)	1.1
	V_{DH} (for CAAC-IGZO FET)	2.5

(b)

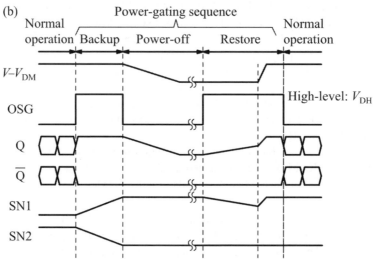

Figure 5.36 (a) Circuit diagram and (b) power-gating sequence of the CAAC-IGZO SRAM. *Source*: Adapted from [6]

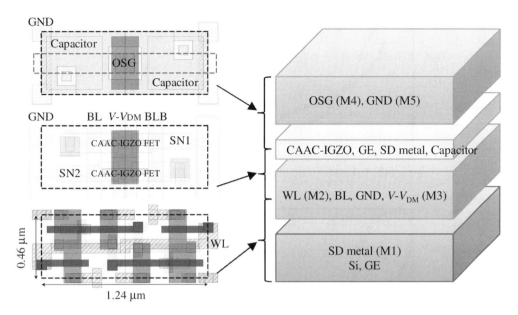

Figure 5.37 Mask layout and layer structure of the CAAC-IGZO SRAM. *Source*: Adapted from [6]

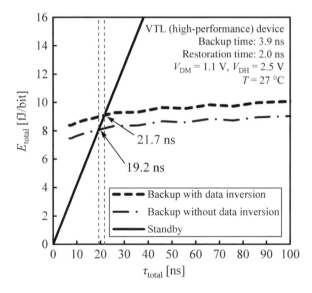

Figure 5.38 Estimation of the BET. *Source*: Reprinted from [6], with permission of IEEE, © 2014

We now replace the standard SRAM cell with the CAAC-IGZO SRAM cell, and examine the effect of the replacement on normal operation of the cell. The results are summarized below.

- After the replacement, the standby power remains unchanged at 419 nW.
- The static noise margin (SNM) during reading, writing, and holding operations is not degraded by the replacement (see Figure 5.39).

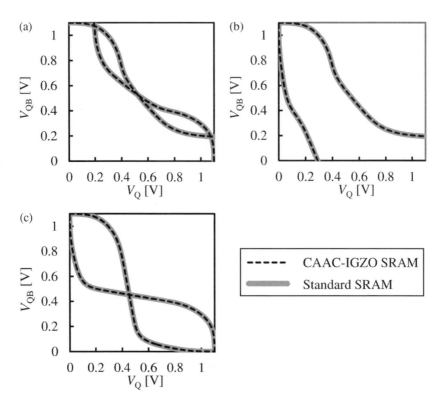

Figure 5.39 Static noise margins of CAAC-IGZO SRAM and standard SRAM during read (a), write (b), and hold (c) operations. The voltages are those in node Q and $\bar{Q}$ shown in Figure 5.36(a). *Source*: Reprinted from [6], with permission of IEEE, © 2014

- Delay time: The inversion time of the bistable nodes in the CAAC-IGZO SRAM cell is 17.7 ps, versus 15.2 ps in the standard SRAM cell. The delay times of word and bit lines were simulated in a CAAC-IGZO SRAM array and a standard SRAM array, each with 256 rows and 128 columns. The rise time of the word line was 39.4 ps and 36.9 ps in the CAAC-IGZO and standard SRAM arrays, respectively. During reading operations, the time between the rise of the word line potential and the fall of the bit line potential to $V_D - V_{BLSENSE}$ was 131 ps in the CAAC-IGZO SRAM array and 126 ps in the standard SRAM array ($V_{BLSENSE}$ is 0.3 V). We assumed that the variation of the differential sense amplifiers (reading circuit) was 0.3 V.
- Dynamic energy: The dynamic energy was measured on test benches identical to those used in the delay simulation. In the CAAC-IGZO and standard SRAMs, the inversion energy of the bistable node was 3.73 and 3.24 fJ, respectively, the charge and discharge energy of the word line was 79.9 and 76.6 fJ, respectively, and the discharge–precharge energy of the bit line was 29.8 and 29.5 fJ, respectively.

According to these results, replacing the standard SRAM cell with a CAAC-IGZO SRAM cell negligibly affects the normal performance, because the CAAC-IGZO FET electrically isolates the bistable nodes used in normal operation from the backup section.

Table 5.8 Simulation results of the CAAC-IGZO SRAM cell. *Source*: Reprinted from [6], with permission of IEEE, © 2014

		CAAC-IGZO SRAM	Standard SRAM
Area		0.5704 (+0%)	0.5704
VTL (HP)	Standby power [nW/bit]	419 (+0%)	419
	Static noise margin	No degradation	—
	Inversion time of bistable nodes [ps]	17.7 (+16.1%)	15.2
	Delay time of word line [ps]	39.4 (+6.7%)	36.9
	Delay time of bit line [ps]	131 (+3.9%)	126
	Inversion energy of bistable nodes [fJ/bit]	3.73 (+15.2%)	3.24
	Delay energy for word line [fJ]	79.9 (+4.3%)	76.6
	Delay energy for bit line [fJ]	29.8 (+1.1%)	29.5
	Backup time [ns]	3.9	—
	Restoration time [ns]	2.0	—
	PG energy [fJ/bit]	9.09	—
	BET [ns]	21.7	—
VTG (LOP)	Standby power [nW/bit]	45.7 (+0%)	45.7
	PG energy [fJ/bit]	7.09	—
	BET [ns]	155	—
VTH (LSTP)	Standby power [nW/bit]	0.589 (+0%)	0.589
	PG energy [fJ/bit]	6.89	—
	BET [ns]	11700	—

Table 5.8 summarizes the characteristics of the CAAC-IGZO SRAM cell. In these results, the VTL device was used for the Si FETs [15]. However, the energies of power gating and BET using VTG (low operating power: LOP) and VTH (low standby power: LSTP) devices are also presented. The VTL model is a transistor parameter in which the threshold voltage is set to a small value and is suitable for high-speed operation. However, leakage current in the model is large. The VTH model is a transistor parameter in which the threshold voltage is set to a large value and aims at reducing leakage current. The VTG model is an intermediate model between VTL and VTH models. The threshold voltage in the model is set to a medium value, and the model is suitable for low-voltage driving. In addition, it can reduce leakage current. VTL, VTG, and VTH are intended for a register file and L1 cache memory, L2 cache memory, and L3 cache memory or low-speed high-capacity SRAM, respectively. The VTG and VTH devices achieved backup and restoration in 3.9 ns and 2.0 ns, respectively, similar to the VTL device.

We now present a fabrication example of a CAAC-IGZO SRAM.

An optical micrograph of a 32-kbit SRAM die is shown in Figure 5.40. The die contains miniaturized CAAC-IGZO FETs with a channel length of 60 nm. The characteristics are listed in Table 5.9.

The standby power is reduced by bit-line floating and power gating. The memory cell array consists of four subarrays, each with 128 word lines and 64 bit lines. The peripheral circuits consist of a backup and restoration driver, power switches, and general driver circuits.

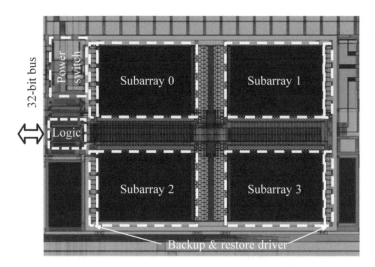

Figure 5.40   Optical micrograph of a 32-kbit CAAC-IGZO SRAM die

Table 5.9   Characteristics of the 32-kbit CAAC-IGZO SRAM. *Source*: Adapted from [16]. Copyright 2015 The Japan Society of Applied Physics

Technology	Si: 180 nm	IGZO: 60 nm
Configuration	32 kbit (1 K words × 32 bits)	
Frequency	85 MHz	
Voltage	Si: 1.8 V/0 V	IGZO: 2.5/−1 V
Power saving technologies	Bit-line floating	
	SRAM cell with backup circuit	
	Power switch	

The circuit configuration of the backup and restoration driver is illustrated in Figure 5.41. To improve the data-retention characteristics, the CAAC-IGZO FET was assigned a high threshold voltage. Because V_{DH} (2.5 V) at the gate (OSG) of the CAAC-IGZO FET exceeds V_{DD} (1.8 V), the backup and restoration driver is installed with level shifters. Furthermore, to prevent unstable states of the OSG line when the backup and restoration driver is powered off, an isolator is inserted.

There are three power domains: the memory cell array (1.8/0 V), the peripheral circuit (1.8/0 V), and the backup and restoration driver (2.5/−1 V). Each of these domains is provided with a power switch (see Figure 5.42). The OSG lines are controlled by a PG signal, and the power switches are controlled by a PS_PERI signal and a PS_MEM signal. The die implements four low-power standby modes: (1) bit-line floating, (2) power gating in the peripheral circuits only (peripheral PG), (3) power gating in the memory array only (array PG), and (4) power gating in all domains (all-domain PG).

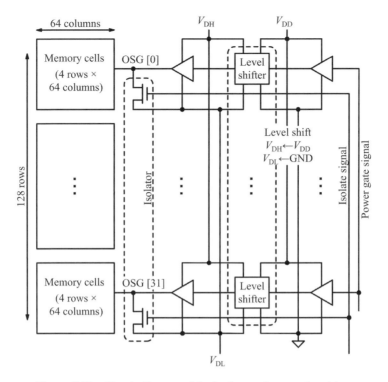

Figure 5.41 Circuit diagram of the backup and restoration driver

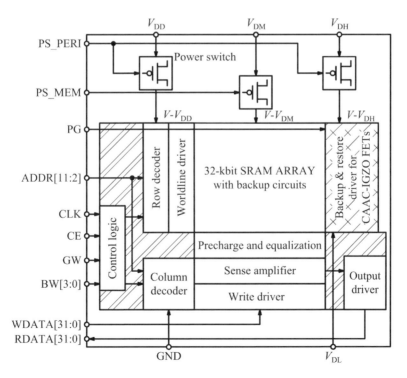

Figure 5.42 Block diagram of 32-kbit CAAC-IGZO SRAM. *Source*: Adapted from [16]. Copyright 2015 The Japan Society of Applied Physics

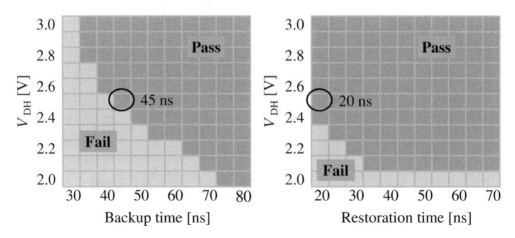

Figure 5.43   Data backup/restoration time

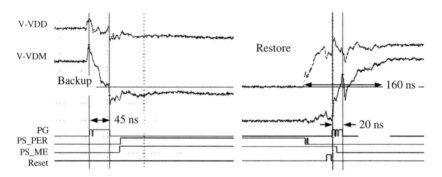

Figure 5.44   Waveforms measured in power gating. *Source*: Adapted from [16]. Copyright 2015 The Japan Society of Applied Physics

Figure 5.43 shows a Shmoo plot of the measured data backup and restoration times, defined as the periods of high OSG. When V_{DH} is 2.5 V, the backup time is 45 ns and the restoration time is 20 ns. Note that to obtain the restoration time of the die, we must add the time for charging the power line to the restoration time. Figure 5.44 shows the oscilloscope waveforms in the power gating. Normal power-gating operation is observed.

Figure 5.45 presents the standby power in each standby mode implemented in the CAAC-IGZO SRAM die. Bit-line floating, peripheral PG, array PG, and all-domain PG reduced the standby power by 3.9%, 0.1%, 95%, and 99.9%, respectively.

During its execution, power gating demands extra power for controlling the power switches, charging the power line, and backing up or restoring the data in the memory cell array. Therefore, if the power-off time is short, power gating may increase the total power consumption.

Figure 5.46 plots the power savings under bit-line floating and the power gating as a function of idle time. The savings are normalized by the standby power in the normal state. At idle times of 700 μs or shorter, the power gating increases the total power consumption. As shown in

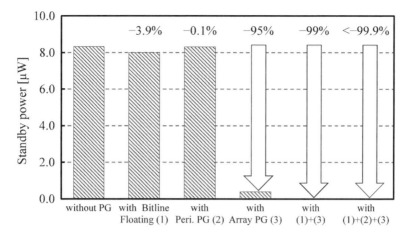

Figure 5.45 Reduction of standby power in 32-kbit CAAC-IGZO SRAM without power gating (w/o PG) and with bit-line floating, power gating in the peripheral circuits (peri. PG), power gating in the memory array (array PG), and power gating in some or all domains. *Source*: Adapted from [16]. Copyright 2015 The Japan Society of Applied Physics

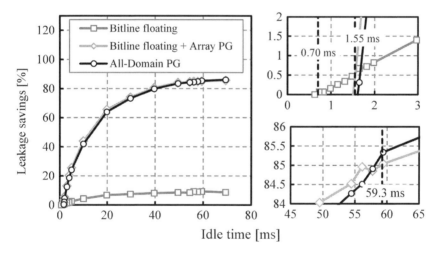

Figure 5.46 Dependence of total power savings on idle time. *Source*: Adapted from [16]. Copyright 2015 The Japan Society of Applied Physics

Figure 5.46, the bit-line floating, array-PG, and all-domain PG modes reduced the total power at idle times between 700 and 1.55 ms, between 1.55 and 59.3 ms, and longer than 59.3 ms, respectively.

Bit-line-floating and power-gating modes are implementable on the fabricated SRAM die. Among these low-standby-power modes, we should optimize the total power consumption. When the memory operates at predetermined time intervals (e.g., when data are sampled),

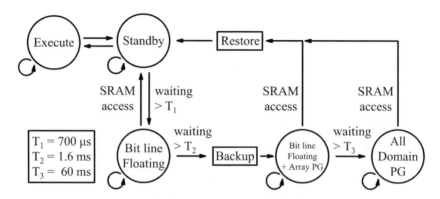

Figure 5.47   State transitions in multiple PG

selecting an appropriate mode will lower the power consumption. However, in applications such as network traffic, the wait times are random or unpredictable, and the appropriate mode is hard to determine in advance. Instead, a state machine might be applied (see Figure 5.47). In a state machine, there is not necessarily any system clock, and the state depends on the input combined with the previous state (i.e., the entire history of the machine is summarized in the current state). When there is no input, the machine is in complete idle and power could be zero if an NVM is employed.

More specifically, a state machine can have multiple power gatings where the domains performing the power gating are gradually increased whenever the idle time exceeds a predetermined value. The effect of multiple-PG was examined by measuring the power with a simple FPGA-based system. The test system chiefly comprised the 32-kbit CAAC-IGZO SRAM, an ARM® Cortex® M0 CPU core, and a power management unit (Figure 5.48). The applied power-gating modes were determined by the length of the SLEEP signal. The active state was then restored by an external interrupt signal. The time to interrupt was assumed to obey a Gamma distribution $\Gamma(\alpha, \lambda)$, which effectively describes Internet traffic behavior (i.e., a Poisson process):

$$\Gamma(\alpha,\lambda) = \frac{x^{\alpha-1}e^{-x/\lambda}}{\lambda^{\alpha}\int\limits_{0}^{\infty} y^{\alpha}e^{-y}dy}. \tag{5.1}$$

In the experiments, the wait times were distributed between 0.1 and 0.25 ms by Equation (5.1), where x represents the wait time. The power consumption of the memory was then measured from the calculation results.

Figure 5.49 shows the total power consumption for various values of (α, λ). In all-domain PG mode, the power consumption was increased by 11% at α and λ of 1 and 0.1 m, respectively. Multiple PG reduced the total power for all α and λ, achieving a 79% reduction at α and λ of 10 and 5 m, respectively.

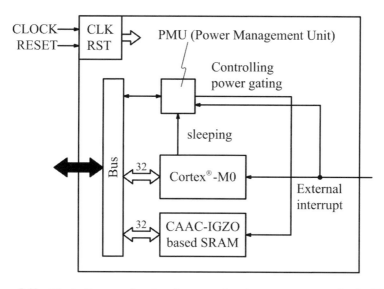

Figure 5.48 Block diagram of system for measuring the power consumption in SRAM

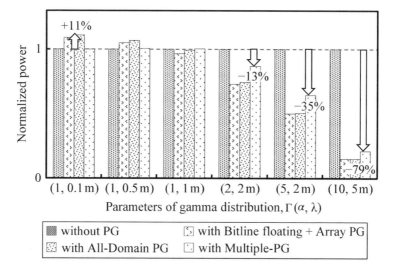

Figure 5.49 Total power consumption of 32-kbit SRAM in various low-standby power modes (the wait time obeys a Gamma distribution)

The structures and performances of various power-gating technologies are summarized in Table 5.10. The prototype CAAC-IGZO SRAM [16] shows superior frequency characteristics in normal operation and dramatically reduced backup and restoration energies. These advantages are obtained by adding the fast-operating backup circuit to SRAM, and employing CAAC-IGZO FETs to realize a low-energy, fast-writing memory operation. The power leakage

Table 5.10 Comparison of various power-gating technologies. *Source*: Adapted from [16]. Copyright 2015 The Japan Society of Applied Physics

	Uesugi [16]	Ishizu [6]	Sakimura [11]	Bartling [10]
Type	CAAC-IGZO	CAAC-IGZO	MTJ	FeCap
Technology	Si: 180 nm	Si: 350 nm	90 nm	130 nm
	IGZO: 60 nm	IGZO: 180 nm		
Frequency	85 MHz	15 MHz	20 MHz	8 MHz
Voltage	Si: 1.8 V	Si: 2.5 V	1.0 V	1.5 V
	IGZO: 2.5/−1 V	IGZO: 2.5/−1 V		
Backup time	45 ns	80 ns	4 ns (2)	320 ns
Restoration time	160 ns (1)	400 ns (1)	5 ns (2)	384 ns (1)
Backup energy	144 fJ/bit	N/A	6 pJ/bit	2.2 pJ/bit
Restoration energy	97 fJ/bit	N/A	0.3 pJ/bit	0.66 pJ/bit

1. Includes charging time of the power supply line.
2. Writing and reading times of MRAM cell.

in the Si die is relatively low when adopting the 180-mm Si technology. As the leakage current of Si FETs increases with scale-down, the power-gating technique becomes even more effective. The CAAC-IGZO-based SRAM also offers promising solutions for high-frequency applications which require SRAM. Although the present prototype was fabricated with 180-nm Si technology, the circuit could be scaled down to realize higher-frequency operation.

References

[1] Ando, K. (2002) "Nonvolatile magnetic memory," *FED Review*, 1 [in Japanese].
[2] Ohmaru, T., Yoneda, S., Nishijima, T., Endo, M., Dembo, H., Fujita, M., *et al.* (2012) "Eight-bit CPU with non-volatile registers capable of holding data for 40 days at 85°C using crystalline In–Ga–Zn oxide thin film transistors," *Ext. Abstr. Solid State Dev. Mater.*, 1144.
[3] Sjökvist, N., Ohmaru, T., Furutani, K., Isobe, A., Tsutsui, N., Tamura, H., *et al.* (2013) "Zero area overhead state retention flip flop utilizing crystalline In–Ga–Zn oxide thin film transistor with simple power control implemented in a 32-bit CPU," *Ext. Abstr. Solid State Dev. Mater.*, 1088.
[4] Sjökvist, N., Ohmaru, T., Isobe, A., Tsutsui, N., Tamura, H., Uesugi, W., *et al.* (2014) "State retention flip flop architectures with different tradeoffs using crystalline indium gallium zinc oxide transistors implemented in a 32-bit normally-off microprocessor," *Jpn. J. Appl. Phys.*, **53**, 04EE10.
[5] Tamura, H., Kato, K., Ishizu, T., Uesugi, W., Isobe, A., Tsutsui, N., *et al.* (2014) "Embedded SRAM and Cortex-M0 core using a 60-nm crystalline oxide semiconductor," *IEEE Micro*, **34**, 42.
[6] Ishizu, T., Kato, K., Onuki, T., Matsuzaki, T., Tamura, H., Ohmaru, T., *et al.* (2014) "SRAM with *c*-axis aligned crystalline oxide semiconductor: Power leakage reduction technique for microprocessor caches," *Proc. IEEE Int. Memory Workshop*, 103.
[7] Yoda, H., Fujita, S., Shimomura, N., Kitagawa, E., Abe, K., Nomura, K., *et al.* (2012) "Progress of STT-MRAM technology and the effect on normally-off computing systems," *IEEE IEDM Tech. Dig.*, 259.
[8] Isobe, A., Tamura, H., Kato, K., Ohmaru, T., Uesugi, W., Ishizu, T., *et al.* (2014) "A 32-bit CPU with zero standby power and 1.5-clock sleep/2.5-clock wake-up achieved by utilizing a 180-nm *c*-axis aligned crystalline In–Ga–Zn oxide transistor," *IEEE Symp. VLSI Circuits Dig. Tech. Pap.*, 49.
[9] Kobayashi, H., Kato, K., Ohmaru, T., Yoneda, S., Nishijima, T., Maeda, S., *et al.* (2013) "Processor with 4.9-μs break-even time in power gating using crystalline In–Ga–Zn-oxide transistor," *IEEE COOL Chips XVI*, Session VI.

[10] Bartling, S. C., Khanna, S., Clinton, M. P., Summerfelt, S. R., Rodriguez, J. A., and McAdams, H. P. (2013) "An 8 MHz 75 µA/MHz zero-leakage non-volatile logic-based Cortex-M0 MCU SoC exhibiting 100% digital state retention at $V_{DD} = 0$ V with < 400 ns wakeup and sleep transitions," *IEEE Int. Solid-State Circuits Conf. Dig. Tech. Pap.*, 432.

[11] Sakimura, N., Tsuji, Y., Nebashi, R., Honjo, H., Morioka, A., Ishihara, K., *et al.* (2014) "A 90 nm 20 MHz fully nonvolatile microcontroller for standby-power-critical applications," *IEEE Int. Solid-State Circuits Conf. Dig. Tech. Pap.*, 184.

[12] Yu, W.-k., Rajwade, S., Wang, S.-E., Lian, B., Suh, G. E., and Kan, E. (2011) "A non-volatile microcontroller with integrated floating-gate transistors," *Proc. IEEE DSNW*, 75.

[13] Shuto, Y., Yamamoto, S., and Sugahara, S. (2012) "Static noise margin and power-gating efficiency of a new nonvolatile SRAM cell based on pseudo-spin-transistor architecture," *Proc. IEEE Int. Memory Workshop*, 233.

[14] Masui, S., Yokozeki, W., Oura, M., Ninomiya, T., Mukaida, K., Takayama, Y., *et al.* (2003) "Design and applications of ferroelectric nonvolatile SRAM and flip-flop with unlimited read/program cycles and stable recall," *Proc. IEEE CICC*, 403.

[15] North Carolina State University (2016) "NCSU Electronic Design Automation (EDA) Wiki, FreePDK45: Contents." Available at: www.eda.ncsu.edu/wiki/FreePDK45:Contents [accessed February 16, 2016].

[16] Uesugi, W., Ishizu, T., Kato, K., Onuki, T., Tamura, H., Isobe, A., *et al.* (2015) "A 32-kb embedded SRAM using 60-nm crystalline oxide semiconductor transistors and power gating with 45-ns 144-fJ/bit data backup," *Ext. Abstr. Solid State Dev. Mater.*, 1146.

6

FPGA

6.1 Introduction

This chapter focuses on a programmable device, namely a field-programmable gate array (FPGA) that includes c-axis-aligned crystalline indium–gallium–zinc oxide (CAAC-IGZO) field-effect transistors (FETs) as non-volatile devices. Non-volatile oxide semiconductor random access memory (NOSRAM) and dynamic oxide semiconductor random access memory (DOSRAM), described in Chapters 3 and 4, respectively, can be used instead of a static random access memory (SRAM) or a flash memory of currently available programmable devices. Not only used as memory, they can also serve as programmable elements in programmable devices, as a substitute for an SRAM or flash memory element. NOSRAM and DOSRAM require much less power and footprint than SRAM to retain charges, and can rewrite data more times with much less power than a flash memory. These characteristics enable the formation of a programmable device with a CAAC-IGZO FET in a programmable element (henceforth CAAC-IGZO programmable device). The CAAC-IGZO programmable device has several attractive features.

Section 6.2 introduces an FPGA with CAAC-IGZO FETs in programmable elements (CAAC-IGZO FPGA) [1–3]. Compared with a common FPGA having a volatile SRAM as a programmable element (SRAM FPGA), a CAAC-IGZO FPGA requires much lower power to retain data and fewer components (e.g., transistors and capacitors) to form the programmable element. Thus, lower power consumption and improved area efficiency can be achieved. While the SRAM FPGA inevitably requires reconfiguration following the return of the source voltage supply, the CAAC-IGZO FPGA does not, thereby allowing high-speed startup. Although high-speed startup is also available in a programmable device (programmable logic device, PLD) using flash memory, the CAAC-IGZO FPGA has a unique characteristic that the PLD lacks: it can improve the operational frequency with a high-speed programmable routing switch (PRS) utilizing a boosting effect.

Physics and Technology of Crystalline Oxide Semiconductor CAAC-IGZO: Application to LSI, First Edition.
Edited by Shunpei Yamazaki and Masahiro Fujita.
© 2017 John Wiley & Sons, Ltd. Published 2017 by John Wiley & Sons, Ltd.

Section 6.3 describes normally-off computing with less power consumption, by effectively utilizing the CAAC-IGZO technology in FPGAs. A CAAC-IGZO FPGA with a fine-granularity, multicontext structure adopts power gating in synchronization with context switching and *in situ* register backup in a normally-off processor introduced in Chapter 5, which results in further-low-power normally-off computing [4–6]. Even with context switching that changes the processing content, such an FPGA can resume processing from the previous end state by reloading the data registered as backup. Such normally-off computing makes the programmable logic element (PLE) sufficient for the processing content to be active, thereby realizing processing with high power efficiency.

In Section 6.4, a method for achieving extremely low-voltage driving (so-called subthreshold driving) is introduced as a way to further reduce the power consumption of the CAAC-IGZO FPGA [7]. In a programmable element comprising a CAAC-IGZO FET, the retainable potential of program data does not depend on the driving voltage; therefore, a voltage higher than the driving voltage can be retained. This feature facilitates overdriving a PRS. In addition, the use of a boosting effect in a programmable power switch (PPS) for power gating enables the PPS to be overdriven without the generation of negative source voltage. These factors enable a CAAC-IGZO FPGA with subthreshold driving, which is expected to be used in sensor networks utilizing power harvesting.

The field of FPGA applications has widened in recent years. For example, a computing system combining the high versatility of a central processing unit (CPU) with the quick response and highly parallel processing performance of an FPGA is becoming popular. With this background, Section 6.5 discusses the possibility of a computing system combining a CPU and a CAAC-IGZO FPGA. While the CAAC-IGZO FPGA applications are not limited to those described here, this chapter will show the high potential of the CAAC-IGZO FPGA.

6.2   CAAC-IGZO FPGA

6.2.1   Overview

An FPGA [8–11] is a device with which the user can change the circuit configuration by changing the program data (configuration data) stored in programmable elements (configuration memory). The basic configuration of an FPGA is shown in Figure 6.1(a). Programmable regions in an FPGA are mainly composed of PLEs and PRSs. Each PLE and PRS has a configuration memory. In response to the configuration data stored in these memories, the circuit configuration can be changed.

Each PLE has a look-up table (LUT), as shown in Figure 6.1(b). In response to the configuration data stored in the configuration memory (CM) of the LUT, the logic of the PLE can be determined. As an example, Figure 6.1(b) shows the configuration of NAND with two inputs. The configuration memories retain the values for the truth table of the NAND function. As shown in Figure 6.1(c), the PRS controls conduction or non-conduction between the PLEs in response to configuration data determining the PRS conduction state stored in the configuration memory. Here, the signal propagates from right to left. A user can freely fabricate a logic circuit by changing the configuration data in the memories of the PLEs and PRSs.

This section introduces a CAAC-IGZO FPGA, an FPGA with a CAAC-IGZO FET as the configuration memory replacing the SRAM element used in conventional FPGAs. The CAAC-IGZO FPGA has the following characteristics: (1) a configuration memory with a small number of transistors, which is advantageous for area reduction; (2) non-necessity of reconfiguration during

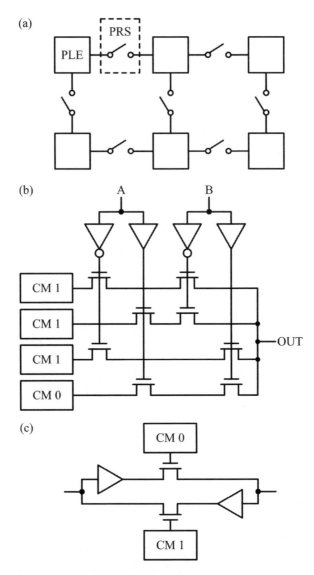

Figure 6.1 Schematic structures of (a) FPGA, (b) PLE, and (c) PRS. CM denotes a configuration memory

power gating because of the non-volatile configuration memory, leading to instant restart; and (3) high-speed switching by a boosting effect in a PRS. The first two characteristics are derived from the characteristics of the non-volatile memory with a CAAC-IGZO FET described in previous chapters. These two characteristics can also be obtained by using other non-volatile memory elements – such as flash memory, magnetoresistive random access memory (MRAM) [12, 13], ferroelectric memory [14], solid-electrolyte switch [15], resistive random access memory (ReRAM) [16], and complementary atom switch [17] elements – for a configuration memory; however, configuration memories with such non-volatile memories do not contribute to the improvement of FPGA operational speed. In contrast, the third characteristic improves the operational speed of the CAAC-IGZO FPGA [3]. The PRS and PLE are described below in detail.

6.2.2 PRS

6.2.2.1 Basic Configuration

User-programmable regions in an FPGA are composed of PLEs, routing fabric (e.g., routing wires and PRSs connecting PLEs), etc. It is recognized that the routing fabric accounts for 50% of the layout area of the user-programmable regions and 60% of delay in critical paths [18]. If the routing fabric, especially PRSs, were to be reduced in an area and improved in response speed, the whole FPGA could be greatly improved in performance.

PRSs widely employ structures in which a signal is applied to a gate of a pass gate (transistor MG) to control conduction and non-conduction between input IN and output OUT (Figure 6.2) [18]. In a PRS where an SRAM element used as configuration memory is connected to the gate of a transistor MG (SRAM pass gate), the high or low of the signals supplied to the gate of the transistor MG is determined by the configuration data stored in the SRAM element. In the pass gate, when the high signal is supplied to the input IN, the voltage of the output OUT decreases from that of IN by the threshold voltage V_{th} of the transistor MG. This is a so-called V_{th} drop [18], which is problematic, particularly in extensive cascading.

6.2.2.2 Improved Configuration

There are several known methods to avoid the V_{th} drop: using a low-threshold-voltage Si FET process; using a keeper latch at OUT; using overdriving [18], that is, supplying signals that drive at a higher voltage (overdrive voltage) than the core source voltage to the gate of the pass-gate transistor MG, as shown in Figure 6.3; and boosting the gate voltage of the transistor MG in a boosting pass gate [19–21] by using the boosting effect shown in Figure 6.4.

Overdriving avoids the V_{th} drop and increases the on-state current of the transistor MG, resulting in improvement of the operational speed compared with a normal pass gate. The

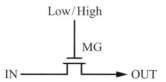

Figure 6.2 Circuit configuration of the pass gate. *Source*: Adapted from [3]

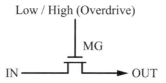

Figure 6.3 Circuit configuration of a pass gate with overdriving. *Source*: Adapted from [3]

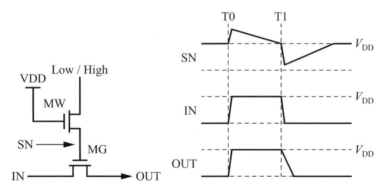

Figure 6.4 Circuit configuration and timing diagram of conventional boosting pass gate. *Source*: Adapted from [3]

circuit configuration of a normal pass gate can be used for overdriving; thus, overdriving can easily be employed, although it requires additional supply voltages. In addition, when the transistor MG is on and a low signal is supplied to IN, the constant high-voltage supply to the gate of the transistor MG affects the reliability of the gate-insulating film [19]. The SRAM element for holding the configuration data in an SRAM pass gate also has to be driven at high voltage, which leads to increased power consumption.

In the conventional boosting pass gate in Figure 6.4, a source voltage is supplied to the gate of a transistor MW and the gate potential of the transistor MG is controlled by a signal supplied to the source of MW. When a PRS with an SRAM element as a configuration memory is connected to the MW source (SRAM boosting pass gate), the high or low of a signal supplied to the MG gate is determined by the configuration data stored in the SRAM element. Supplying a high signal to the MW source sets node SN to a potential corresponding to the high signal, turning on the transistor MG and turning off the transistor MW. That is, the gate of the transistor MG (node SN) is in a weakly floating state. When the signal to IN rises (i.e., at time T0 in Figure 6.4), the gate potential of the transistor MG (the potential of the node SN) is boosted to improve the drive capability; thus, the signal transmission speed to OUT is high.

The effect of boosting the potential of the transistor MG gate (node SN) is called a boosting effect. However, the off-state current of the transistor MW acts as a leakage current from the node SN, and the boosted potential of SN drops to the former potential in a maximum of several seconds. Thus, at time T1, the transistor MG has low drive capability. Even worse, the signal supplied to IN falls at this time, whereby the MG gate potential is further decreased by negative boosting and the drive capability becomes very low. Therefore, it is difficult to use the boosting pass gate in a possible critical path. One report [19] indicates that large limitations are imposed on the use of the boosting pass gate when flexibility in changing circuit configuration (as in FPGA) is required.

6.2.2.3 CAAC-IGZO PRS

The problems with the conventional pass gates are derived from the effects of the incomplete insulation property of the node SN. If SN is in a *complete* floating state, the boosted gate

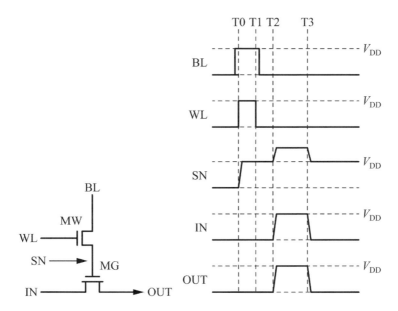

Figure 6.5 Circuit configuration and timing diagram of the CAAC-IGZO boosting pass gate. *Source*: Adapted from [3]

potential can be maintained and the drive capability of the transistor MG can be retained, even when the signal to IN falls in potential. In order to obtain such a complete floating state, the structure shown in Figure 6.5 is proposed, in which the transistor MW in the boosting pass gate uses a CAAC-IGZO FET (CAAC-IGZO boosting pass gate) [10]. The CAAC-IGZO boosting pass gate can be regarded as a memory cell with the same structure as that of the NOSRAM memory element in Chapter 3 or the sensor pixel of the CAAC-IGZO image sensor in Chapter 7. In other words, it can be treated as a PRS where a configuration memory storing configuration data is combined with a pass gate. The CAAC-IGZO boosting pass gate is characterized by a smaller number of transistors to constitute a PRS than the SRAM pass gate or SRAM boosting pass gate.

The configuration data in the CAAC-IGZO boosting pass gate are written in the following manner: a positive logic signal is supplied to a signal line, WL, to turn on the transistor MW, and configuration data are supplied to the source of MW, so as to be stored in the node SN (at time T0 in Figure 6.5). After writing, the transistor MW is turned off (at time T1). Writing configuration data in the configuration memory of the CAAC-IGZO boosting pass gate is controlled in the same way as writing control of the configuration memory of the SRAM pass gate (SRAM element). The CAAC-IGZO boosting pass gate does not require a high-voltage signal, unlike flash memory, and can have a similar structure to that of the SRAM pass gate for control circuits such as decoders.

The CAAC-IGZO FET has an extremely low off-state current, as described in previous chapters. In the CAAC-IGZO boosting pass gate, the off state of the transistor MW therefore makes the node SN into an *almost completely* floating state. Of course, like the conventional boosting pass gate, the driving capability of the transistor MG is heightened when the signal to IN rises (at time T2). In addition, even when the signal to IN falls in potential, the drive capability can be maintained because the node SN potential remains boosted (at time T3).

To achieve an almost completely floating state of node SN, the following conditions should be satisfied: a very low off-state current of the transistor MW, a very low gate-leakage current from the transistor MG, and very high translation properties of the interlayer films. When these conditions are satisfied sufficiently well, a *complete* floating state can be obtained. Refreshing (i.e., rewriting) the configuration data at a very low frequency effectively compensates for any deficiency in the floating state. As described in Subsection 6.2.4.2, the pass gate does not require refreshing for at least 8 days, meaning that the power consumption for refreshing is practically negligible. To improve the retention characteristics of node SN, the addition of a storage capacitor to SN is also effective; however, too much storage capacitance makes it difficult to obtain the boosting effect and lowers the response speed of the PRS. Thus, the tradeoff between the charge-retention period of node SN and the response speed should be considered.

6.2.3 PLE

6.2.3.1 Block Configuration

The PLE in the SRAM FPGA includes, for example, a four-input LUT and register, as shown in Figure 6.6(a), and functions as carry logic or scan logic for grouping adjacent PLEs to make a carry chain or register chain [22]. The CAAC-IGZO FPGA can have a similar PLE.

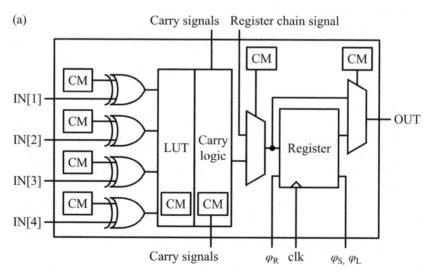

Figure 6.6 (a) Block diagram of the PLE composed of an LUT, carry logic, scan logic, and a register; (b) configuration memory in the PLE; and (c) non-volatile register. The "register" in (a) is a non-volatile register in the CAAC-IGZO FPGA. The configuration memory is denoted by CM. *Source*: Adapted from [6]

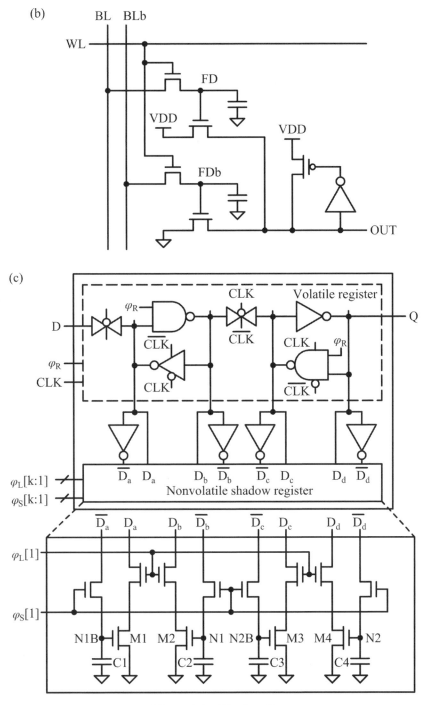

Figure 6.6 (*Continued*)

6.2.3.2   Configuration Memory

The circuit configuration shown in Figure 6.6(b) can be used for the configuration memory in the PLE of the CAAC-IGZO FPGA. The circuit configuration of the PLE can be set by storing complementary data corresponding to the configuration data in the configuration memory (nodes FD and FDb hold data corresponding to the configuration data and inverse of the configuration data, respectively).

6.2.3.3   Non-volatile Register

A register with a non-volatile memory composed of CAAC-IGZO FETs (non-volatile register) shown in Figure 6.6(c) enables the non-volatile operation described below. The non-volatile register contains a volatile register of Si FETs and a non-volatile memory of CAAC-IGZO and Si FETs (non-volatile shadow register). In normal operation, the non-volatile register can operate like a common one by utilizing the volatile register. When control signals φ_S and φ_L to the shadow register are set to be active asynchronously with clock signals, the data of the volatile register can be stored in/loaded from the non-volatile shadow register. After storing the data, the power can be shut off without risk of data loss. By loading the data after power return, the process immediately before the store operation can be resumed. In contrast, when an SRAM is used for the shadow register, the power to the shadow register cannot be shut off, and thus the above resuming operation cannot be executed.

6.2.4   Prototype

6.2.4.1   Entire Configuration

A prototype CAAC-IGZO FPGA with the above-introduced structure has been fabricated with a hybrid of the processes used for the 1.0-μm CAAC-IGZO and 0.5-μm Si FETs [3]. The details of the fabricated CAAC-IGZO FPGA are described below.

In the CAAC-IGZO FPGA, the CAAC-IGZO boosting pass gate from Subsection 6.2.2 is used as a PRS, the configuration memory with CAAC-IGZO FETs in Subsection 6.2.3 is used as the configuration memories in a PLE and a user I/O, and the non-volatile register with CAAC-IGZO FETs is used as a register in the PLE. Such a structure offers non-volatility of the FPGA operations.

The schematic structure of the fabricated FPGA is described with reference to Table 6.1 and Figure 6.7. The FPGA comprises a logic array block, user I/Os, a word driver, a bit driver, and a configuration controller. The logic array block comprises 20 PLEs in two columns (10 PLEs per column) and routing fabrics with PRSs that connect PLEs to one another or a PLE to a user I/O.

The configuration of the PLE can be set by configuration data stored in the configuration memory with CAAC-IGZO FETs in the PLE [circuit configuration as in Figure 6.6(b); 64 bits in each PLE].

The PRS has a CAAC-IGZO boosting pass gate consisting of a configuration memory with a CAAC-IGZO FET integrated with a pass gate. The routing fabric has a multiplexer circuit configuration with a CAAC-IGZO boosting pass gate (see Figure 6.8). A PRS exists between two PLEs or between a PLE and a user I/O. The routing fabrics include 5760 CAAC-IGZO boosting pass gates (i.e., there is a 5760-bit configuration memory). In the CAAC-IGZO boosting pass

Table 6.1 Specifications of CAAC-IGZO FPGA

Process	1.0-µm CAAC-IGZO/0.5-µm Si
Die size	5.5 mm × 4.5 mm
Number of PLE	20
Number of IO	20
LUT inputs	4
Total bit number in CM	1760 bits
Total bit number in PRS	5760 bits
PLE size	390 µm × 717 µm
CM size	45 µm × 12 µm
PRS size	45 µm × 5 µm
Structure of CM	2 CAAC-IGZO FETs, 4 Si FETs, 2 capacitors (4 fF)
PRS structure	1 CAAC-IGZO FET, 2 Si FETs, 1 capacitor (4 fF)
Supply voltage	3.3 V

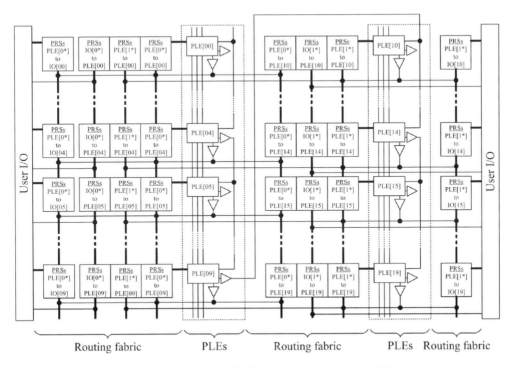

Figure 6.7 Entire block diagram of prototype FPGA. PRSs denote groups of PRSs selecting output from PLE to user I/O, from user I/O to PLE, and from PLE to PLE. *Source*: Reprinted from [3], with permission of IEEE, © 2015

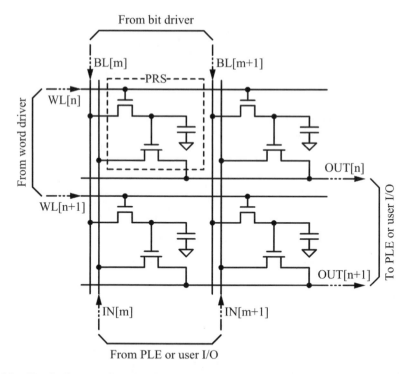

Figure 6.8 Circuit diagram of routing fabric in the prototype FPGA, where a CAAC-IGZO boosting pass gate is used for PRS. *Source*: Adapted from [3]

gate, the gate capacitance of the transistor MG is 15 fF and the storage capacitance with parasitic capacitance is 4 fF; that is, the combined capacitance in the node SN is 19 fF. The CAAC-IGZO FPGA has a 7520-bit configuration memory in total.

There are 10 user I/Os in each of the left and right columns (i.e., 20 in total). Each user I/O can select input/output (logic output/open drain output). The configuration of the user I/O can be set by configuration data stored in the configuration memory with CAAC-IGZO FETs in the user I/O [where the circuit is configured as in Figure 6.6(b) and there are 8 bits in each user I/O].

The word and bit drivers control the writing processing of configuration data in the configuration memories. The word driver controls the gate potential of the transistor MW in the CAAC-IGZO boosting pass gate. The bit driver supplies configuration data to the transistor MW source. The configuration controller controls the configuration and generates system clocks. The source voltage at each block is V_{DD} and ground; the word driver further includes the source voltages of V_{DDH} and V_{SSH}. The word driver has a level shifter for converting signals of V_{DD} and ground into those of V_{DDH} and V_{SSH}, and generates a signal to be applied to the gate of the CAAC-IGZO transistor MW in the CAAC-IGZO boosting pass gate. To lower the off-state current of the transistor MW, ground is set to be equal to or larger than V_{SSH}, so as to apply negative bias to the gate.

For comparison, an FPGA with an SRAM element as a configuration memory (SRAM FPGA) is also fabricated entirely in Si and with the same process node (0.5 μm). The circuit block configurations other than the configuration memory are the same as those of the CAAC-IGZO FPGA.

6.2.4.2 Measurement

Basic Operation
The basic functions of the CAAC-IGZO FPGA operate correctly under operational testing; specifically, the configuration as an up/down counter or shift circuit with all PLEs works as intended. Figure 6.9(a) and (b) shows the operational waveforms of the up/down counter configuration and those of the shift circuit configuration, respectively. They are the waveforms under $V_{DD} = V_{DDH} = 2.5$ V, $V_{SSH} = -1.0$ V, and a frequency of 1 MHz.

Area Efficiency
One major feature of the CAAC-IGZO FPGA is its small-area programmable elements. Figure 6.10(a) and (c) shows micrographs of the fabricated CAAC-IGZO and SRAM FPGAs,

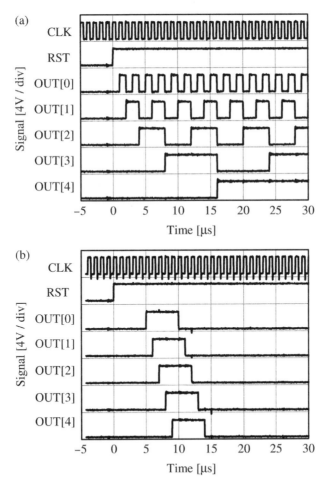

Figure 6.9 Operational waveforms at 1 MHz of (a) up/down counter and (b) shift circuit. Each circuit is configured by a CAAC-IGZO FPGA. $V_{DD} = V_{DDH} = 2.5$ V, $V_{SSH} = -1.0$ V. *Source*: Reprinted from [3], with permission of IEEE, © 2015

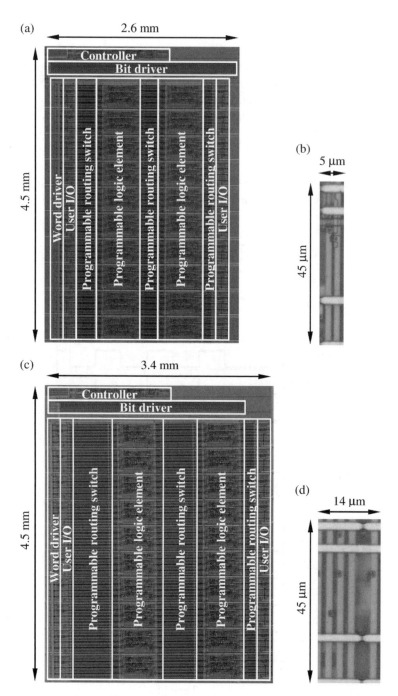

Figure 6.10   CAAC-IGZO FPGA micrographs of (a) core and (b) PRS, and SRAM FPGA micrographs of (c) core and (d) PRS.

respectively [3]. Figure 6.10(b) and (d) shows enlarged micrographs of the PRSs of the CAAC-IGZO and SRAM FPGAs, respectively. In the CAAC-IGZO FPGA, the layout areas of the PRS, routing fabric, PLE, and core are 61%, 54%, 6%, and 22% smaller than those of the SRAM FPGA, respectively. The CAAC-IGZO boosting pass gate can be composed of a smaller number of transistors than that of the SRAM pass gate, as explained in Section 6.2, which contributes to the reduction in routing fabric layout area. In addition, the reduction in the placement area of power lines contributes to the reduction in the layout areas of the routing fabric and PLE.

The configuration memories with CAAC-IGZO FETs in PRSs and PLEs offer the benefit of layout area reduction.

Non-volatility

To confirm that the CAAC-IGZO FPGA can perform non-volatile operations, its operation with and without a source voltage is examined. The data for a volatile register are stored in a shadow register before stopping the source voltage supply, and the stored data are loaded from the shadow register after resupplying the source voltage. The configuration is again an up/down counter. Here, $V_{DD} = V_{DDH} = 2.5$ V, $V_{SSH} = -0.7$ V, and the frequency is 20 kHz.

The waveforms of Figure 6.11 show that the output value after power return is continuous with that before power shut-off. In addition, the operation resumes after the power return without reconfiguration. Thus, non-volatile operation is available, which is the second major feature of the CAAC-IGZO FPGA. Reconfiguration immediately after source voltage supply is unnecessary, which allows instant startup.

Operational Speed

To confirm the acceleration of the PRS in the CAAC-IGZO FPGA, the relationship between the source voltage and the maximum operational frequency is examined with the CAAC-IGZO and

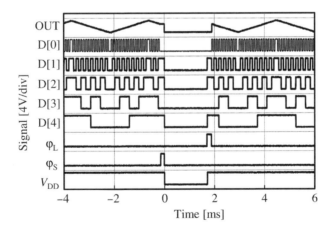

Figure 6.11 Demonstration of performing the power (V_{DD}) on/off for the CAAC-IGZO FPGA in an up/down counter configuration. A store/load signal controls the register data store in/load from the non-volatile block in the non-volatile register, and D[4:0] corresponds to outputs of the up/down counter. The OUT is the output of an A/D converter whose inputs are D[4:0]. *Source*: Reprinted from [2], with permission of IEEE, © 2013

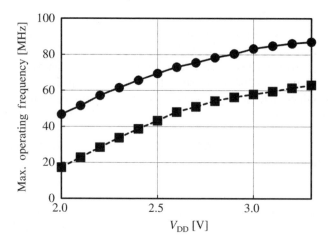

Figure 6.12   Relationship between the source voltage and the highest operational frequency in counter circuits of the CAAC-IGZO FPGA (solid line) and SRAM FPGA (dotted line). $V_{DD} = V_{DDH} = $ 2.5 V to 3.3 V and $V_{SSH} = -1.0$ V. *Source*: Reprinted from [3], with permission of IEEE, © 2015

SRAM FPGAs. The results are shown in Figure 6.12. The configuration is set to the counter circuit, and the measurement is performed under conditions of $V_{DD} = V_{DDH} = 2.5$ V to 3.3 V and $V_{SSH} = -1.0$ V.

As shown in Figure 6.12, the CAAC-IGZO FPGA is superior to the SRAM FPGA in terms of its operational frequency. For example, at V_{DD} of 2.0, 2.5, and 3.0 V, the operational frequency of the CAAC-IGZO FPGA is, respectively, 2.7 times, 1.6 times, and 1.4 times higher than that of the SRAM FPGA. This is because the PRS is composed of the above-described CAAC-IGZO boosting pass gate and the switching speed is improved by the boosting effect. This is the third major feature of the CAAC-IGZO FPGA.

Retention Characteristics
To discuss the non-volatile operations of the CAAC-IGZO FPGA, the data-retention characteristics in the CAAC-IGZO configuration memory are examined. Here, in addition to continuous configuration preservation, high-speed operation should be maintained. Figure 6.13 shows a long-term operational test with the CAAC-IGZO FPGA. In the test, 13 PLEs, each with an inverter structure, are cascaded to create the configuration of a ring oscillator. The oscillation frequency in the ring oscillator is monitored. Here, $V_{DD} = V_{DDH} = 2.5$ V and $V_{SSH} = -1.0$ V. According to Figure 6.13, the decrease in the oscillation frequency after 8 days is less than 3%. The operational frequency of the CAAC-IGZO FPGA is 1.6 times higher than that of the SRAM FPGA at $V_{DD} = 2.5$ V in Figure 6.12. In light of these results, an oscillation frequency higher than that of the SRAM FPGA can be maintained for at least 8 days. In other words, the boosting effect can be maintained in the CAAC-IGZO boosting pass gate for at least 8 days while holding the configuration data in memory.

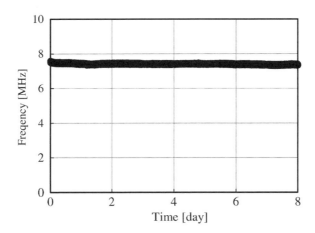

Figure 6.13 Change in oscillation frequency in the long-term operation test. The CAAC-IGZO FPGA has the configuration of a 13-stage ring oscillator circuit with 13 PLEs. $V_{DD} = V_{DDH} = 2.5$ V and $V_{SSH} = -1.0$ V. *Source*: Reprinted from [3], with permission of IEEE, © 2015

6.3   Multicontext FPGA Realizing Fine-Grained Power Gating

6.3.1   Overview

The CAAC-IGZO FPGA brings greater advantages when it has a fine-grained structure with a relatively large number of PRSs and a multicontext structure capable of fine-grained power gating in synchronization with context switching. Such a structure enables fine-grained power control, by which power is supplied only to PLEs contributing to effective operation. In addition, normally-off computing can be achieved by loading (storing) registered data from (into) the non-volatile register of a PLE in synchronization with context switching [5, 6].

6.3.2   Normally-Off Computing

6.3.2.1   Fine-Grained Multicontext Method

Dynamic reconfiguration is a technique whereby the circuit configuration of an FPGA can be switched during system operation. This technique is advantageous because a large-scale system can be composed of a small number of hardware resources. One of the dynamic reconfiguration techniques is a multicontext method. In this method, an FPGA holds multiple sets of configuration data corresponding to multiple circuit configurations, and switches these sets to select the objective circuit configuration dynamically [23]. Figure 6.14 illustrates the conceptual diagram of a multicontext-type dynamic reconfiguration FPGA.

In the multicontext method, switching among internal configuration memory sets [see Figure 6.15(a)] allows short-term reconfiguration. There is no need to read any dataset from an external memory to store the data in an internal configuration memory; thus, the method is excellent in terms of responsiveness and power consumption. However, multiple configuration memory sets are required for storing multiple configuration datasets. In particular, the

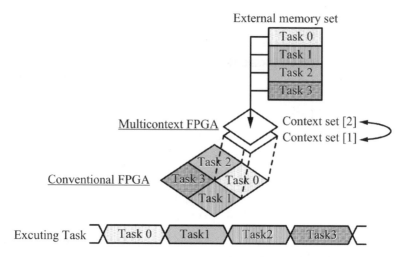

Figure 6.14   Conceptual diagram of dynamic reconfiguration FPGA

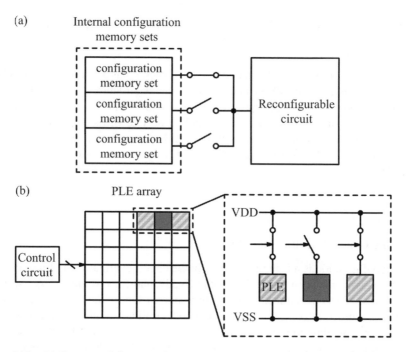

Figure 6.15   (a) Conceptual diagram of the multicontext method; (b) fine-grained power gating

relative increase in PRS area, which accounts for a large part of the configuration memory, becomes noticeable. Accordingly, the multicontext method is said to be suitable only for a coarse-grained FPGA, which has a relatively small number of PRSs. In other words, it is appropriate for the coarse-grained FPGA with relatively large-scale PLEs.

Power gating, which reduces power consumption by shutting off the power supply to circuits currently having no contribution to the operation, is effective in an FPGA. For effective power gating, the power supply is controlled for each small block so that only the necessary processing portions of circuits consume power; therefore, the total power consumption can be reduced. The contribution by power gating to system efficiency increases with PLE granularity – i.e., with small PLEs [Figure 6.15(b)]. However, such fine-grained power gating generates circuit area overhead due to additional complicated control circuits, and thus requires more consideration for application to FPGA.

One idea for facilitating power gating in a multicontext FPGA is to change the power-gating setting in each PLE at the time of changes in circuit configuration at context switching. However, as described above, the coarse-grained structure is advantageous in light of area efficiency, while the fine-grained structure is advantageous in light of power consumption. It is challenging to find a structure that satisfies both requirements. Here, a CAAC-IGZO FPGA has merits in terms of PRS area and power consumption. Employment of a fine-grained structure with a relatively large number of PRSs in the CAAC-IGZO FPGA has few disadvantages, and has the advantage of fine-grained power gating. Thus, a fine-grained multicontext method can fully utilize the characteristics of the CAAC-IGZO FPGA [4, 24]. Furthermore, normally-off computing for an FPGA becomes possible in the fine-grained multicontext CAAC-IGZO FPGA, which further reduces power consumption. The driving architecture is described below.

6.3.2.2   Proposed Driving Architecture

The architecture is achieved with a fine-grained multicontext CAAC-IGZO FPGA in the following manner: the power gating is performed on PLEs in response to context switching, and in synchronization with context switching, data store/load is conducted between a volatile register and a non-volatile shadow register in a PLE register.

In order to achieve the above, the configuration data for a PLE corresponding to each context includes data determining whether a power supply to the PLE is required (i.e., whether power gating should be performed). Accordingly, power gating is applied in those PLEs that are not contributing to the operation of the present context task.

The register of a PLE has a normal volatile register as well as a register with non-volatile shadow registers corresponding to respective contexts. The data of the volatile register immediately before context switching are transferred into the non-volatile shadow register corresponding to the present context. Immediately after switching, the saved data in the non-volatile shadow register corresponding to the presently selected context is loaded to the volatile register. In other words, by using the saved data in the non-volatile register, it is possible to return to the end state of the previous execution of the context.

Such a mechanism allows *in situ* data storage/loading in transition to an arbitrary context, which reduces power consumption. After data storing, the power gating can be conducted in PLEs without risk of data loss; thus, the degree of configuration freedom increases with reduced power consumption.

An operational sequence in Figure 6.16 is given as a specific example of the proposed architecture. Figure 6.16 illustrates an example of a fine-grained multicontext CAAC-IGZO FPGA with k contexts, j shadow registers, and p PLEs. Let us suppose that the contexts are switched in the following order: context [1] at the first step, context [2] at the second step, context [1] at the

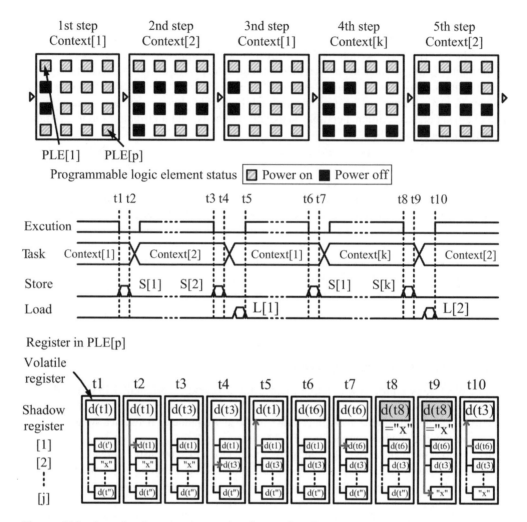

Figure 6.16 Operational sequence example of normally-off computing in FPGA. *Source*: Adapted from [6]

third step, context [k] at the fourth step, and context [2] at the fifth step. The p PLEs are classified into (1) PLEs to which power is supplied, that is, PLEs without power gating (gray blocks) and (2) PLEs to which power is not supplied, that is, PLEs where power gating is conducted (black blocks). The number of PLEs of both types varies depending on the context.

A typical store-operation example is introduced to understand the sequence, with reference to Figure 6.16. The operation from the first step (context [1]) to the second step (context [2]) is described. Immediately before the context switching (in time t1 to t2), in each PLE, data in a volatile register are transferred into the corresponding non-volatile shadow register [1] corresponding to context [1]. A more detailed explanation is given below, focusing on data retained in a register in the PLE [p]. In the PLE [p], a volatile register and non-volatile shadow registers [1],

[2], …, and [j] have data d(t1), d(t'), "x", …, and d(t"), respectively, before context switching (t1). Here, d(t1), d(t'), and d(t") are some values and "x" is an undefined value. Immediately before the context switching (t1 to t2), d(t1) in the volatile register is stored in the shadow register [1] corresponding to context [1]. The store operation from the second step (context [2]) to the third step (context [1]) can be explained in a similar manner; the terms t1, t2, shadow register [1], context [1], d(t1), and d(t') here correspond to t3, t4, shadow register [2], context [2], d(t3), and d(t1), respectively.

Next, a typical load operation example is introduced with reference to Figure 6.16. The operation from the second step (context [2]) to the third step (context [1]) is described. Immediately after context switching (in time t4 to t5) in each PLE, data are loaded from the non-volatile shadow register [1] corresponding to context [1] to the volatile register. A more detailed explanation is given below, focusing on data retained in the register of PLE [p]. In this register, immediately after context switching, d(t1) in the non-volatile shadow register [1] corresponding to context [1] is loaded to the volatile register. Thus, the task of context [1] at the third step can be started from the end state of the task of context [1] at the first step. The load operation from the fourth step (context [k]) to the fifth step (context [2]) can be explained in a similar manner; the terms t4, t5, shadow register [1], context [1], and d(t1) correspond to t9, t10, shadow register [2], context [2], and d(t3), respectively.

Context switching requires only store operation in some cases. As a specific example, the operation from the first step (context [1]) to the second step (context [2]) is described. If the task of context [2] has never been performed before the second step, an indefinite value ("x") is stored in the non-volatile shadow register at time t2. Thus, it is not necessary to resume the task from the end state of the previous execution. It is unnecessary to load the data from the non-volatile shadow register [2], and loading can be skipped. Then, the power consumption for the load can be reduced. Without loading, the selected context [2] can be executed with continuous use of the data of context [1] immediately before switching.

Similarly, context switching requires only load operation in some cases. As a specific example, the operation from the fourth step (context [k]) to the fifth step (context [2]) is described. In a register of a PLE to which power is not supplied in the former context (e.g., PLE [p]), the volatile register has data of an indefinite value ("x"). In addition, power is not supplied to the register when that task is restarted, and thus the data are not necessarily stored in the non-volatile register [k]. Therefore, storing can be omitted, which leads to a reduction in power consumption for the storing.

As mentioned above, the present normally-off computing can schedule the processor to selectively conduct storing and loading, only storing, or only loading.

As seen from the above, only the PLEs required for the current operation are supplied with power, which is called "normally off computing." It enables interruption processing – e.g., suspending the first context task, processing the second context task, and then resuming the first context task. Fine-grained power gating by the PLE unit in accordance with the context allows power to be supplied only to necessary PLEs at the necessary time. Since the same number of non-volatile shadow registers as contexts exist in the register of each PLE, switching from any context to any other context is feasible; thus, a complicated schedule for context switching can be made available.

In the multicontext FPGA with the proposed normally-off computing, the following goals can be achieved at once: (1) resuming the task of the context from the end state at the previous execution, i.e., improving the operational efficiency; (2) improving the use efficiency of PLEs, i.e., heightening the flexibility of circuit configuration; and (3) conducting fine-grained power gating for each PLE, that is, reducing the power consumption.

6.3.2.3 Required Hardware

PRS
Figure 6.17(a) illustrates a PRS set. This set has k routing switches corresponding to k contexts and a multiplexer. Control signal lines BL and WL for configuration data writing are connected to a bit driver and a word driver, respectively. In the PRS set, a PRS is selected by a context signal and the on/off value of the selected PRS determines the conduction/non-conduction from input IN to output OUT. The input IN of the PRS set is connected to an output signal line of a PLE or user I/O, while the output OUT of the PRS is connected to an input signal line of another PLE or user I/O. One PRS set is present between two PLEs or between a PLE and a user I/O.

Figure 6.17(b) shows the circuit diagram of a PRS. This circuit has the same configuration as that introduced in the previous section (Figure 6.8). Compared with the SRAM-based routing switch, the circuit in Figure 6.17(b) has the advantages of small area, low power consumption, and high-speed driving.

PLE
The PLE in Figure 6.18 has the same function as that in Section 6.2. In the multicontext CAAC-IGZO FPGA proposed in the present section, the configuration memory set contains k configuration memories (configuration memory [1] to configuration memory [k]) for contexts [1] to [k]. The configuration memory is also used for power switching that controls the PLE-by-PLE power supply [4, 24]. The power supply to a PLE is determined by the configuration data stored in the configuration memory, which enables PLE-by-PLE power gating without complicated control circuits [4, 24]. By switching a selection signal, $\varphi_{CTX[k:1]}$, in response to the task schedule, the circuit configuration can be changed together with power control switching for the PLEs.

Configuration Memory
Figure 6.19(a) shows a configuration memory set used in a PLE. This memory set comprises two configuration memories and a multiplexer. One configuration memory corresponds to 1-bit

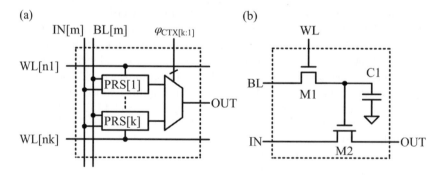

Figure 6.17 Circuit configurations of (a) PRS set and (b) PRS. *Source*: Adapted from [6]

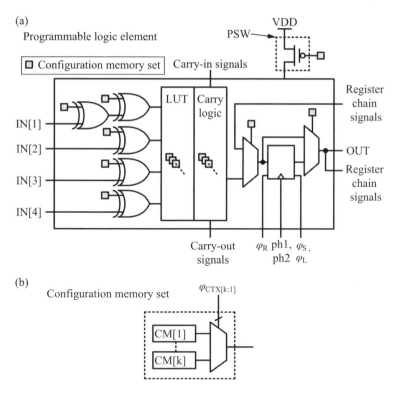

Figure 6.18 (a) PLE block diagram and (b) configuration memory set diagram. The configuration memory is denoted by CM. *Source*: Adapted from [6]

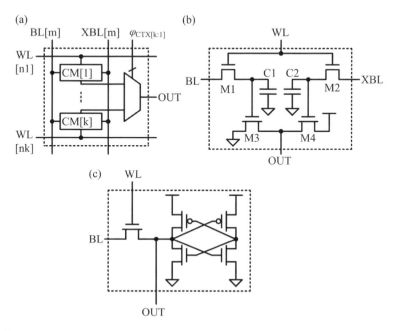

Figure 6.19 Circuit configurations of (a) configuration memory set, (b) configuration memory, and (c) SRAM-based configuration memory. *Source*: Adapted from [6]

configuration data. In the configuration memory set, a context signal, $\varphi_{CTX[k:1]}$, selects one of the k configuration memories, and then the configuration data in the selected configuration memory are output. The control signal lines BL, XBL (BL-inversed signal line), and WL for writing the configuration data are connected to the bit and word drivers.

Figure 6.19(b) is the circuit diagram of the configuration memory. In the CAAC-IGZO configuration memory, the CAAC-IGZO FET M1 (M2) of the configuration memory to be written is turned on by the WL, and the configuration data are written in the storage capacitor C1 (C2) from BL (XBL) through M1 (M2). When either Si FET M3 or M4 is turned on by the storage capacitor C1 or C2, an output corresponding to the configuration data can be obtained. Since the off-state current of the CAAC-IGZO FETs M1 and M2 is extremely low, the potential of C1 and C2 can be kept constant by switching off M1 and M2. That is, the configuration memory is non-volatile.

The CAAC-IGZO configuration memory has four transistors (two Si FETs and two CAAC-IGZO FETs) and two capacitors, and the number of elements is almost equivalent to the SRAM-based configuration memory having five Si FETs [Figure 6.19(c)]. However, through the hybrid process, CAAC-IGZO FETs can be stacked over Si FETs. With this structure, the CAAC-IGZO configuration memory can have a smaller layout area than the SRAM-based configuration memory. Note that the area reduction by the stack structure can be achieved to a greater degree in the PRS mentioned above, and thus it is enhanced in the fine-grained structure, which has a relatively large number of PRSs.

Non-volatile Shadow Register

To store (load) from (to) the volatile register at high speed with low power consumption (Figure 6.16) without loss of the advantages of fine power gating, it is effective to adopt a shadow register structure in which a non-volatile memory is included in the volatile component of a register.

A register with a shadow register component, as shown in Figure 6.20, is proposed for the PLE of the FPGA. Figure 6.20(a) is a circuit diagram of the register and Figure 6.20(b) is its timing diagram. The register comprises a volatile register of Si FETs and a non-volatile shadow register with CAAC-IGZO FETs and Si FETs. In normal operation, the volatile register is used as a common register. The register uses a 2-phase non-overlap clock, which is effective for normally-off computing. When a store-control signal, $\varphi_{S[i]}$, and a load-control signal, $\varphi_{L[i]}$, corresponding to the shadow register [i] are activated, the data of the volatile register can be stored in/loaded from the non-volatile shadow register.

In the shadow register, the logic of the data stored in node N1 (N2) is always inverted from that in node N1B (N2B); that is, they are complementary to each other. Thus, at loading, either M1 (M3) or M2 (M4) serves as a pull-down circuit that supplies the ground potential to an inverter latch in the volatile block.

6.3.3 Prototype

6.3.3.1 Entire Structure

The fine-grained multicontext CAAC-IGZO FPGA has been fabricated with the hybrid process to evaluate the effect of the proposed normally-off computing. The processing technology is the

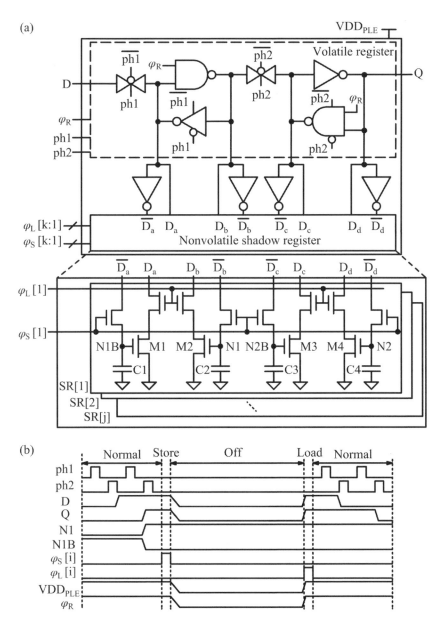

Figure 6.20 (a) Register circuit configuration and (b) register timing diagram. The timing diagram shows the timing of storage and loading with the *i*th shadow register. *Source*: Adapted from [6]

same as that used in Section 6.2, that is, the hybrid process of 1.0-μm CAAC-IGZO and 0.5-μm Si FETs.

Figure 6.21 shows the entire structure of the fabricated fine-grained multicontext CAAC-IGZO FPGA, and Table 6.2 shows the specifications. The PLE array comprises 20 PLEs.

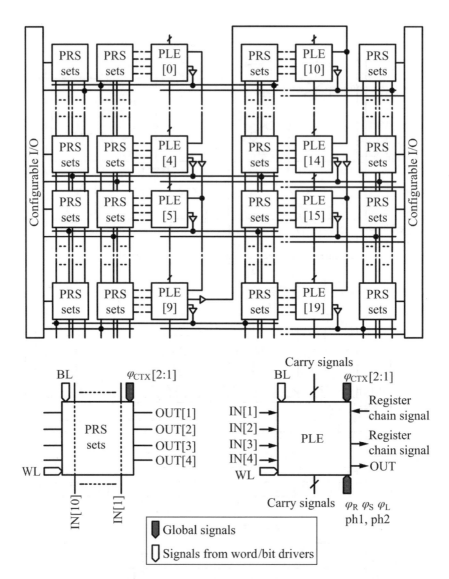

Figure 6.21 Entire block diagram of prototype FPGA. *Source*: Reprinted from [6], with permission of IEEE, © 2015

The number of contexts, k, is 2. The configuration memory uses 3760 bits for each context, that is, 7520 bits in total. The number of non-volatile shadow registers, j, is 1, which is the simplest register structure needed to examine the effects of normally-off computing. The non-volatile shadow register has a storage capacitance of 32.7 fF under the assumption of a storage time of 1 year.

The time required for data storage/loading in the register in the case of overdriving the store control signal φ_S is shown in the Shmoo plot of Figure 6.22. The CAAC-IGZO FET has an

Table 6.2 Specifications of CAAC-IGZO FPGA

Process	1.0-μm CAAC-IGZO/0.5-μm Si
Die size	5.5 mm × 4.5 mm
Number of PLE	20
Number of IO	20
LUT inputs	4
Total bit number in CM	1760 bits
Total bit number in PRS	5760 bits
PLE size	390 μm × 765 μm
CM size	45 μm × 18 μm
PRS size	45 μm × 12 μm
Structure of CM	2 CAAC-IGZO FETs, 4 Si FETs, 2 capacitors (184 fF)
PRS structure	1 CAAC-IGZO FET, 2 Si FETs, 1 capacitor (184 fF)
Supply voltage	3.3 V
Power gating control	Individual PLE

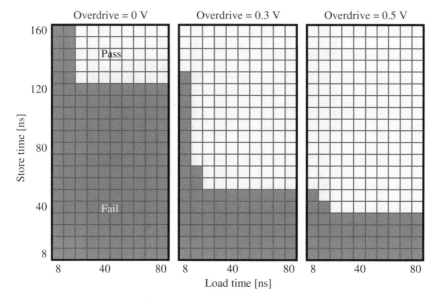

Figure 6.22 Store time and load time of register data in non-volatile shadow register. *Source*: Adapted from [6]

extremely low off-state current region at a gate voltage of 0 V, and a threshold voltage higher than that of an enhancement-type Si FET. The store time is longer than the load time, but can be shortened by overdriving the store control signal φ_S. From Figure 6.22, the store time and load time are 128 ns and 24 ns, respectively, at a V_{DD} of 2.5 V and a φ_S overdrive voltage of 0 V. The store operation is charge to/discharge from the capacitors C1 to C4 in Figure 6.20; therefore, the writing speed of the high potential can be heightened by overdriving the φ_S signal. At an overdrive voltage of 0.5 V, the storage time and loading time are 40 ns and 8 ns, respectively, which are clearly shortened from the case without overdriving. In this overdriving case, storage consumes 1.6 pJ and loading consumes 17.4 pJ. While the power consumption at an overdrive voltage of 0.5 V increases by 3%, the storage time is a third of that without overdriving.

A micrograph of the fabricated CAAC-IGZO FPGA is shown in Figure 6.23 [6]. The use of a non-volatile component in the register of the PLE causes a 40% area overhead of the register, a 0.6% area overhead of the PLE, and a 0.3% area overhead of the core, having little influence on the core area. Such a bit-area increase enables the addition of the normally-off computing function to the CAAC-IGZO FPGA and the reduction in power consumption.

When the number of shadow registers is increased by 1, the register, PLE, and core increase in area by 40%, 0.6%, and 0.3%, respectively. The area increase due to the addition of one context (without the addition of shadow registers) is approximately 30% for the configuration memory, 25% for the PRS, and 19.7% for the core. These values indicate that it is possible to form a CAAC-IGZO FPGA capable of further effective normally-off computing with increased shadow registers and contexts.

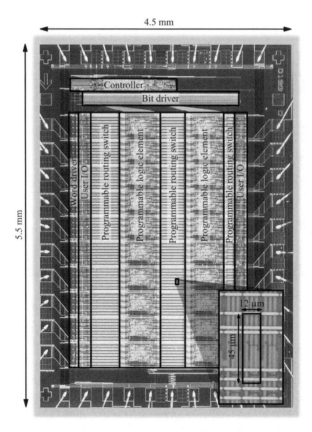

Figure 6.23 Micrograph of a prototype fabricated by a hybrid process. *(For color detail, please see color plate section)*

6.3.3.2   Prototype Measurement

Test Sequence

The behavior of normally-off computing on the prototype is examined. To evaluate the advantages of the fabricated multicontext FPGA, the fabricated FPGA is compared with FPGAs where the essential hardware for obtaining normally-off computing (i.e., a non-volatile shadow register, fine-grained PLE-by-PLE power gating, and multicontexts) is somewhat limited or disabled. The following four types of hardware structure (Figure 6.24) are used: (A) the fabricated FPGA with a non-volatile shadow register, fine-grained power gating, and multicontexts; (B) an FPGA with a volatile shadow register, fine-grained power gating, and multicontexts; (C) an FPGA with a volatile shadow register and multicontexts; and (D) an FPGA with multicontexts without shadow register.

Type A has a non-volatile shadow register, while type B has a volatile shadow register; the shadow-register difference will cause a difference in the performance of the fine-grained power gating. Since data are stored in the volatile shadow register in type B, power gating is not applicable to non-used PLEs in some cases. On the contrary, in type A, the number of PLEs subjected to fine-grained power gating is expected to increase, owing to the non-volatile shadow register. The performance difference will increase if the number of used PLEs differs largely between contexts. A comparison between types A and C will show the effectiveness of the fine-grained power gating with the non-volatile shadow register. A comparison between types A and D will show the effectiveness of normally-off computing with the power gating and non-volatile shadow register. A comparison between types B and C will show the effectiveness of fine-grained power gating with the volatile shadow register. The performance difference between types B and C will increase if the number of used PLEs in each context is small with respect to the total number of PLEs. A comparison between types B and D will show the effectiveness of power gating and the volatile shadow register. A comparison between types C and D will show the effectiveness of the volatile shadow register. The effectiveness will increase if the number of used PLEs in each context and the number of contexts increase.

The purpose of the experiment is to examine the following effects of the normally-off computing task processing compared with that without normally-off computing: (1) the continuousness of task processing, (2) the improvement in the use efficiency of the PLEs, and (3) the reduction in power consumption. The experimental conditions are as follows. The task schedule has the first step of context [1], the second step of context [2], and the third step of context [1]. Context [1] and context [2] correspond to task [1] and task [2], respectively. An 8-stage shift circuit is configured by the 8 PLEs in task [1], and a 4-stage binary counter circuit is configured by the 5 PLEs in task [2]. To improve the switching characteristics, the source voltage in the word and bit drivers is 2.6 V, while that in the other parts is 2.5 V, and the operational frequency varies from 1 to 20 MHz.

In context [1], the first to eighth stages of the shift circuit are configured by PLE [0] to PLE [7], and outputs from the respective PLEs are called OUT [0] to OUT [7]. In types A and B, PLE [8] to PLE [19] can be subjected to power gating. In context [2], the least significant bit (LSB) to the most significant bit (MSB) of the counter circuit are configured by PLE [0] to PLE [3] in types A to C, while they are configured by PLE [10] to PLE [13] in type D. Outputs from the respective PLEs are called OUT [0] to OUT [3]. The control logic of the counter circuit is configured by PLE [4] in types A to C and by PLE [14] in type D. In types A and B, PLE [8] to PLE [19] can be subjected to power gating. In addition, in type A, PLE [5] to PLE [7] can also be subjected to power gating.

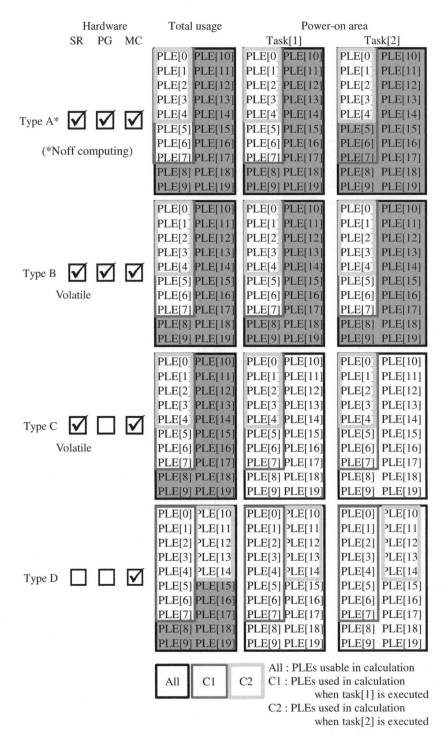

Figure 6.24 Normally-off computing and other driving methods of executing tasks. A gray area in the "Total usage" column represents a region including unused PLE. A gray area in the "Power-on area" columns represents a region including PLEs subjected to power gating. SR: shadow register, PG: power gating, MC: multicontexts. *Source*: Reprinted from [6], with permission of IEEE, © 2015

Continuousness of Task Processing

First, the operation result in type A along the above-mentioned task schedule (prototype multi-context FPGA) is described. As shown in Figure 6.25, the output waveforms corresponding to the first step of context [1], the second step of context [2], and the third step of context [1] are obtained. In particular, the values of OUT [0] to OUT [7] at store start time (2.7 μs), immediately before context switching from the first step to the second step, are, respectively, equivalent to the values of OUT [0] to OUT [7] at the load end time (7.6 μs) immediately after context switching from the second step to the third step. This is true, of course, for OUT [5] to OUT [7] corresponding to PLE [5] to PLE [7], which are subjected to power gating at the second step of context [2]. Thus, the first effect of the proposed normally-off computing, the continuousness of task processing, has been confirmed. The task can be resumed from the end state of the previous processing. Types B, C, and D operate similarly; however, type A is preferable to the others for achieving an improvement in the use efficiency of PLEs and a reduction in power consumption in addition to the continuous task processing, as described below.

Use Efficiency of PLEs

To examine the second effect of normally-off computing, the improvement in efficiency on using PLEs, the used PLEs are compared among types A to D in the "Total usage" column in Figure 6.24. The PLEs not used in task [1] and task [2] are shown in gray, and those used for the tasks are shown in white.

Since type D is a multicontext FPGA without a shadow register, exclusive assignment of the PLEs to tasks [1] and [2] is necessary (that is, the PLEs used in task [1] should be different from those used in task [2]) to resume a task from the end state of the previous processing. Thus, the efficiency in usage of the PLEs cannot be heightened. The total number of necessary PLEs is

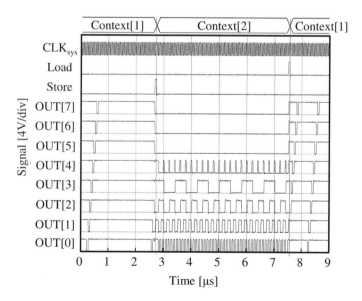

Figure 6.25 Operational waveforms and event schedule in testing the fabricated FPGA. *Source:* Reprinted from [6], with permission of IEEE, © 2015

the sum of the PLEs used in task [1] and those used in task [2]. The type-D FPGA does not receive the benefit of the multicontexts.

In contrast, types A to C have non-volatile or volatile shadow registers. Thus, even if the PLEs are shared in tasks [1] and [2], a task can be resumed from the end state of the previous processing. The high use efficiency of the PLEs can be obtained in these types.

In the proposed normally-off computing, a shadow register, regardless of whether it is volatile or non-volatile, is effective in improving the use efficiency of the PLEs. In the case of executing task [1] and task [2] in accordance with the above task schedule, the necessary number of PLEs with types A, B, and C is 8; thus, compared with type D with the 13 necessary PLEs, their use efficiency improves by $(1-8/13) \times 100 = 38\%$.

Reduction in Power Consumption

To examine the third effect of normally-off computing with the proposed FPGA, reduction in power consumption, the power consumption needed to execute the tasks in the above four types of FPGA along the above-mentioned schedule is analyzed (see Figure 6.26). For each type, the measured power consumption of the core is scaled to the calculated power consumption of the core using the Simulation Program with Integrated Circuit Emphasis (SPICE). The power consumption of each PLE is then calculated and shown by task (task [1] and task [2]). The clock frequency is set to 1 MHz or 20 MHz. In task [1], a pulse of 1 toggle per 10 clocks is the input to the shift register, and the operation activation rate of PLE [0] to PLE [7] is 10%.

In the type-C FPGA, the power consumption at a clock frequency of 1 MHz is 98.7 μW in task [1] and 176 μW in task [2], while that at 20 MHz is 1.96 mW in task [1] and 3.46 mW in task [2]. In the type-D FPGA, the power consumption at a clock frequency of 1 MHz is 111 μW in task [1] and 191 μW in task [2], while that at 20 MHz is 2.24 mW in task [1] and 3.79 mW in task [2]. The power-consumption difference between types C and D is due to the effect of the volatile shadow register. Thanks to the volatile shadow register, the power consumption by type C is reduced from that of type D by 11.1% in task [1] and by 7.85% in task [2] at a clock frequency of 1 MHz, and by 12.5% in task [1] and by 8.71% in task [2] at a clock frequency of 20 MHz.

In the type-B FPGA, the power consumption at a clock frequency of 1 MHz is 83.0 μW in task [1] and 154 μW in task [2], while that at 20 MHz is 1.63 mW in task [1] and 3.07 mW in task [2]. The difference between types B and C expresses the effect of the fine-grained power gating with the volatile shadow register. Thanks to the power gating, the power consumption of type B is reduced from that of type C; the decrease in power relative to that consumed by type D accounts for 14.1% in task [1] and 11.5% in task [2] at a clock frequency of 1 MHz, and 14.7% in task [1] and 10.3% in task [2] at a clock frequency of 20 MHz.

In the type-A FPGA, the power consumption at a clock frequency of 1 MHz is 82.6 μW in task [1] and 151 μW in task [2], while that at 20 MHz is 1.62 mW in task [1] and 2.98 mW in task [2]. The power-consumption difference between types A and D expresses the effect of normally-off computing by the multicontext and non-volatile shadow register. Owing to normally-off computing, the power consumed in type A is decreased from that in type D by 25.6% in task [1] and by 20.9% in task [2] at a clock frequency of 1 MHz, and by 27.7% in task [1] and by 21.4% in task [2] at a clock frequency of 20 MHz. The difference between types A and C expresses the effect of fine-grained power gating with a non-volatile shadow register. Because of the power gating, the power consumed by the type-A FPGA is reduced from that of type C; the power decrease from type C to type A relative to the power consumed by type D is

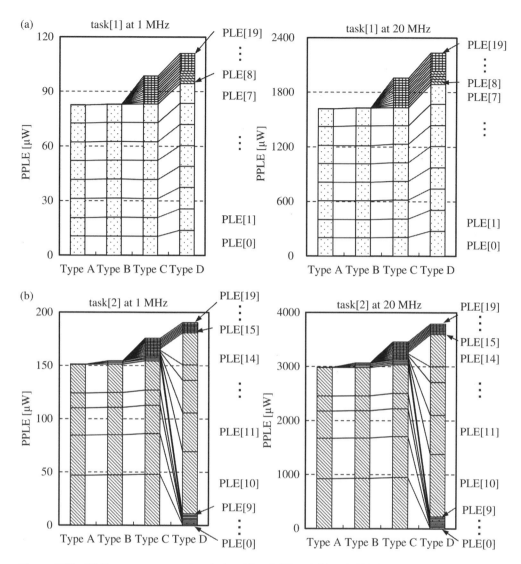

Figure 6.26 PLE power consumption during (a) task [1] and (b) task [2]. *Source*: Reprinted from [6], with permission of IEEE, © 2015

14.5% in task [1] and 13.1% in task [2] at a clock frequency of 1 MHz, and 15.2% in task [1] and 12.7% in task [2] at a clock frequency of 20 MHz. The difference between types A and B expresses the increased effect of fine-grained power gating by use of the non-volatile shadow register instead of the volatile register. Such an increased power-gating effect reduces the power consumption of type A from that of type B; the decrease in power relative to that consumed by type D accounts for 0.360% in task [1] and 1.57% in task [2] at a clock frequency of 1 MHz, and 0.446% in task [1] and 2.37% in task [2] at a clock frequency of 20 MHz.

These results show that each of the non-volatile shadow register, fine-grained PLE-by-PLE power gating, and multicontexts required for normally-off computing, as well as the synergetic effects due to combinations thereof, contribute to the reduction in power consumption.

Discussion

It has been confirmed that the increase in use efficiency of the PLEs and reduction in power consumption can be effectively achieved in the proposed normally-off computing by the non-volatile shadow register, fine-grained PLE-by-PLE power gating, and multicontexts.

As to the reduction in power consumption, there are two types of overhead to be considered: power consumption in power-gating operation and power consumption in store in/load from operation of the non-volatile shadow register.

Power consumption in the case where power gating is conducted for 8 PLEs (in context [1]) and 12 PLEs (in context [2]) as in type A at an operational frequency of 20 MHz is calculated, with use of the calculation method in the reports [4, 24]. Power consumption by power gating is advantageous when the task operation periods in contexts [1] and [2] are longer than 15 μs and 12 μs, respectively. Conversely, if the contexts are switched in less than such time periods, the scheduling without power gating is effective.

The power consumptions of other non-volatile memories are discussed below. The energy required for 1-bit data storage is 1.48 pJ in MRAM [25] and 100 pJ in a ferroelectric random access memory (FeRAM) [26]. The proposed non-volatile shadow register with CAAC-IGZO FETs requires an energy of 1.6 pJ for 1-bit data storage, which is sufficiently competitive with that of the MRAM or FRAM, given the fact that MRAM density is lower than that of CAAC-IGZO FET non-volatile memory.

6.4   Subthreshold Operation of FPGA

6.4.1   Overview

To further improve the low-power characteristics of the CAAC-IGZO FPGA in Section 6.3, an FPGA architecture capable of driving at an extremely low voltage (subthreshold voltage) is now discussed. Utilizing the ideal floating gate with excellent charge-retention characteristics formed by CAAC-IGZO FETs, the CAAC-IGZO FPGA proposed here overdrives PPSs for power gating and PRSs. This enables the overdrive of PPSs without generating negative potentials and that of PRSs with low power and a high I_{on}/I_{off} ratio.

An FPGA is expected to be the optimal device for sensor networks if it can flexibly change its circuit configurations to operate at low enough voltage to use energy harvesting in the standby mode and to perform high-performance processing in the active mode [27]. The CAAC-IGZO FPGA has the functions (fine-grained power gating, normally-off driving, and context switching) that achieve low power consumption, as described in the above section. Therefore, a CAAC-IGZO FPGA driven at low voltage from energy harvesting with natural energy, such as solar or wind power, would be a battery-less device free from maintenance and suitable for a sensor network application.

6.4.2   Subthreshold Operation

6.4.2.1   Design Problem

Low voltage sometimes leads to non-operation; in general, this malfunction is caused mainly by decreasing I_{on}/I_{off} ratio in the transistors. In cases using the same size of transistors, a gate circuit with a large number of transistor stacks has a lower I_{on}/I_{off} ratio than one with a small number of transistor stacks. To use low voltage without the ratio decrease, the number of transistor stacks is limited in an application-specific integrated circuit (ASIC). Such a limitation is generally adopted to achieve subthreshold operation [28, 29].

Unlike an ASIC, an FPGA, which has a high degree of freedom for circuit configuration, has the following unique problems:

1. non-transmission of logic signals due to threshold voltage drop in a pass gate;
2. decrease of the I_{on}/I_{off} ratio along with an increase of the static leakage current source [27].

The first problem leads to a decrease in the voltage amplitude of logic signals. When the source voltage is sufficiently high compared to the threshold voltage, the decrease in voltage amplitude by the threshold voltage drop reduces the circuit performance but does not completely disable logic transmission. However, when the source voltage is as low as the threshold voltage (e.g., in a subthreshold voltage region), the voltage amplitude decrease due to the threshold voltage drop makes it difficult to obtain the necessary I_{on}/I_{off} ratio for output from the latter-stage gate circuit, disabling the correct logic transmission. Therefore, a structure that does not decrease the voltage amplitude of logic signals in a pass gate (even at low voltage) is required.

The second problem is inevitable in FPGAs. The circuit configuration of an FPGA can be freely changed by the user, which means that the number of elements connected to a node can be large. In other words, an FPGA is a circuit with a large static leakage-current source that inevitably has a circuit region with a lower I_{on}/I_{off} ratio (typically, a PRS) than that of an ASIC. As the number of parallel-connected PRSs increases, it becomes more difficult to lower the used voltage.

Furthermore, an additional unique problem occurs when maintaining the characteristics of the CAAC-IGZO FPGA [4, 6]. As introduced in the previous section, the fine-grained power gating per PLE is a key technology for reducing power consumption; however, the number of transistor stacks in the PLEs increases in practice by adding the PPSs necessary for power gating, leading to a decrease in the I_{on}/I_{off} ratio and difficulty in low-voltage operation. Techniques that solve this problem are necessary to lower the voltage in PLEs.

6.4.2.2   Design Guides

To solve the above problems, first, the number of transistor stacks in the CAAC-IGZO FPGA is reduced in a manner similar to that employed in an ASIC. Specifically, a logic circuit is composed of a gate circuit with two or fewer transistor stacks (NOT, 2-input NAND, 2-input NOR, or the register with the gate circuit), whereby the I_{on}/I_{off} ratio increases. In addition, the overdrive utilizing a floating node with a charge-retention function formed by a CAAC-IGZO FET is conducted in PRSs and PPSs to solve the unique problem with FPGAs.

The principles of overdriving PRSs and PPSs with CAAC-IGZO FETs (in the structure of the CAAC-IGZO FPGA in Section 6.3) are explained specifically with reference to Figure 6.27 and Figure 6.28. This overdriving becomes available by supplying a high potential only when updating configuration and context data.

The PRSs are overdriven in the following sequence (see Figure 6.27). The data of high or low (V_{DDH}/ground) are written in N_{cfg} at configuration and in N_{ctx} at context switching (the data to be written in N_{cfg} and N_{ctx} are denoted by D_{H_cfg} and D_{H_ctx}, respectively). Subsequently, the CAAC-IGZO transistor MO_{cfg} (MO_{ctx}) is turned off, whereby N_{cfg} (N_{ctx}) becomes a floating

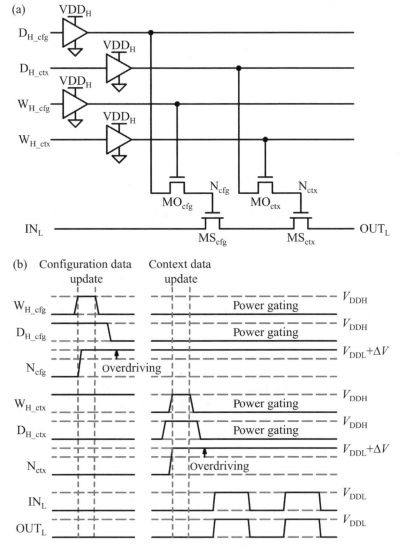

Figure 6.27 Principles of overdriving PRS: (a) circuit configuration and (b) timing diagram. *Source:* Adapted from [7]

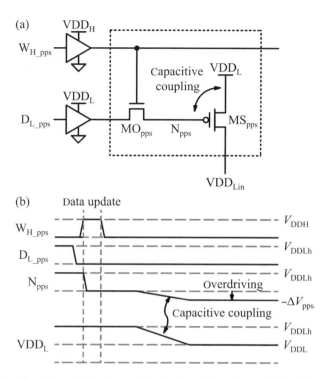

Figure 6.28 Principles of overdriving PPS: (a) circuit configuration and (b) timing diagram. *Source:* Adapted from [7]

node to retain the potential. In the PRSs, the pass transistors MS_{cfg} and MS_{ctx} are controlled by high-potential logics, even at a subthreshold voltage operation; therefore, the voltage amplitude drop due to the threshold voltage drop at signal transmission from IN_L to OUT_L can be avoided and a high I_{on}/I_{off} ratio can be obtained, regardless of transmission signal potentials. The high potential can be maintained with an extremely small current due to the excellent current characteristics of CAAC-IGZO FETs [3, 4]. In this operation, the high potential of V_{DDH} should be higher than the high potential of V_{DDL}, at least by the threshold voltage of the pass transistor.

The PPSs are overdriven in the following sequence (see Figure 6.28). First, the VDD_L domain voltage is set to a high potential (V_{DDLh}) at context switching. The CAAC-IGZO transistor MO_{pps} is turned off after writing configuration and context data, whereby N_{pps} becomes a floating node. Next, the capacitive coupling via parasitic capacitance of a PPS gate and VDD_L source wiring is utilized. The VDD_L domain voltage is reduced from a high potential (V_{DDLh}) to a low potential (V_{DDL}), whereupon the N_{pps} potential decreases. If the potential of N_{pps} is low when the transistor MO_{pps} is turned off, the potential of N_{pps} becomes lower than ground. For example, writing low (a potential of 0 V) in the PPS gate can generate a negative potential, which enables negative-potential overdrive. In other words, a decrease in the gate potential of the PPS leads to an increase in the on-state current. It is possible to avoid a decrease in the on-state current in the PPS because of the overdrive; thus, the fine-grained power gating

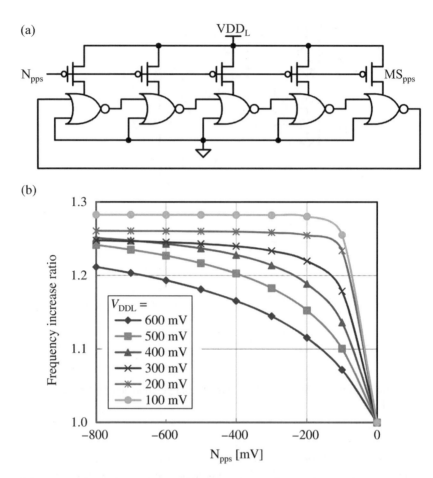

Figure 6.29 (a) Five-stage ring oscillator (RO5) and (b) dependence of RO5 operational frequency ratio on source voltage V_{DDL} and N_{pps}. The operational frequency ratio is normalized by the frequency at an N_{pps} of 0 mV. *Source*: Adapted from [7]. Copyright 2015 The Japan Society of Applied Physics

necessary for reducing the power consumption by an FPGA can be maintained even at a subthreshold voltage.

The contribution of the PPS overdrive to low-voltage operation can be verified by a SPICE simulation. A 5-stage ring oscillator (RO5), in which NOR2s each having a PRS are serially connected to each other in 5 stages, is used [see Figure 6.29(a)]. In Figure 6.29(b), the vertical axis represents the operational frequency ratio normalized by the operational frequency without overdrive, and the horizontal axis represents the overdrive voltage (N_{pps}) for the PPS. The drive voltage, V_{DDL}, of RO5 is changed in the range of 100 to 600 mV. The simulation result in Figure 6.29(b) shows that the operation frequency of RO5, that is, the on-state current of the PPS, can be increased by applying a negative potential of approximately −200 mV to the PPS gate, especially at a drive voltage of 300 mV or lower. The PPS overdrive becomes more effective as the drive voltage is lowered.

6.4.2.3 CAAC-IGZO FPGA

The structure of the CAAC-IGZO FPGA designed in accordance with the above guidelines is explained below with reference to a micrograph in Figure 6.30 and specifications in Table 6.3. The CAAC-IGZO FPGA has peripheral circuits necessary for controlling the overdrive of the PRSs and PPSs (see Figure 6.31), and has the following three source domains: a (high-potential) VDD_H domain that contains a configuration controller (including bit and word drivers) for controlling the configuration of the CAAC-IGZO FPGA and a context controller for controlling multicontext switching; a (low-potential) VDD_L domain that contains a logic array block (including PLEs and PRSs) that is a key circuit of the FPGA constantly supplied with power, and a low-potential region of a programmable I/O that is an input/output circuit for users; and an I/O domain that contains a high-potential region of the programmable I/O.

The programmable I/O has both high- and low-potential regions because external signals and internal signals belonging to the VDD_L domain are transmitted or received through a level shifter (LS). The subindexes in signal names, "$_H$" and "$_L$," indicate that the signals are generated by the VDD_H and VDD_L domains, respectively.

The PRS has a multicontext structure in which the PRS_{cfg} controlled by configuration data and the PRS_{ctx} controlled by context data are connected serially and in parallel (see

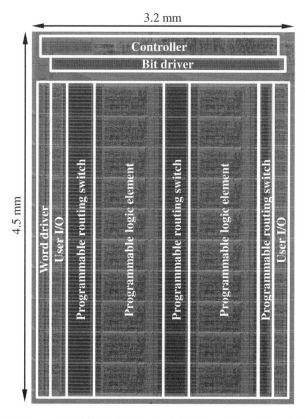

Figure 6.30 Micrograph of CAAC-IGZO FPGA designed for subthreshold operation.

Table 6.3 Specifications of CAAC-IGZO FPGA designed for subthreshold operation

Process	0.8-μm CAAC-IGZO/0.18-μm Si
Die size	5.5 mm × 4.5 mm
Number of PLE	20
Number of IO	20
LUT inputs	4
Total bit number of CM	1760 bits
Total bit number in PRS	5760 bits
PLE size	360 μm × 753 μm
CM size	45 μm × 18 μm
PRS size	45 μm × 7.5 μm
Structure of CM set	5 CAAC-IGZO FETs, 8 Si FETs, 2 capacitors (3.8 fF)
PRS structure	1 CAAC-IGZO FET, 2 Si FETs, 1 capacitor (3.8 fF)
Supply voltage (programmable area)	180 mV–1000 mV
Power gating control	Individual PLE

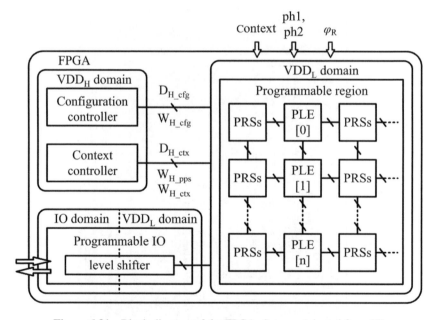

Figure 6.31 Block diagram of the FPGA. *Source*: Adapted from [7]

Figure 6.32). To overdrive the PRSs, writing in the N_{cfg} and N_{ctx} nodes is controlled by CAAC-IGZO FETs so that the overdrive potential is maintained. The multiple input/output signals are connected to the PRSs (see Figure 6.33). The input/output signal wirings are arranged in a matrix. The input signal $IN_L[n]$ is the input to the nth column and the output signal $OUT_L[m]$ is the output to the mth row.

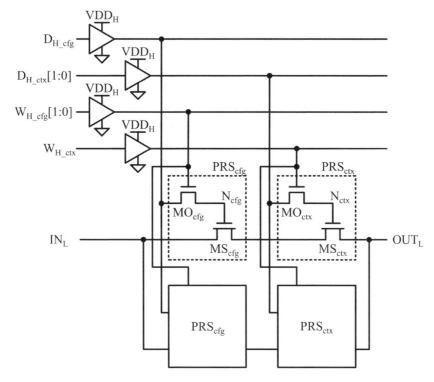

Figure 6.32 Circuit diagram of PRS. *Source*: Adapted from [7]

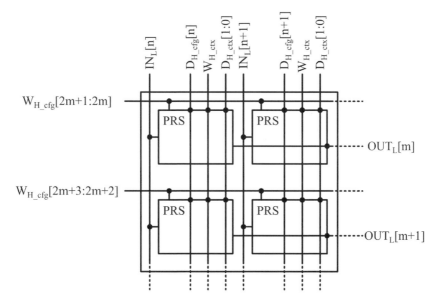

Figure 6.33 Connection relationship between PRSs. *Source*: Adapted from [7]. Copyright 2015 The Japan Society of Applied Physics

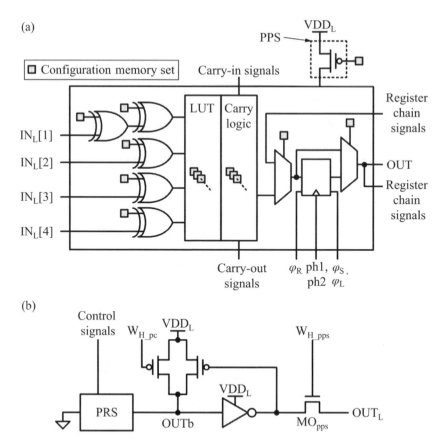

Figure 6.34 Circuit diagrams of (a) PLE and (b) configuration memory. *Source*: Adapted from [7]

The PLE here is different from that in Section 6.3 with respect to the PPS overdrive and the circuit of the configuration memory (see Figure 6.34). In the configuration memory, a PRS for controlling the conduction with the ground potential and a dynamic logic circuit are connected to a floating node. After precharging OUTb, in a manner depending on the configuration and context data in the PRS, the configuration memory maintains the V_{DDL} potential of OUTb when the PRS is off, or discharges that potential when the PRS is on (see Figure 6.35). The transistor MO_{pps} is on during context switching, whereby the output data from the configuration memory are supplied to the output node to update (refresh) the configuration data.

6.4.3 Prototype

6.4.3.1 Entire Structure

To confirm that the CAAC-IGZO FPGA based on the above design guidelines is capable of subthreshold operation, a CAAC-IGZO FPGA is fabricated with a hybrid of the processes used for the 0.8-μm CAAC-IGZO and 0.18-μm Si FETs. In addition, a test element group (TEG) of

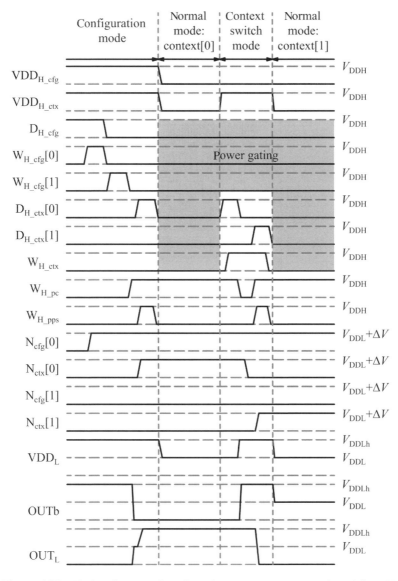

Figure 6.35 Timing diagram of configuration memory. *Source*: Adapted from [7]

a PLE is fabricated to measure directly the overdrive effect of the PPS and the subthreshold operation voltage signals.

6.4.3.2 Low-Voltage Driving

To confirm whether the PPSs can be effectively overdriven, the subthreshold operation is examined with the TEG of a PLE. The PLE has a multicontext structure ($k = 2$) and the

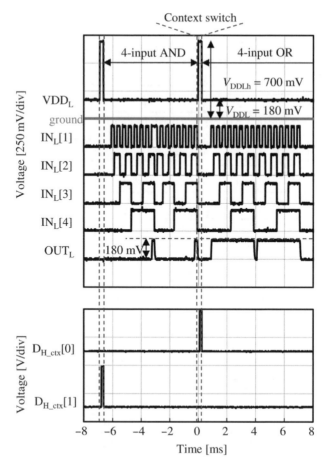

Figure 6.36   Input/output waveforms of PLE TEG with 4-input AND or 4-input OR. *Source*: Adapted from [7]

configuration of a 4-input AND or a 4-input OR gate. To check the PLE operation, the PLE is operated as the 4-input AND first, and then is switched to operate as the 4-input OR by context switching. As a result, it operates at a minimum operational voltage of 180 mV in this sequence (see Figure 6.36). In updating the configuration data, the V_{DDLh} (a boosted potential of V_{DDL}) necessary for overdriving the PPS is boosted to 700 mV.

6.4.3.3   Comparison of Power Delay Product

Next, how much overdriving PPS and PRS contribute to voltage lowering is examined with the prototype CAAC-IGZO FPGA. In order to understand the dynamic characteristic differences due to the circuit configuration, the CAAC-IGZO FPGA with the configuration of a 3-stage ring oscillator (RO3) and that with the configuration of a 4-bit counter (CNT4) are used.

The operational frequency (F), power consumption, and dependence of power delay product (PDP) on V_{DDL} are measured in the case of RO3 (see Figure 6.37); the maximum operational frequency (F_{max}), power consumption, and dependence of PDP on V_{DDL} are measured in the case of CNT4 (see Figure 6.38). The power supply to non-active PLEs is stopped by

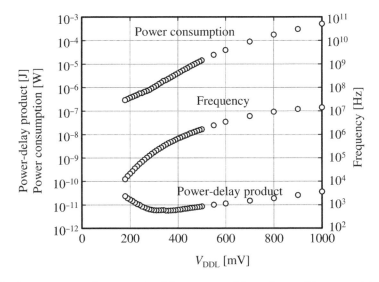

Figure 6.37 Power consumption, operational frequency, and dependence of the power delay product on operational voltage in a CAAC-IGZO FPGA with the RO3 configuration. *Source*: Adapted from [7]. Copyright 2015 The Japan Society of Applied Physics

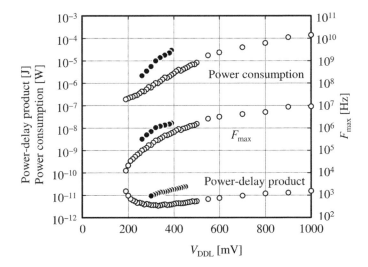

Figure 6.38 Dependence of power consumption, maximum operational frequency, and PDP on operational voltage in a CAAC-IGZO FPGA (O) with the configuration of a CNT4 and SRAM FPGA (●). *Source*: Adapted from [7]. Copyright 2015 The Japan Society of Applied Physics

Table 6.4 Characteristic comparison between FPGAs. *Source*: Adapted from [7]. Copyright 2015 The Japan Society of Applied Physics

Specifications	CAAC-IGZO FPGA designed for subthreshold operation	CAAC-IGZO FPGA presented in Subsection 6.3.3	SRAM-based FPGA [30]
Process node	0.8-μm CAAC-IGZO/ 0.18-μm Si	1.0-μm CAAC-IGZO/ 0.5-μm Si	0.18-μm Si
Die size	5.5 mm × 4.5 mm	5.5 mm × 4.5 mm	4.4 mm × 4.1 mm
Configuration memory	Non-volatile (CAAC-IGZO FET + C)	Non-volatile (CAAC-IGZO FET + C)	Volatile (SRAM)
Number of PLEs	20	20	148
PRS implementation	Pass transistor + CAAC-IGZO FET	Pass transistor	Multiplexers
Routing tracks	32	32	48
Configuration	4-bit counter	10-stage shifter	4-bit counter
Minimum operating voltage	190 mV	900 mV	260 mV
Frequency	12.5 kHz at 190 mV 28.6 kHz at 330 mV (8.6 MHz at 900 mV)	33.3 kHz at 900 mV	332 kHz at 260 mV
Minimum PDP	3.40 pJ at 330 mV (12.9 pJ at 900 mV)	13.5 pJ at 900 mV	6.72 pJ at 260 mV

a fine-grained power gating. The source voltages are set as follows: $V_{DDLh} = 1.2$ V and $V_{DDH} = 2.5$ V. The minimum operational voltage is 180 mV in the case of RO3. The minimum PDP is 3.40 pJ at $V_{DDL} = 330$ mV and $F_{max} = 28.6$ kHz in the case of CNT4, which is lower than the SRAM FPGA [30] by approximately 49% (see Table 6.4). Compared with the CAAC-IGZO FPGA in Section 6.3 [4, 6] with an F_{max} of 33.3 kHz at a minimum operational voltage of 900 mV, the F_{max} of the FPGA in this section increases to 8.6 MHz at 900 mV. This means that the challenges in voltage reduction faced by the previous FPGA can be solved by this section's FPGA. Thus, it has been confirmed that the employment of the subthreshold operational structure can achieve both low-power driving and high-performance processing, albeit not simultaneously.

To confirm the contribution of overdriving PPS to the FPGA characteristics, the difference in the PDP caused by the use or non-use of overdriving is examined with the FPGA of CNT4 (see Figure 6.39). Overdriving the PPS can lower the minimum operational voltage from 390 mV to 180 mV. In addition, the overdrive improves the minimum PDP by 24%, from 4.48 pJ at 390 mV to 3.40 pJ at 330 mV.

6.4.3.4 Context Switching

Under the condition where PDP has a minimum value (330 mV and 28.6 kHz) in CNT4, the context is switched from CNT3 to CNT4 in 1 clock period (see Figure 6.40). The output

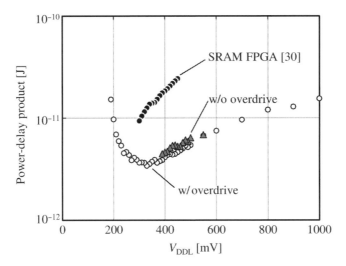

Figure 6.39 Difference in dependence of PDP on operational voltage between the cases with and without overdriving in the CAAC-IGZO FPGA of CNT4. *Source*: Adapted from [7]. Copyright 2015 The Japan Society of Applied Physics

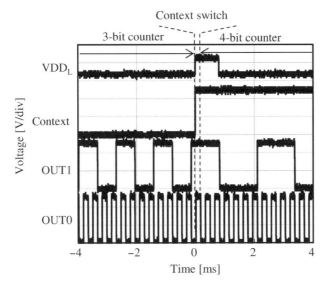

Figure 6.40 Operational waveforms of context switching from CNT3 to CNT4. *Source*: Adapted from [7]. Copyright 2015 The Japan Society of Applied Physics

signals are boosted to 2.5 V by the level shifter to obtain the waveforms. The energy required for the context switching in the CAAC-IGZO FPGA is calculated by SPICE simulation, being 6.42 nJ. The average power of CNT3 is 3.86 µW. Even if the contexts are switched once per second, the power consumption will be 0.17% of that under CNT3 operation or lower. Thus, a

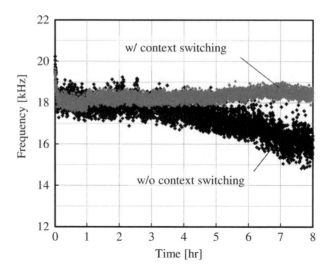

Figure 6.41 Time change in RO3 oscillation frequency at a source voltage V_{DDL} of 180 mV. V_{DDLh} is 1200 mV only at the instance of context switching. *Source*: Adapted from [7]. Copyright 2015 The Japan Society of Applied Physics

high source voltage supplied only when the configuration data are updated will cause negligible overhead.

6.4.3.5 Retention Characteristics

To confirm the duration of the overdrive effect, the time change in the oscillation frequency is measured with the configuration of RO3 (see Figure 6.41). Without context switching, the operational frequency decreases only by 4.5%, on average, in 4 h. With a refresh operation (context switching) once per hour, the overdrive can be maintained within a decrease rate of the operational frequency of 1.0%. The duration can be adjusted by the intentional provision of a capacitor for a floating node for overdriving.

6.5   CPU + FPGA

6.5.1   Overview

Thanks to circuit scaling and promotion of the FPGA development environment, FPGAs are becoming larger and larger, and FGPA applications are widening. In particular, it is recognized that an FPGA can perform tasks more efficiently than a CPU when processing continuous large data and/or short latency of computation is required. While there have been conventional computing systems with high-end CPUs or a combination of a CPU and a graphic processing unit (GPU), a computing system combining a versatile CPU and a quick-response, high-operation-performance FPGA is apparently gaining popularity [31].

Because of the above background, the feasibility of a computing system that combines a CPU and a CAAC-IGZO FPGA is discussed in this section. We describe the bottleneck for achieving versatility and high-performance computing with a CPU; a computing system combining a CPU and a GPU and its limitations; a computing system combining a CPU and an FPGA, its characteristics, and a possible field of application; and the possibility of a computing system combining a CPU and a CAAC-IGZO FPGA.

6.5.2   CPU Computing

Most common CPUs adopt the von Neumann architecture, a stored-program system [32]. A CPU with this architecture sequentially decodes and executes instructions. In addition to the instructions, the data necessary for executing an instruction and the data generated from execution of the instruction are also stored in the memory (see Figure 6.42). Each instruction has high versatility, and the CPU has high flexibility for achieving any function as long as a desired program is made by combining instructions [33].

The performance of a CPU can be learned from the number of executed instructions per unit time [34]. Early CPUs executed one instruction in several clocks in many cases. Thus, memory was accessed once in several clocks. To improve this structure, a structure adopting a pipeline and a cache memory was proposed to execute one instruction per clock [35]. Its memory was accessed once in each clock. In addition, the superscalar was proposed to execute several instructions per clock [36]. The memory was accessed several times (the number of times is equivalent to the number of instructions) per clock. Such improvement requiring larger-scale hardware was effective during the period over which the frequency of the clock smoothly increased by miniaturization in accordance with Moore's Law (2000s), and the performance of CPUs continued to increase exponentially.

However, beginning in the middle of the first decade of the 21st century, the rate of increase in the clock frequency started to slow down because of power-consumption limitations. The progress of miniaturization for high-speed operation caused an increase in leakage current, leading to an explosive increase in power consumption [37]. The performance increase of

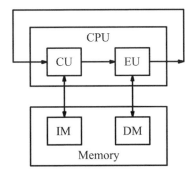

Figure 6.42   Operational procedure of a CPU. The control unit (CU) reads an instruction from the instruction memory (IM). After decoding the instruction, the execution unit (EU) reads the data to be processed from the data memory (DM), and the processed data are stored in the memory. The operation is repeated

CPUs then depended on scaling up the hardware structure. A multithread structure [38] executing more instructions in parallel in the same CPU core and a multicore structure [39] having more CPU cores in the same die were proposed.

To fully educe the (theoretical) performance of the multiple threads and CPU cores, the existing program cannot be used as it is. With a compiler corresponding to the multithread or multicore, the source code should be compiled again. Furthermore, a change in the source code (e.g., indication of a part capable of parallel processing) is required. Thus, the existing software resources cannot be fully utilized.

6.5.3   CPU + GPU Computing

CPUs can have the desirable function by combining versatile instructions, and are capable of achieving various required functions flexibly. However, it is not effective to use CPUs for data-intensive routine processing. Historically, the use of an external circuit (co-processor) for floating-point arithmetic [40] and a special unit for that arithmetic [41] have been proposed. After that, a CPU specializing in arithmetic processing appeared, namely the GPU [42].

A GPU has a very large number of computing units. While a CPU has only several tens of computing units, even with a multithread or multicore structure, a GPU has several hundreds to several thousands of computing units in many cases. The GPU can execute massively parallel multithread processing on single-instruction multiple data (SIMD); that is, it can execute the processing on the mass data with a single instruction (see Figure 6.43) [42].

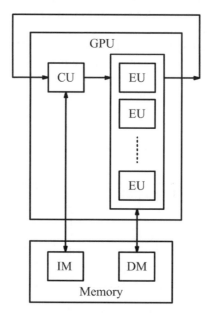

Figure 6.43   Operational procedure of a GPU. The CU reads an instruction from the IM. After decoding the instruction, EUs read multiple types of data to be processed from the DM. After processing, the data are stored in the DM. The operation is repeated

For graphics processing, product-sum operation (as matrix operation) on massive data should be repeatedly executed. Thus, it is effective to collectively obtain an enormous volume of data to be processed from the memory and execute a product-sum operation on the whole dataset with a large number of computing units. When comparing a single-core GPU and a single-core CPU, the power efficiency of the GPU is said to be several to several dozen times that of a CPU in terms of the product-sum operational performance. In recent years, computing systems using a GPU as co-processor to a CPU have rated highly in the rankings of supercomputers [43].

To fully educe the (theoretical) performance of a GPU, a source code for the GPU should be newly developed, together with one for the CPU [44], which requires additional work by software engineers. Meanwhile, the performance increase by the use of the GPU is very large. The development environment for GPU software is constructed on the basis of the CPU software-development environment, with the addition of special instructions or libraries for data processing. Thus, a CPU software engineer can deal with GPU software development relatively easily [44]. This is one factor in the spread of the computing systems combining CPUs and GPUs.

6.5.4   CPU + FPGA Computing

6.5.4.1   Background

The benefit of FPGAs, the flexibility in hardware change, has been recognized since it first appeared on the market. However, it had drawbacks in those days. For example, the mass production cost was much higher than that for ASICs, and users could not use an FPGA unless they learned the dedicated hardware description language. Such drawbacks disturbed the spread of FPGAs [45].

FPGAs are highly integrated thanks to the development of fine processing technology, and have become capable of having dedicated peripheral circuits, such as a large-scale memory and a high-speed communication interface in addition to the programmable logics. Even one FPGA can configure a large-scale system on chip (SoC). The cost, including development, of FPGAs is becoming relatively advantageous compared with that of an ASIC, whose photomask cost has risen along with device miniaturization. In addition, a development environment similar to that for GPUs has become available for FPGAs, and thus software engineers can easily enter into the development of FPGAs [46]. Further progress of the development environment in the future will enable an engineer in a data server company to rewrite the circuit directly, for example. In this way, FPGAs are being used by increasing numbers of users.

Furthermore, applications that enjoy a performance advantage when using systems constructed from FPGAs (killer applications for FPGAs) have been appearing – e.g., a data center, a high-speed database, big data processing, a network search engine, and high-speed automated stock trading, and bit coin mining. They have the computing system using an FPGA as co processor of the CPU [47], utilizing the higher flexibility in circuit configuration and quicker response of the FPGA than those of a GPU. In routine processing, the power efficiency of an FPGA is said to be several dozen times as high as that of a CPU and several times as high as that of a GPU. With the successes in these applications, FPGAs are gradually being regarded as the third computing device, following CPUs and GPUs.

6.5.4.2   High Throughput

An FPGA is effective for applications needing large data processing and emphasizing high throughput, such as network searching, image searching, big data processing, machine learning for artificial intelligence, and a server accelerator [48, 49]. These have been regarded as effective FPGA applications after the development of large-scale FPGAs.

    The circuit configuration can be changed flexibly in an FPGA, so that a series of deep pipelines corresponding to multiple operations can be formed (see Figure 6.44). If the pipeline process is executed for a data-processing series, memory access will be needed only at the start and finish of the series. In other words, processing with high power efficiency becomes possible with a CPU or GPU.

    In the above applications, the structure where the plural FPGAs each forming deep pipelines are connected by a high-speed interface to form deeper pipelines or pipelines with higher parallelism is also expected to be effective.

6.5.4.3   Quick Response

An FPGA is also utilized effectively in applications where real-time response characteristics are important, such as high-frequency trading (HFT) achieving high-speed automated stock trading, real-time bidding (RTB) for Internet competitive tendering, complex event processing (CEP) for processing big data in real time, and a network interface card (NIC) [50]. These

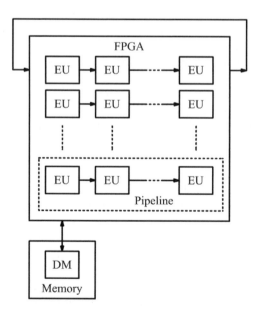

Figure 6.44 Operational procedure in a deep-pipeline FPGA. The FPGA comprises multiple EUs. The EUs are connected to form a pipeline. There are multiple pipelines. When data to be processed, which are read out from the DM, are input into the head of each pipeline, the pipeline executes the data processing. After processing, the data are stored in the DM. The operation is repeated

applications utilize the FPGA's flexibility to change the hardware configuration into a configuration capable of high-speed response.

In case of a CPU, when an event is detected and responsive processing is executed as in the above applications, the CPU should determine the content of the event after detection and execute a responsive program. To determine the event content, and execute a responsive program, several instructions should be executed. In addition, access to external data is required to detect the event, and thus data acquisition is expected to take a very long time.

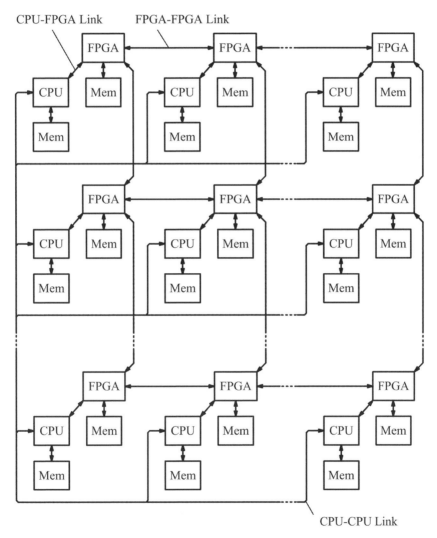

Figure 6.45 Example of a CAAC-IGZO CPU + CAAC-IGZO FPGA computing system. CPUs are connected by a common link, a CPU and an FPGA forming a pair are connected by a high-speed link, and adjacent FPGAs are connected by a dedicated link. Mem denotes memory

The FPGA where an event is converted into a trigger signal in real time and the responsive processing is included as hardware can respond immediately. In HFT and the like, the optimal event-detection algorithm changes every day; therefore, an FPGA whose circuit configuration can be changed in accordance with the algorithm change is effective. Note that not every processing route is suitable for conversion into hardware. Processing by CPU can be utilized additionally in such cases.

6.5.5   CAAC-IGZO CPU + CAAC-IGZO FPGA Computing

If the existing FPGA is replaced by a CAAC-IGZO FPGA having low power consumption and high power efficiency, a low-power, high-power-efficiency computing system can be constructed. In particular, a computing system combining a CAAC-IGZO CPU and CAAC-IGZO FPGA should reduce the whole power consumption and facilitate the construction of larger systems. In such a system, normally-off computing by the CAAC-IGZO CPU and CAAC-IGZO FPGA would also be available.

An example of the computing system combining CAAC-IGZO CPUs and CAAC-IGZO FPGAs is illustrated in Figure 6.45. The CPUs are connected by a common link, whereby versatile parts of programs are executed dispersedly. A CPU and an FPGA forming a pair are connected by a high-speed link, whereby the heavy-load processing is executed by the FPGA. Adjacent FPGAs are connected by dedicated links; in large data processing, a deep pipeline

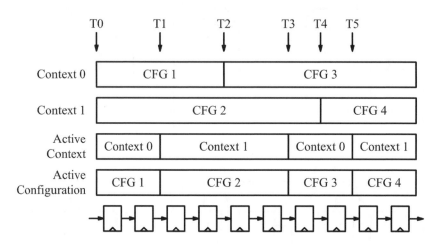

Figure 6.46   Example of dynamic reconfiguration in multicontext FPGA, having two contexts. CFG1, CFG2, CFG3, and CFG4 denote configuration data. At time T0, the operation starts from the initial state with context 0 of CFG1, context 1 of CFG2, active context of context 0, and active configuration of CFG1. At time T1, the active context is changed from context 0 to context 1, and accordingly the active configuration is changed from CFG1 to CFG2. At time T2, non-active configuration data of context 0 are changed to CFG3. At time T3, the active context is changed from context 1 to context 0, and accordingly the active configuration is changed from CFG2 to CFG3. At time T4, non-active configuration data of context 1 are changed to CFG4. At time T5, the active context is changed from context 0 to context 1, and accordingly the active configuration is changed from CFG3 to CFG4

is formed at each FPGA and pipelines parallel to adjacent FPGAs are made to improve the throughput of the whole data processing.

Easy construction of the fine-grained multicontext structure features the CAAC-IGZO FPGA. This feature is advantageous for forming a flexible deep pipeline. The multicontext structure enables the dynamic reconfiguration in response to processing executed by the FPGA, as in the startup of software in a CPU system in response to an application.

For example, as shown in Figure 6.46, changing the configuration data of the non-selected context allows for the construction of an apparently very deep pipeline. When the configuration data corresponding to plural operations are prepared and the operations are executed with the circuit configuration changed bit by bit, the processing can be conducted as if different operations proceed simultaneously.

References

[1] Okamoto, Y., Nakagawa, T., Aoki, T., Ikeda, M., Kozuma, M., Osada, T., *et al.* (2013) "Novel application of crystalline indium–gallium–zinc-oxide technology to LSI: Dynamically reconfigurable programmable logic device based on multi-context architecture," *ECS Trans.*, **54**, 141.

[2] Kurokawa, Y., Okamoto, Y., Nakagawa, T., Aoki, T., Ikeda, M., Kozuma, M., *et al.* (2013) "Applications of crystalline indium–gallium–zinc-oxide technology to LSI: Memory, processor, image sensor, and field programmable gate array," *Proc. Asia Symp. Quality Electronic Design*, 66.

[3] Okamoto, Y., Nakagawa, T., Aoki, T., Ikeda, M., Kozuma, M., Osada, T., *et al.* (2015) "A boosting pass gate with improved switching characteristics and no overdriving for programmable routing switch based on crystalline In–Ga–Zn–O technology," *IEEE Trans. Very Large Scale Integr. (VLSI) Syst.*, **23**, 422.

[4] Kozuma, M., Okamoto, Y., Nakagawa, T., Aoki, T., Ikeda, M., Osada, T., *et al.* (2014) "Crystalline In–Ga–Zn–O FET-based configuration memory for multi-context field-programmable gate array realizing fine-grained power gating," *Jpn. J. Appl. Phys.*, **53**, 04EE12.

[5] Aoki, T., Okamoto, Y., Nakagawa, T., Ikeda, M., Kozuma, M., Osada, T., *et al.* (2014) "Normally-off computing with crystalline InGaZnO-based FPGA," *IEEE Int. Solid-State Circuits Conf. Dig. Tech. Pap.*, 502.

[6] Aoki, T., Okamoto, Y., Nakagawa, T., Kozuma, M., Kurokawa, Y., Ikeda, T., *et al.* (2015) "Normally-off computing for crystalline oxide semiconductor-based multicontext FPGA capable of fine-grained power gating on programmable logic element with nonvolatile shadow register," *IEEE J. Solid-State Circuits*, **50**, 2199.

[7] Kozuma, M., Okamoto, Y., Nakagawa, T., Aoki, T., Kurokawa, Y., Ikeda, T., *et al.* (2015) "180-mV Subthreshold operation of crystalline oxide semiconductor FPGA realized by overdriving programmable power switch and programmable routing switch," *Ext. Abstr. Solid State Dev. Mater.*, 1174.

[8] Brown, S. D., Francis, R. J., Rose, J., and Vranesic, Z. G. (1992) *Field-Programmable Gate Array*. Boston, MA: Kluwer Academic.

[9] Trimberger, S. M. (1994) *Field-Programmable Gate Array Technology*. Boston, MA: Kluwer Academic.

[10] George, V. and Rabaey, J. M. (2001) *Low-Energy FPGAs – Architecture and Design*. Boston, MA: Kluwer Academic.

[11] Lemieux, G. and Lewis, D. (2004) *Design of Interconnection Networks for Programmable Logic*. Boston, MA: Kluwer Academic.

[12] Suzuki, D., Natsui, M., Ohno, H., and Hanyu, T. (2010) "Design of a process-variation-aware nonvolatile MTJ-based lookup-table circuit," *Ext. Abstr. Solid State Dev. Mater.*, 1146.

[13] Suzuki, D., Natsui, M., Ikeda, S., Hasegawa, H., Miura, K., Hayakawa, J., *et al.* (2009) "Fabrication of a non-volatile lookup-table circuit chip using magneto/semiconductor-hybrid structure for an immediate-power-up field programmable gate array," *IEEE Trans. Very Large Scale Integr. (VLSI) Syst.*, 80.

[14] Masui, S., Ninomiya, T., Oura, M., Yokozeki, W., Mukaida, K., and Kawashima, S. (2003) "A ferroelectric memory-based secure dynamically programmable gate array," *IEEE J. Solid-State Circuits*, **38**, 715.

[15] Miyamura, M., Nakaya, S., Tada, M., Sakamoto, T., Okamoto, K., Banno, N., *et al.* (2011) "Programmable cell array using rewritable solid-electrolyte switch integrated in 90nm CMOS," *IEEE Int. Solid-State Circuits Conf. Dig. Tech. Pap.*, 228.

[16] Liauw, Y. Y., Zhang, Z., Kim, W., Gamel, A. E., and Wong, S. S. (2012) "Nonvolatile 3D-FPGA with mono-lithically stacked RRAM-based configuration memory," *IEEE Int. Solid-State Circuits Conf. Dig. Tech. Pap.*, 406.

[17] Tada, M., Sakamoto, T., Miyamura, M., Banno, N., Okamoto, K., Iguchi, N., *et al.* (2011) "Highly reliable, com-plementary atom switch (CAS) with low programming voltage embedded in Cu BEOL for nonvolatile program-mable logic," *IEEE IEDM Tech. Dig.*, 689.

[18] Lewis, D. and Chromczak, J. (2012) "Process technology implications for FPGAs," *IEEE IEDM Tech. Dig.*, 565.

[19] Eslami, F. and Sima, M. (2011) "Capacitive boosting for FPGA interconnection networks," *Proc. IEEE FPL*, 453.

[20] Fujii, K. and Douseki, T. (2003) "A sub-1 V bootstrap pass-transistor logic," *IEICE Trans. Electron.*, **E86-C**, 604.

[21] Chun, K. C., Jain, P., Lee, J. H., and Kim, C. H. (2011) "A 3T gain cell embedded DRAM utilizing preferential boosting for high density and low power on-die caches," *IEEE J. Solid-State Circuits*, **46**, 1495.

[22] Altera Corporation (2006) *Stratix Device Handbook*, Vol. 1. Available at: www.altera.com/literature/hb/stx/stra-tix_handbook.pdf [accessed February 11, 2016].

[23] Trimberger, S., Carberry, D., Johnson, A., and Wong, J. (1997) "A time-multiplexed FPGA," *Proc. IEEE FCCM*, **22**.

[24] Kozuma, M., Okamoto, Y., Nakagawa, T., Aoki, T., Ikeda, M., Osada, T., *et al.* (2013) "Crystalline In–Ga–Zn–O FET-based configuration memory for multi-context field-programmable gate array realizing fine-grained power gating," *Ext. Abstr. Solid State Dev. Mater.*, 1096.

[25] Sakimura, N., Sugibayashi, T., Nebashi, R., and Kasai, N. (2008) "Nonvolatile magnetic flip-flop for standby-power-free SoCs," *Proc. IEEE CICC*, 355.

[26] Masui, S., Ninomiya, T., Oura, M., Yokozeki, W., Mukaida, K., and Kawashima, S. (2003) "A ferroelectric mem-ory-based secure dynamically programmable gate array," *IEEE J. Solid-State Circuits*, **38**, 715.

[27] Calhoun, B. H., Ryan, J. F., Khanna, S., Putic, M., and Lach, J. (2010) "Flexible circuits and architectures for ultralow power," *Proc. IEEE*, **98**, 267.

[28] Lotze, N. and Manoli, Y. (2012) "A 62 mV 0.13 μm CMOS standard-cell-based design technique using Schmitt-trigger logic," *IEEE J. Solid-State Circuits*, **47**, 47.

[29] Zimmermann, R. and Fichtner, W. (1997) "Low-power logic styles: CMOS versus pass-transistor logic," *IEEE J. Solid-State Circuits*, **32**, 1079.

[30] Grossmann, P. J., Leeser, M. E., and Onabajo, M. (2012) "Minimum energy analysis and experimental verification of a latch-based subthreshold FPGA," *IEEE Trans. Circuits Syst.*, **59**, 942.

[31] Putnam, A., Caulfield, A. M., Chung, E. S., Chiou, D., Constantinides, K., Demme, J., *et al.* (2014) "A recon-figurable fabric for accelerating large-scale datacenter services," *Proc. IEEE ISCA*, 13.

[32] Godfrey, M. D. and Hendry, D. F. (1993) "The computer as von Neumann planned it," *IEEE Ann. Hist. Comput.*, **15**, 11.

[33] Backus, J. (1978) "Can programming be liberated from the von Neumann style? A functional style and its algebra of programs," *Commun. ACM*, **21**, 613.

[34] Faggin, F., Shima, M., Hoff Jr., M. E., Feeney, H., and Mazor, S. (1972) "The MCS-4 – an LSI micro computer system," *Proc. IEEE Region 6 Conf.*, 8.

[35] Patterson, D. A. and Ditzel, D. R. (1980) "The case for the reduced instruction set computer," *ACM SIGARCH Comput. Architect. News*, **8**, 25.

[36] McGeady, S. (1990) "The i960CA SuperScalar implementation of the 80960 architecture," *Proc. IEEE COMP-CON*, 232.

[37] Danowitz, A., Kelley, K., Mao, J., Stevenson, J. P., and Horowitz, M. (2012) "CPU DB: Recording micropro-cessor history," *Commun. ACM*, **55**, 55.

[38] Preston, R. P., Badeau, R. W., Bailey, D. W., Bell, S. L., Biro, L. L., Bowhill, W. J., *et al.* (2002) "Design of an 8-wide superscalar RISC microprocessor with simultaneous multithreading," *IEEE Int. Solid-State Circuits Conf. Dig. Tech. Pap.*, **1**, 334.

[39] Vangal, S., Howard, J., Ruhl, G., Dighe, S., Wilson, H., Tschanz, J., *et al.* (2007) "An 80-tile 1.28TFLOPS network-on-chip in 65nm CMOS," *IEEE Int. Solid-State Circuits Conf. Dig. Tech. Pap.*, 98.

[40] Nave, R. and Palmer, J. (1980) "A numeric data processor," *IEEE Int. Solid-State Circuits Conf. Dig. Tech. Pap.*, 108.

[41] Schutz, J. (1991) "A CMOS 100 MHz microprocessor," *IEEE Int. Solid-State Circuits Conf. Dig. Tech. Pap.*, 90.

[42] Tarditi, D., Puri, S., and Oglesby, J. (2006) "Accelerator: Using data parallelism to program GPUs for general-purpose uses," *ACM SIGARCH Comput. Architect. News*, **34**, 325.

[43] Mittal, S. and Vetter, J. S. (2015) "A survey of CPU-GPU heterogeneous computing techniques," *ACM Comput. Surv.*, 47.

[44] Du, P., Weber, R., Luszczek, P., Tomov, S., Peterson, G., and Dongarra, J. (2012) "From CUDA to OpenCL: Towards a performance-portable solution for multi-platform GPU programming," *Parallel Comput.*, **38**, 391.

[45] Bacon, D. F., Rabbah, R., and Shukla, S. (2013) "FPGA programming for the masses," *Commun. ACM*, **56**, 56.

[46] Chen, D. and Singh, D. (2012) "Using OpenCL to evaluate the efficiency of CPUS, GPUS and FPGAS for information filtering," *Proc. IEEE FPL*, 5.

[47] Herbordt, M. C., VanCourt, T., Gu, Y., Sukhwani, B., Conti, A., Model, J., *et al.* (2007) "Achieving high performance with FPGA-based computing," *Computer*, **40**, 50.

[48] Chen, X.-W., and Lin, X. (2014) "Big data deep learning: Challenges and perspectives," *IEEE Access*, **2**, 514.

[49] Zhang, C., Li, P., Sun, G., Guan, Y., Xiao, B., and Cong, J. (2015) "Optimizing FPGA-based accelerator design for deep convolutional neural networks," *Proc. ACM/SIGDA Int. Symp. on FPGAs*, 161.

[50] Leber, C., Geib, B., and Litz, H. (2011) "High frequency trading acceleration using FPGAs," *Proc. IEEE FPL*, 317.

7

Image Sensor

7.1 Introduction

This chapter introduces an application of the c-axis-aligned crystalline indium–gallium–zinc oxide (CAAC-IGZO) technology to image sensors. As described in Chapters 3 and 4, the non-volatile oxide semiconductor random access memory (NOSRAM) and dynamic oxide semiconductor random access memory (DOSRAM) elements, each comprising a capacitor and a CAAC-IGZO field-effect transistor (FET) with an extremely low off-state current, can have excellent charge-retention characteristics. Also, complementary metal–oxide semiconductor (CMOS) image sensors [1] (CIS for short) rely on charge retention, so implementing CAAC-IGZO FETs is expected to give the CIS excellent image-capturing characteristics.

Section 7.2 explains how CAAC-IGZO FETs can be used to realize an image sensor with a global shutter mode [2], avoiding some of the disadvantages of Si-based CIS with a global shutter [3]. A common CIS adopts a rolling shutter mode, in which image distortion occurs due to the non-simultaneous image capture of moving objects. Accordingly, an image taken by a rolling-shutter CIS often has a lower quality compared with images captured by traditional silver halide film cameras or charge-coupled devices (CCD) [4]. The image distortion problem caused in a rolling shutter can be resolved by a global shutter, which can be implemented both mechanically and electrically. In the former case, the exposure time will be shorter and the sensitivity therefore lower. In the latter case, power consumption and sensor pixel density will suffer [5].

Developed with a global shutter in mind, a CAAC-IGZO image sensor with multiple storage nodes in a sensor pixel [6] is described in Section 7.3. This image sensor can employ a capturing method by which several continuously captured images are retained in the respective storage nodes in a sensor pixel and are read out sequentially. The method does not necessarily require a special high-speed analog-to-digital converter (ADC), and enables continuous image

Physics and Technology of Crystalline Oxide Semiconductor CAAC-IGZO: Application to LSI, First Edition.
Edited by Shunpei Yamazaki and Masahiro Fujita.
© 2017 John Wiley & Sons, Ltd. Published 2017 by John Wiley & Sons, Ltd.

capturing at extremely short time intervals. The image sensor is expected to be very effective at performing trajectory tracking for a high-speed moving object – i.e., obtaining a so-called optical flow [7].

The combination of a capacitor and a CAAC-IGZO FET with an extremely low off-state current enables construction of an ideal analog memory, which also allows construction of an analog arithmetic circuit. Section 7.4 introduces a CAAC-IGZO image sensor with a differential circuit comprising an analog arithmetic circuit in the sensor pixels [8, 9]. The image sensor can retain captured data from a reference frame in a storage node in each sensor pixel and generate difference data between the reference frame and the current frame. In addition to normal imaging, the image sensor can conduct difference determination between frames using a very simple analog arithmetic circuit without the help of any ADC, that is, it serves as an effective motion sensor [10]. It is expected to have significant important in surveillance applications, particularly where low power is important. In this way, value-added functional image sensing with not only a simple image-capturing function but also new additional functions such as analog arithmetic processing will be an attractive application of CAAC-IGZO image sensor technology.

7.2   Global Shutter Image Sensor

This section shows that a sensor pixel in the CAAC-IGZO image sensor (CAAC-IGZO sensor pixel) has excellent charge-retention characteristics, and that the image sensor can be driven by an electronic global shutter [3]. In the sensor pixel, a CAAC-IGZO FET with an extremely low off-state current offers a storage node with very high electrical insulation properties. This allows the use of an electronic global shutter and hence higher-quality image capturing of moving objects without distortion.

7.2.1   Sensor Pixel

Chapters 3 and 4 covered NOSRAM and DOSRAM with memory elements comprising a CAAC-IGZO FET for controlling access to a storage node, and this subsection will demonstrate how a sensor pixel of the image sensor can be designed in a similar way. The DOSRAM memory element (configuration: 1Tr1C, where "Tr" and "C" denote a transistor and a capacitor, respectively) described in Chapter 4, the NOSRAM memory element (2Tr1C) described in Chapter 3, and a CIS sensor pixel circuit with CAAC-IGZO FETs (4Tr1C1PD, where "PD" denotes a photodiode) are illustrated in Figure 7.1(a), (b), and (c), respectively. The storage node FD in (c) is formed by a capacitor and a CAAC-IGZO FET, similar to the node SN in the memory cells in (a) and (b). Thus, the sensor pixel can have a function similar to that of the memory.

The operation of the circuit in Figure 7.1(c) is illustrated in Figure 7.2. The operation includes the following: (1) reset [initialization of charges in the storage node FD by a reset signal (RST)]; (2) accumulation [accumulation control of charges generated by the photodiode with an accumulation control signal (TX)]; (3) retention; and (4) readout [output by a readout control signal (SE)]. The period after accumulation to readout corresponds to (3) the retention period. The accumulated charges in FD should be retained until the readout.

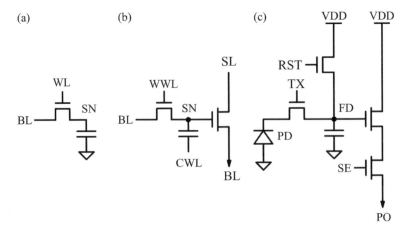

Figure 7.1 Relationship between circuit configuration and storage node in (a) DOSRAM memory element, (b) NOSRAM memory element, and (c) CIS sensor pixel with a CAAC-IGZO FET

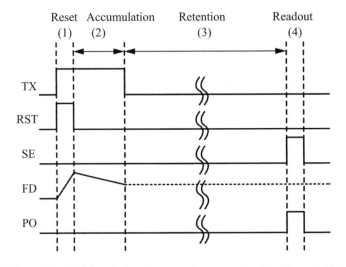

Figure 7.2 Driving timing diagram of a sensor pixel in Figure 7.1(c)

7.2.2 Global and Rolling Shutters

To understand the relationship between the retention characteristics of a sensor pixel and the captured image, the operation of the whole CIS is described here. In a common CIS, captured data in the sensor pixels of each frame are converted into digital values row by row by an ADC, as shown in Figure 7.3. The major image-capturing methods are (a) a rolling shutter mode adopted widely in CISs and (b) a global shutter mode. The timing diagrams of the two image-capturing modes are shown in Figure 7.4. A major difference between the two is the timing of image capture at the sensor pixels. With the rolling shutter, image capture and readout

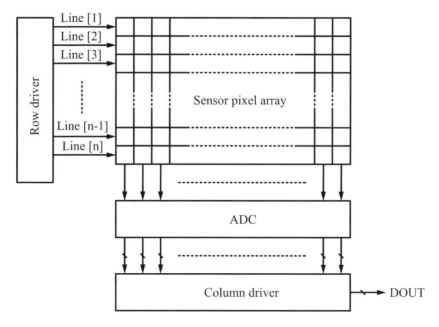

Figure 7.3 Block diagram of a general CIS

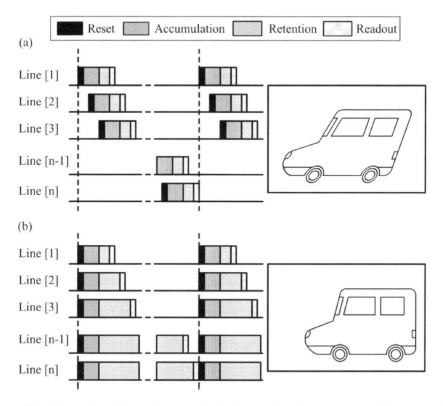

Figure 7.4 Comparison between image capturing by a rolling shutter (a) and a global shutter (b)

of the obtained data are performed row by row. That is, the timing of image capture is different between rows of the sensor pixels. In contrast, with the global shutter, images are concurrently captured in all sensor pixels, and then the obtained data are read out, row by row. Such a timing difference clearly affects the image quality, specifically the distortion degree of the image of a moving object. For example, when the image of an object (car) moving at high speed from the right to the left is captured with a rolling shutter, the sensor pixels on the upper rows, where images are captured at an early time, obtain the data of the object when it is still at the right, whereas the sensor pixels on the lower rows, where image capture is performed later, obtain data on the object when it has moved to the left. As a result, the captured image of the object is distorted, as shown in Figure 7.4(a). In contrast, with the global shutter, when an image of the car moving at high speed from the right to the left is captured, the sensor pixels from the uppermost and lowermost rows simultaneously obtain the data of the object at a certain position; thus, the captured image is not distorted, as shown in Figure 7.4(b).

7.2.3 Challenges Facing Adoption of Global Shutter

The global shutter mode is preferable to the rolling shutter mode in terms of image quality, as described above. However, it is difficult to employ a global shutter in the CIS because the sensor pixels have insufficient data-retention characteristics. With the global shutter, the period between the simultaneous image capturing in all sensor pixels and the readout (the retention period of the captured data) differs depending on the rows. In the sensor pixels in the row from which the data are read out last, the data-retention time is the longest; therefore, it is necessary to have additional mechanisms in order to avoid a loss of captured data due to charge leakages from storage nodes. With the rolling shutter mode, since the period from image capture to readout is constant among all sensor pixels, the data-retention period is within an acceptable range. Consequently, the rolling shutter mode is widely used in CISs. Note that it is ensured that complete images can be captured simultaneously in silver-halide photography and CCD image sensors; therefore, image distortion of a moving object does not occur in principle. Thus, it has been said for a long time that a CCD image sensor can take a high-quality image, while a CIS has the advantage of low cost (but takes a low-quality image). There are still great challenges facing the adoption of a global shutter mode in a CIS.

Several methods have been proposed to achieve the global shutter mode in a CIS. One is to use a mechanical shutter to realize a global-shutter-like mode (a mechanical global shutter mode). Exposure with a photodiode is mechanically controlled, during which image capture (i.e., accumulation of charges when the mechanical shutter is opened) is simultaneously performed among the sensor pixels, even though the sensor pixels are driven with a rolling shutter mode [Figure 7.5(a)]. However, this mechanism increases the cost, due to the addition of a mechanical shutter, and vibration arising from switching the shutter can also be problematic. Another proposed method is an electronic global shutter mode, where the image-capturing period is controlled by the transistors in the sensor pixels. The high-speed readout relieves the limitation of the retention characteristics [Figure 7.5(b)], and the degradation of the captured data is suppressed with the use of a special charge-retention structure [11–13]. However, such a structure complicates the circuit configuration and driving method compared with a CIS with the rolling shutter. The classification of the image sensors discussed here is shown in Table 7.1.

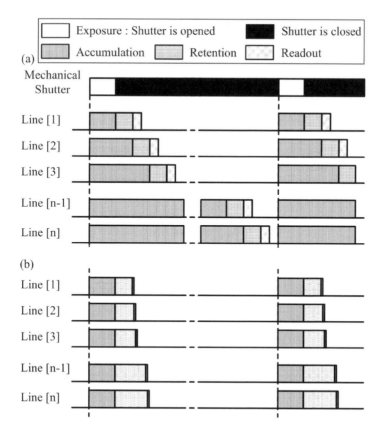

Figure 7.5 Timing diagrams of mechanical global shutter mode (a) and electronic global shutter mode with high-speed readout (b)

Table 7.1 Classification of image sensors

Process	Power	Shutter mode	Moving object	Retention time in sensor pixel data	Challenge
CCD	High	Global shutter	Good	Short	—
CIS	Low	Rolling shutter	Bad		—
		Mechanical global shutter	Good		Cost, vibration
				Long	Off-state current
		Electronic global shutter			

7.2.4 CAAC-IGZO Image Sensor

7.2.4.1 Configuration

To solve the above problem with the electronic global shutter, a proposal to utilize a CAAC-IGZO FET is given. When a CAAC-IGZO FET is used in a sensor pixel, a storage node can be used for its substantially complete electrical insulation properties, even without a special charge-retention circuit or driving method. As a result, the charge-retention characteristics improve, and an image sensor with an electronic global shutter mode without increased readout frame rate can be formed.

In order to understand the charge-retention characteristics of a sensor pixel, simulation of the captured image has been carried out with an assumed structure consisting of three transistors, one capacitor, and one photodiode (3Tr1C1PD) without a reset transistor (Figure 7.6). The assumed structure is a simplified version of the sensor pixel circuit in Figure 7.1(c). In the sensor pixel circuit, the photodiode is a horizontal Si diode; the three transistors are a Si-amplifying transistor (AMP), a transfer transistor [T; a CAAC-IGZO FET in (a) and a Si FET in (b)], and a Si-selection transistor (S). Figure 7.6(c) shows the timing diagram of the sensor pixel in

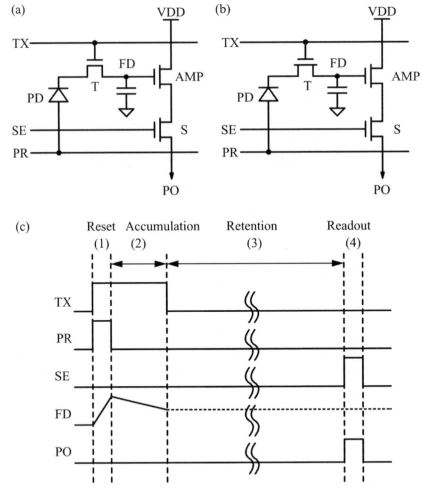

Figure 7.6 Sensor pixel circuit: (a) T is a CAAC-IGZO FET; (b) T is a Si FET. The transistors other than T are Si FETs. (c) Driving timing diagram of the sensor pixel. *Source*: Adapted from [3]

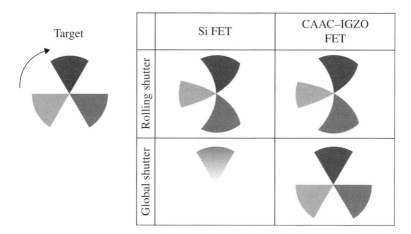

Figure 7.7   Target to be captured and simulated captured images

Figure 7.6(a) and (b). The target to be captured is a fan ("Target" in Figure 7.7), rotating in a clockwise direction at 640 rpm. The number of sensor pixels is 640 × 480 and the frame rate is 60 Hz. The rolling shutter mode and the global shutter mode are separately assumed as the image-capturing method; to compare the influences of the off-state current, a CAAC-IGZO FET and a Si FET are separately assumed as transfer transistor. Thus, four combinations are measured. A captured data-retention period from image capture to readout [Retention (3) in Figure 7.6(c)] is 30 μs at all rows with the rolling shutter, whereas the period is 30 μs (at the first readout row) to 16.7 ms (at the last readout row) with the global shutter.

The simulated captured images in the table of Figure 7.7 show that the quality depends on the shutter mode and transistor. With the rolling shutter, since the captured data-retention period is uniformly short in all sensor pixels, the influence of leakage current on the image quality is negligible regardless of whether Si FETs or CAAC-IGZO FETs are used; however, since image capture is not performed simultaneously among the sensor pixels, the captured image of the target is distorted. In contrast, with the global shutter, since the image capture is performed simultaneously among all the sensor pixels, image distortion does not occur, but the influence of leakage current increases. Specifically, with the Si FET, the captured data are lost from the sensor pixels selected in late time, which need to hold the captured data longer (the image at lower rows went white in the drawing). With the CAAC-IGZO FET, however, the captured data have been retained, even when the retention period is long.

7.2.4.2   Prototype 1

In order to confirm the above-simulated effect of retaining captured data in a CAAC-IGZO sensor pixel for a long time, a CAAC-IGZO image sensor was fabricated [3]. Figure 7.8 illustrates the device structure of the fabricated CAAC-IGZO image sensor. The hybrid structure of CAAC-IGZO and Si FETs is achieved by stacking these FETs (see Chapter 2 for detailed stacking techniques). The circuit configuration of the CAAC-IGZO sensor pixel is the same as that simulated in Figure 7.6(a). With such a sensor circuit configuration, the storage node FD is

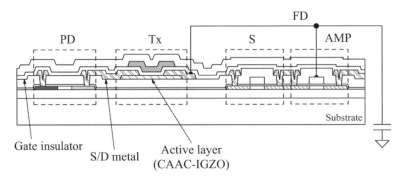

Figure 7.8 Device structure. *Source*: Adapted from [3]

completely surrounded by insulating films (including a gate insulator and interlayer insulator) and the CAAC-IGZO FET. Therefore, FD is a completely insulated region when the CAAC-IGZO FET is off, so its charge-retention characteristics are improved. In addition, high-speed readout is realized by the Si FET. The signal timing of the driving is shown in Figure 7.6(c).

To evaluate the charge-retention characteristics, a test element group (TEG) of the CAAC-IGZO sensor pixel shown in Figure 7.6(a) and a TEG of a sensor pixel circuit only with Si FETs (a CMOS image sensor) in Figure 7.6(b) were fabricated and compared. For the measurement, the voltage value of the output of the captured data (PO) is continuously monitored to examine the charge-retention characteristics of FD. The selection transistor is always on so that the voltage of PO might be monitored, and the other timings are the same as those shown in Figure 7.6(c).

The dependence on illuminance (0 to 1000 lx) was measured and is shown in Figure 7.9. Figure 7.9(a) shows the TEG of the CAAC-IGZO sensor pixel, and Figure 7.9(b) shows the TEG of the CMOS sensor. In the CMOS sensor pixel, the voltage value of PO started to decrease after several milliseconds of charge retention, as shown in the graph. In contrast, since the leakage current is much smaller in the CAAC-IGZO sensor pixel than in the CMOS sensor pixel, a sufficient PO voltage value is retained so as not to affect the sensor pixel output even after approximately 16.7 ms or longer [corresponding to 1 frame at 60 frames per second (fps)]. These results suggest that the CAAC-IGZO sensor pixel allows the achievement of an electronic global shutter even without any special charge-retention circuit.

The photograph and specifications of the fabricated CAAC-IGZO image sensor are shown in Figure 7.10 and Table 7.2, respectively [3].

The fabricated CAAC-IGZO image sensor can capture a clear image of a rotating object with a global shutter. An image of the rotating object at approximately 640 rpm is captured, and the captured images are shown in Figure 7.11. Figure 7.11(a) is the result taken by the rolling shutter, Figure 7.11(b) is the reference result taken by the global shutter in the case where the transfer transistor is always weakly opened, assuming a large gate leakage current of the transfer transistor (i.e., supposing a Si FET), and Figure 7.11(c) is the result taken by the global shutter. The image captured by the rolling shutter is distorted because of the non-simultaneous image capture among the sensor pixels, while the images captured by the global shutter are not distorted. In addition, since charges are retained from image capture to readout of the captured data with the global shutter, the captured image (c) shows an accurate shape of the rotating object

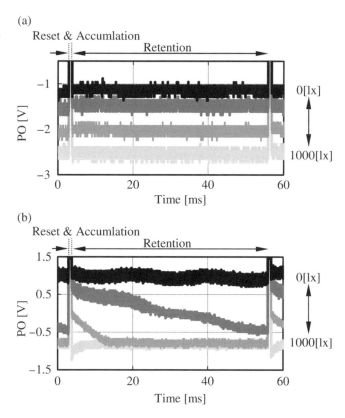

Figure 7.9 Charge-retention characteristics of floating diffusion (FD) at 0 to 1000 lx in a structure with (a) a CAAC-IGZO FET [Figure 7.6(a)] and (b) only Si FETs [Figure 7.6(b)]. *Source*: Adapted from [3]. Copyright 2011 The Japan Society of Applied Physics

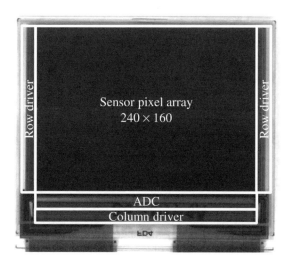

Figure 7.10 Photograph of the global shutter image sensor

Table 7.2 Specifications of the global shutter image sensor [3]

Process	2.0-μm CAAC-IGZO/2.0-μm Si
Chip size	96 mm × 78 mm
Number of sensor pixels	640 × 480
Sensor pixel size	126 μm × 126 μm
Sensor pixel configuration	3 transistors, 1 capacitor
Fill factor	3.8%
Supply voltage	3.3 V
ADC	6-bit ring oscillator

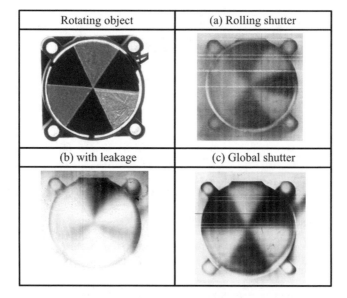

Figure 7.11 Image of a rotating object. *Source*: Adapted from [3]. Copyright 2011 The Japan Society of Applied Physics

without decrease in output value from the sensor pixels shown in image (b). These results demonstrate that the CAAC-IGZO image sensor enables the global shutter mode.

7.2.4.3 Prototype 2

As an example of a further scaled-down prototype utilizing the effectiveness of the CAAC-IGZO image sensor described above, a sub-micrometer fabrication process is used. The micrograph and specifications of the fabricated CAAC-IGZO image sensor are shown in Figure 7.12 and Table 7.3, respectively. The channel length L of the CAAC-IGZO FET is 0.35 μm. The image sensor has the circuit configuration of Figure 7.1(c), and the transistors and photodiodes of the sensor pixels are CAAC-IGZO FETs and horizontal PIN photodiodes, respectively. The peripheral circuits are composed of CMOS circuits of 0.18-μm Si FETs.

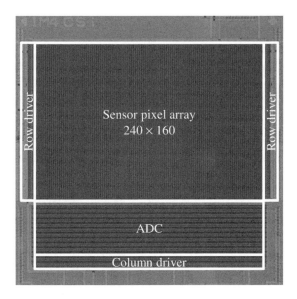

Figure 7.12   Micrograph of the global shutter image sensor

Table 7.3   Specifications of the global shutter image sensor

Process	0.35-μm CAAC-IGZO/0.18-μm Si
Die size	6.5 mm × 6.0 mm
Number of sensor pixels	240 × 160
Sensor pixel size	20 μm × 20 μm
Sensor pixel configuration	4 transistors, 1 capacitor
Fill factor	30%
Supply voltage	1.8 V
ADC	8-bit single slope

Figure 7.13 shows the results of imaging of an object rotating at 400 rpm with use of the fabricated CAAC-IGZO image sensor. Figure 7.13(a) is the reference imaging result with the rolling shutter, taken with a commercial smartphone camera, and Figure 7.13(b) is the result with the global shutter, taken with the CAAC-IGZO image sensor. As shown in Figure 7.13(a), the image of the object is distorted because the capture is performed with the rolling shutter in the commercial smartphone and the exposure is not conducted simultaneously among the sensor pixels; in contrast, as shown in Figure 7.13(b), the image of the object obtained with the global shutter is not distorted. This suggests that, as in the case of the prototype in Figure 7.10, the image sensor with an electronic global shutter can be formed even with sub-micrometer CAAC-IGZO FETs.

Rotating object	(a) Rolling shutter

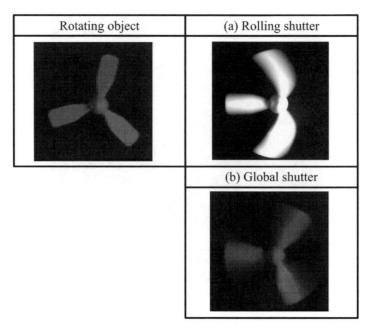

(b) Global shutter

Figure 7.13   Images of a rotating object [above left: image of resting object, (a) image taken with rolling shutter using a commercial smartphone and (b) image taken with global shutter using a fabricated prototype]

7.3   Image Sensor Conducting High-Speed Continuous Image Capture

7.3.1   Overview

Based on the CAAC-IGZO image sensor with excellent charge-retention characteristics allowing the electronic global shutter described in the previous section, a high-speed continuous-capturing image sensor with short shutter time [6] can be realized. In this section, "high speed" refers to short shutter time. Although one storage node with excellent charge-retention characteristics is positioned in each sensor pixel in the previous section, it is possible to provide multiple similar storage nodes in each sensor pixel. In that case, pieces of data captured continuously can be, respectively, stored in the storage nodes in the sensor pixel. After that, the pieces of captured data can be read out sequentially, without synchronizing with the image capture. In this image-capturing method, capture intervals depend on the sensitivity of the sensor pixel, not the operational speed of peripheral circuits such as an ADC. That is, without the massive A/D conversion that is forcibly conducted in a general high-speed camera with large power consumption, this method can continuously capture images at high speed. In addition, high-speed peripheral circuits are not essential to the method, and thus peripheral circuits of the image sensor can be formed with p-channel Si (Pch-Si) and CAAC-IGZO FETs. Furthermore, the use of the method will lead to trajectory tracking of a high-speed moving object (i.e., a so-called optical flow system) with low power consumption.

7.3.2   Conventional High-Speed Continuous-Capturing Image Sensor

In a common CIS, data captured in each sensor pixel at each frame are generally converted into digital data in an ADC [14]. One method to directly realize high-speed continuous image capturing at short time intervals is to increase the frame rate [15,16]. In that case, short-time capturing and fast readout by the sensor pixels are required – i.e., highly sensitive and high-speed sensor pixels are required. In addition, a high-speed ADC is needed. The use of such an ADC will cause problems, such as an increase in circuit area, larger power consumption, and heat generation. The solutions to these problems are technically challenging.

Methods by which the problems with a high-speed ADC can be avoided have been proposed. First, an image sensor is proposed in which DRAM (an analog memory) is provided adjacent to the sensor pixel region, and captured data in the sensor pixels at each frame are read out and stored in the analog memory as analog data without A/D conversion [17]. This image sensor does not require a high-speed ADC; however, captured data must be transferred at high speed from the sensor pixels to the analog memory, which may lead to high power consumption. In addition, the image quality may be degraded because of the off-leakage current of Si FETs in the analog memory. Another image sensor is that in which captured data in the sensor pixels are stored by the CCD method and then transferred [18,19]. Although the problems with a high-speed ADC can be avoided in this method also, a high driving voltage causing high power consumption is required to transfer captured data between the CCDs. In addition, a CCD manufacturing process is required, and thus the cost advantage of the CIS process is cancelled out.

7.3.3   High-Speed Continuous-Capturing CAAC-IGZO Image Sensor

7.3.3.1   Configuration

The above problems can be solved by utilizing a CAAC-IGZO FET with an extremely low off-state current. Specifically, multiple storage nodes, each like the one in Figure 7.1(c), are included in sensor pixels of a CAAC-IGZO image sensor, and captured data are continuously stored in the storage nodes and then sequentially read out. Such a configuration allows high-speed continuous image capture without a high-speed ADC [6].

A CAAC-IGZO image sensor for performing high-speed continuous image capturing with a short shutter time is described below. A circuit diagram of a sensor pixel of the image sensor is shown in Figure 7.14. The sensor pixel has k sub-sensor pixels, each including a storage node. Each sub-sensor pixel comprises four transistors and one photodiode. The photodiode is shared among the sub-sensor pixels with the use of a sharing transistor, which enables charges to be accumulated in any one of the storage nodes (FD) by the photocurrent from k photodiodes. Such a structure offers superior sensitivity to that in a general method by which charges are accumulated in one FD with the use of one photodiode. This compensates for the increased necessary sensitivity due to short-time-interval image capture. When the transistors in the sensor pixel are CAAC-IGZO FETs, the charge-retention characteristics of the FD improve, and thus degradation due to leakage of the captured data hardly occurs. Figure 7.15 is a timing diagram of the image sensor. When accumulation control signals TX1, TX2, ..., TXk are sequentially activated, image capture is performed continuously at short time intervals. Next, the captured data are read out row by row and subjected to A/D

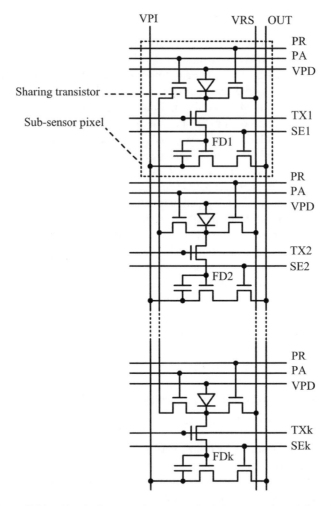

Figure 7.14 Circuit diagram of a sensor pixel. *Source*: Adapted from [6]

conversion with an ADC. That is, high-speed continuous image capture with a short shutter time can be obtained without high-speed ADC. Therefore, lower power consumption is achieved. The complete block diagram of the image sensor is shown in Figure 7.16. The image sensor comprises a sensor pixel array, a row driver, a column driver, and an ADC. The image sensor can operate with a circuit configuration similar to that of a common image sensor.

7.3.3.2 Prototype 1

In order to examine the CAAC-IGZO image sensor conducting high-speed continuous image capture, an image sensor formed by the hybrid process of 0.35-μm CAAC-IGZO and 0.18-μm

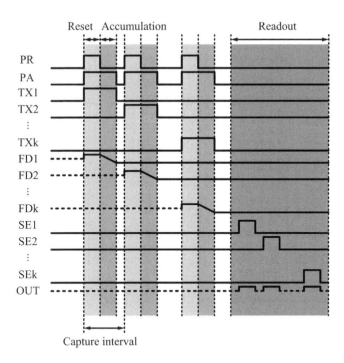

Figure 7.15 Timing diagram of the high-speed continuous-capture method. *Source*: Adapted from [6]

Si FETs was fabricated [6]. Table 7.4 and Figure 7.17 show the specifications and micrograph, respectively. Here the sensor pixel has the configuration shown in Figure 7.14 with $k = 2$, which is the simplest configuration. All FETs of the sensor pixels contain CAAC-IGZO. The readout frame rate in Table 7.4 indicates the number of readouts per second.

To analyze whether the image quality is affected by the sensor pixel configuration, a uniform planar light source is used as an object to be captured. Images captured at sub-sensor pixels 1 (controlled by TX1) and 2 (controlled by TX2) with the global shutter are called TX1G and TX2G images, respectively, and those captured at sub-sensor pixels 1 and 2 with high-speed continuous, non-global shutter image capture are called TX1C and TX2C images, respectively. TX1G, TX2G, TX1C, and TX2C images are captured under various conditions, and the TX1G image sampled by the global shutter at 60 fps is used as a reference image. Each pixel value [digital output value by ADC with respect to the captured data at each (sub-)sensor pixel] difference between the reference image and the TX1G/TX2G/TX1C/TX2C image is obtained in the manner shown in Figure 7.18, and the average values and standard deviations of the differences are shown in Figures 7.19 and 7.20, respectively. Figure 7.19 shows the results under readout frame rates of 0.1, 0.3, 1, 3, 10, 30, and 60 fps (note that the capture interval is 100 μs under all conditions). Figure 7.20 shows the results with high-speed continuous image capture at a readout frame rate of 60 fps and capture intervals of 100, 200, 500, and 1000 μs. In Figures 7.19 and 7.20, a least significant bit (LSB) is used as a unit of pixel value to quantitatively discuss the difference between the two images.

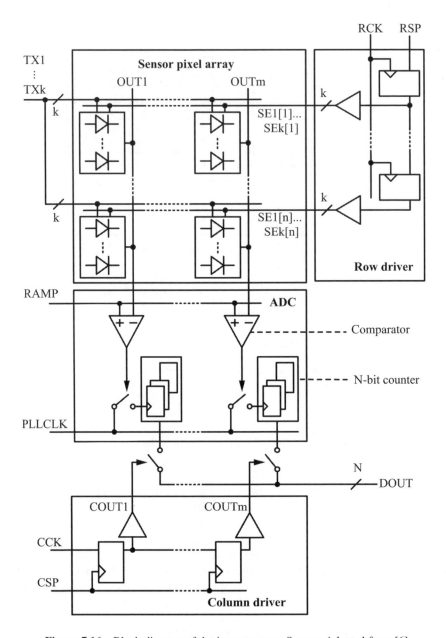

Figure 7.16 Block diagram of the image sensor. *Source*: Adapted from [6]

In Figure 7.19(a) and (b), within the readout frame rate of 1 to 60 fps, the average and standard deviation of pixel value differences between the reference and TX1G/TX2G/ TX1C/TX2C images are distributed in the range ±0.5 LSB. Therefore, within this range of the readout frame rate, there is no significant difference between the results under the

Table 7.4 Specifications of the image sensor. *Source*: Adapted from [6]

Process	0.35-μm CAAC-IGZO/0.18-μm Si	
Die size	6.5 mm × 6.0 mm	
Number of sensor pixels	240 × 80 (each sensor pixel has 2 sub-sensor pixels)	
Sub-sensor pixel size	20 μm × 20 μm	
Sub-sensor pixel configuration	4 transistors, 1 capacitor	
Fill factor	31%	
Supply voltage	Nominal	1.8 V
	Sensor pixel, comparator in ADC	3.3 V
ADC	8-bit single slope	
Clock frequency	Row driver	2.76 kHz
(readout frame rate: 60 fps)	Column driver	718 kHz
	ADC	1.77 MHz
Reset time/accumulation time	45 μs/45 μs	

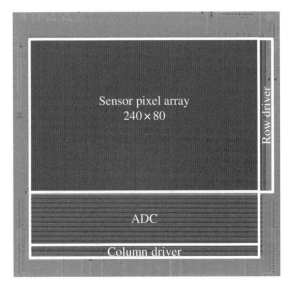

Figure 7.17 Micrograph of the image sensor

global shutter and high-speed continuous image capturing, and the latter can therefore be effectively employed. In particular, high-speed continuous image capturing with a short shutter time can be used even at a readout frame rate as low as 1 fps, leading to a later-described proposal for a low-power image sensor with peripheral circuits comprising Pch-Si and CAAC-IGZO FETs. In Figure 7.20(a) and (b), within the interval of 100–1000 μs under high-speed continuous image capturing, the average and standard deviation of the pixel value differences between the reference image and the TX1C/TX2C images are distributed in the

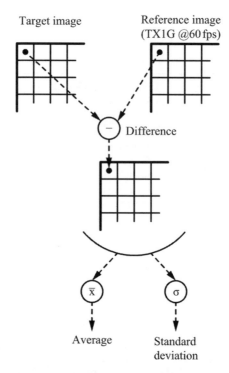

Figure 7.18 Method for calculating average value and standard deviation

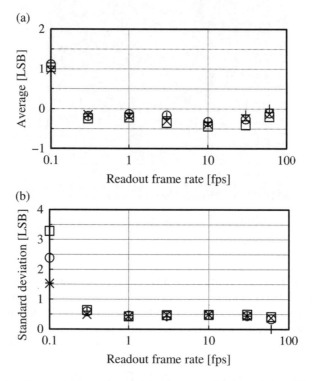

Figure 7.19 Readout frame rate dependence of the image quality degradation: (a) average pixel value difference between the target and reference images; (b) standard deviation of the pixel value difference between target and reference images (○, TX1C; □, TX2C; +, TX1G; ×, TX2G)

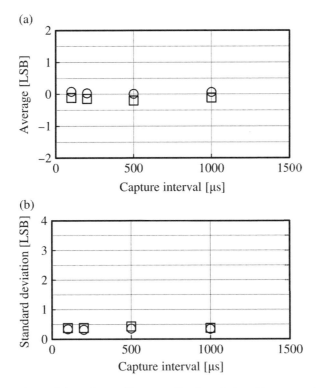

Figure 7.20 Capture interval dependence of image quality degradation: (a) average pixel value difference between the target and reference images; (b) standard deviation of the pixel value difference between target and reference images (O, TX1C; □, TX2C)

range ±0.5 LSB. Thus, within this range of capture interval, there is no significant difference between the results under global shutter and high-speed continuous image capturing, and the latter can be effectively used even at a capture interval as short as 100 μs. This leads to a later-described proposal for an optical flow system.

High-speed continuous image capturing was used to capture two fans rotating at 6500 rpm and 10000 rpm in the counterclockwise direction. The images are captured at a readout frame rate of 1 fps and capture intervals of 100 μs and 1000 μs. The driving signal waveforms at a capture interval of 100 μs are shown in Figure 7.21. In the drawing, the waveforms of PR, PA, TX1, and TX2 correspond to those in Figure 7.15 in the case of $k = 2$. Figure 7.22 shows the captured images. Figure 7.22(a) and (b) are a TX1C image and TX2C image, respectively, captured at an interval of 100 μs, while (c) and (d) are a TX1C image and TX2C image, respectively, captured at an interval of 1000 μs. From Figure 7.22, the rotation speed of each fan can be calculated. For example, the angle between the right fan positions in Figure 7.22(a) and (b) is 6.0°. Since the capture interval is 100 μs, the rotation speed of the right fan is calculated to be 6.0° / 100 μs × 60 s / 360° = 10000 rpm. Similarly, the angle between the left fan positions in (a) and (b) is 3.9°, and the rotation speed of the left fan is calculated to be 3.9° / 100 μs × 60 s / 360° = 6500 rpm. These calculated rotation speeds are coincident with

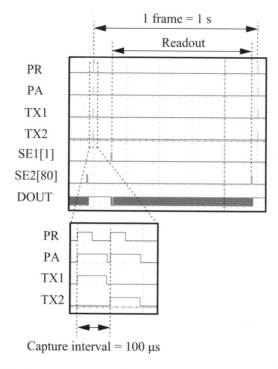

Figure 7.21 Waveforms at a readout frame rate of 1 fps and capture interval of 100 μs

Figure 7.22 Images of rotating fans (the left-hand fan rotates at 6500 rpm and the right-hand fan rotates at 10000 rpm): (a) TX1C image (capture interval 100 μs); (b) TX2C image (capture interval 100 μs); (c) TX1C image (capture interval 1000 μs); (d) TX2C image (capture interval 1000 μs)

the actual rotation speeds. Furthermore, the same rotation speeds are obtained in the case of 1000 μs from a right fan angle of 60° and a left fan angle of 39° in Figure 7.22(c) and (d).

The captured image quality is hardly degraded in the CAAC-IGZO image sensor, even at a low readout frame rate, as described above, because of the excellent retention characteristics of the sensor pixels with CAAC-IGZO FETs. The captured data can be read out slowly; therefore, even when the ADC speed is lowered in accordance with the low readout frame rate, there is no influence on image quality. This allows low-power driving of the ADC, leading to a reduction in power consumption by the whole image sensor. However, the short shutter time requires high illuminance and/or a sensor with high sensitivity. By increasing the number of sub-sensor pixels, it is possible to capture sequences of more than two images per readout.

To confirm the above-mentioned effect of reducing power consumption, the power consumption and energy necessary to obtain one image are measured under the three conditions shown in Table 7.5. The first is a reference condition. In the second condition, the readout frame rate is lowered from 60 fps of the reference to 1 fps. In the third condition, the source voltages of the peripheral circuits such as the ADC are lowered from the second condition, while the other conditions are kept the same. Under all conditions, the capture interval is 100 μs. Table 7.6 shows the results. The power consumption under the second condition is 93.3% of that under the first condition; however, the energy required to obtain one image increases. This is because the possible number of images to be captured is decreased by lowering the readout frame rate and because static power consumption accounts for the majority of power consumption in the ADC. In contrast, the consumed power under the third condition is as low as 0.71% of that under the first condition, and the energy for acquiring an image is 9.2 μJ, lower than

Table 7.5 Measurement condition of power consumption by the image sensor

		1st cond.	2nd cond.	3rd cond.
Readout frame rate		60 fps	1 fps	1 fps
Supply voltage	ADC (comparator/counter)	3.3 V/1.8 V	3.3 V/1.8 V	2.5 V/1.8 V
	Comparator bias	−0.07 V	−0.07 V	−0.6 V
	Row driver	3.3 V	3.3 V	2.5 V
	Column driver	3.3 V	3.3 V	2.5 V

Table 7.6 Power consumption by the image sensor

		1st cond.	2nd cond.	3rd cond.
Power (μW)	Comparator	1216.9	1204.0	7.4
	Counter	19.1	0.9	1.0
	Column driver	56.8	1.0	0.5
	Row driver	0.2	0.0	0.0
	Sensor pixels	0.3	0.3	0.3
	Total	**1293.2**	**1206.2**	**9.2**
Energy (μJ)		21.6	1206.2	9.2

21.6 µJ under the first condition. Thus, it has been confirmed that combining high-speed continuous image capturing with the lowered readout frame rate and lowered voltage in the peripheral circuits including ADCs can effectively offer reduced power consumption. Although lowering the comparator bias and ADC voltages reduces the dynamic range of the image sensor, it is still sufficient for some machine vision applications with high contrast images, such as the fan in the example above.

7.3.3.3 Prototype 2

The proposed method of high-speed continuous image capturing does not require any special high-speed operation of the peripheral circuits unless a high readout frame rate is necessary, as described above. The $I_d - V_g$ characteristics [Figure 7.23(a)] and noise characteristics [Figure 7.23(b) and (c)] of the CAAC-IGZO and Si FETs formed by the hybrid process indicate that the CAAC-IGZO FET tends to be superior to the n-channel Si FET (Nch-Si FET) with

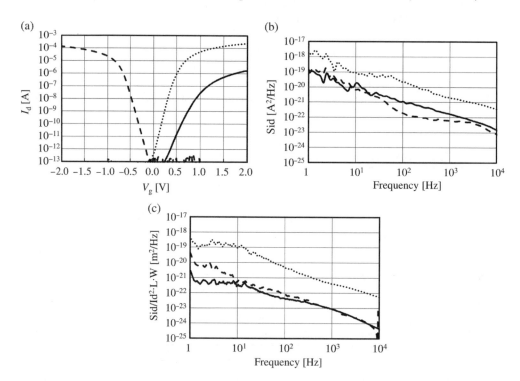

Figure 7.23 Characteristics of FETs formed by hybrid process: (a) $I_d - V_g$ curve ($V_d = 1.9$ V) (solid line: CAAC-IGZO, $L/W = 0.36$ µm / 1.0 µm; dotted line: Nch-Si, $L/W = 0.19$ µm / 1.74 µm; dashed line: Pch-Si, $L/W = 0.19$ µm / 1.74 µm); (b) 1/f noise (solid line: CAAC-IGZO, $L/W = 0.8$ µm / 10 µm; dotted line: Nch-Si, $L/W = 0.35$ µm / 1.6 µm; dashed line: Pch-Si, $L/W = 0.35$ µm / 3.2 µm); and (c) 1/f noise in case of using scaled-down CAAC-IGZO FET (solid line: CAAC-IGZO, $L/W = 30$ nm / 30 nm; dotted line: Nch-Si, $L/W = 0.8$ µm / 10 µm; dashed line: Pch-Si, $L/W = 0.8$ µm / 10 µm). W denotes the channel width. *Source*: Adapted from [6]

respect to noise levels. These factors may allow an image sensor with a hybrid structure of CAAC-IGZO and Pch-Si FETs where no Nch-Si FET is used. In such an image sensor, all sensor pixel circuits can be implemented entirely by CAAC-IGZO FETs (except for the photo-diode, which is Si), and the peripheral circuits such as drivers and ADCs can be implemented by Pch-Si FETs and CAAC-IGZO FETs [6]. Thus, the steps for forming Nch Si FETs are possibly eliminated, leading to a reduction in the process cost. To verify this, a prototype image sensor was fabricated, and the high-speed continuous image capturing operation is examined.

The specifications and micrograph of the fabricated image sensor are shown in Table 7.7 and Figure 7.24, respectively. The peripheral circuits are formed of CAAC-IGZO and Pch-Si FETs.

Table 7.7 Specifications of image sensor. *Source*: Adapted from [6]

Process	0.35-μm CAAC-IGZO/0.18-μm Si (Pch)	
Die size	6.5 mm × 6.0 mm	
Number of sensor pixels	240 × 80	
Sub-sensor pixel size	20 μm × 20 μm	
Sub-sensor pixel configuration	4 transistors, 1 capacitor	
Fill factor	31%	
Supply voltage	Nominal	1.8 V
	Sensor pixel, comparator in ADC	3.3 V
ADC	8-bit single slope	
Clock frequency	Row driver	46.0 Hz
(readout frame rate: 1 fps)	Column driver	12.0 kHz
	ADC	29.5 kHz
Reset time/accumulation time	45 μs/45 μs	

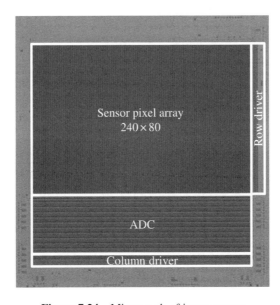

Figure 7.24 Micrograph of image sensor

All transistors of the sensor pixels are CAAC-IGZO FETs. Figure 7.25 illustrates the hybrid structure of the fabricated image sensor. Figure 7.26 shows the layout of the sub-sensor pixels. In such a structure, the area of the peripheral circuits, such as drivers and ADCs, can be reduced. In addition, since the sensor pixel transistors can all be CAAC-IGZO, an arrangement of Si FETs on a Si substrate is not required; therefore, the photodiode can be increased in size up to that of the sub-sensor pixel. The number of sub-sensor pixels is $k = 2$, i.e., the same as shown in Figure 7.14.

First, the operation of the peripheral circuits of the fabricated image sensor is examined. Figure 7.27 shows the signal waveforms of the column driver. The circuits operate at a low readout frame rate of 1 fps. As shown in the figure, the column driver outputs an image data output enable signal (COUT) in synchronization with a clock signal (CCK). With the sensor, high-speed continuous image capturing is performed under the following conditions: the read-out frame rate is 1 fps, the capture interval is 100 μs with 45 μs accumulation, and the objects to be captured are two fans rotating at 6500 rpm and 10000 rpm in a counterclockwise direction. The captured images are shown in Figure 7.28. The photographs show the rotation angles of the fans (3.9° and 6.0°) during the capture interval. In the same manner as the calculation for the image sensor of Prototype 1, the rotation speeds of the left and right fans are calculated to be 6500 rpm and 10000 rpm, respectively, which are coincident with the actual rotation speeds.

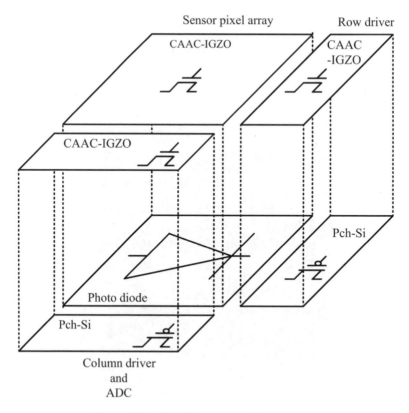

Figure 7.25   Hybrid structure of image sensor

CAAC-IGZO FETs   Photo diode

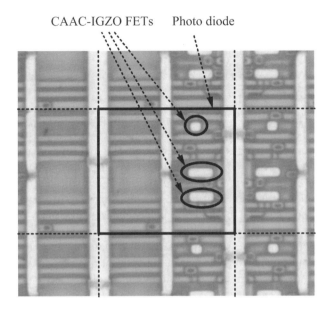

Figure 7.26   Micrograph of sub-sensor pixel in image sensor. *Source*: Adapted from [6]. Copyright 2014
The Japan Society of Applied Physics

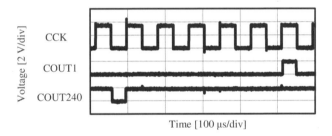

Figure 7.27   Measured waveforms of the column driver in the image sensor. *Source*: Adapted from [6]

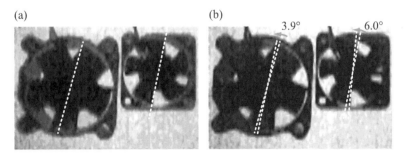

Figure 7.28   Images of a rotating fan: (a) TX1C image; (b) TX2C image

Thus, it has been confirmed that high-speed continuous image capturing can be performed normally even in the test image sensor.

7.3.4 Application to Optical Flow System

A low-power optical flow system [20] is an example of an application of the image sensor operated with high-speed continuous image capturing. An optical flow expresses the visual movement of an object as a vector, and is an important concept in fields where motion capture, such as image recognition or machine vision, is the target. While a short shutter time is required to accurately obtain an optical flow, a high sampling rate is not required in many applications where flow calculation is carried out. A conventional image sensor with high frame rate may not be the optimum solution, because it captures unnecessarily many images and requires high power for a fast readout. In contrast, the image sensor with high-speed continuous image capturing captures only the necessary images for an accurate optical flow calculation. The images can be read out slowly from the start of the optical flow calculation without degrading the image quality, and the system power consumption can be reduced.

An optical flow system includes, for example, an image sensor, a controller, a digital signal processor (DSP), and a display (Figure 7.29). The image sensor uses high-speed continuous image capturing. The image sensor captures two images during short shutter times at a short time interval in response to the controller, and then reads out the images at a low readout frame rate (e.g., 1 fps) depending on the application. The DSP creates an optical flow from the two images, and instructs the controller to conduct additional image capturing if necessary. The optical flow and captured images are then synthesized and displayed.

To examine the feasibility of the optical flow system, an optical flow is calculated with the use of the images in Figure 7.22 captured by the image sensor in Subsection 7.3.3.2. As for the calculation algorithm, the common Lucas–Kanade method [7] is used. The result in the case of a 100-μs capture interval [i.e., the optical flow obtained from the images in Figure 7.22(a) and (b)] is shown in Figure 7.30(a). The result in the case of a 1000-μs capture interval [i.e., the optical flow obtained from the images in Figure 7.22(c) and (d)] is shown in Figure 7.30(b). Figure 7.30(a) shows that an accurate optical flow indicating two fans rotating in a counterclockwise direction is obtained at a capture interval of 100 μs. In contrast, Figure 7.30(b) shows that the flows of the left fan are varied and the rotating direction is not calculated at all, and that it even seems as if the right fan was calculated to rotate in a clockwise direction (the inverse direction) at a capture interval of 1000 μs. That is, when the capture interval is long, such as 1000 μs, it is difficult to obtain the optical flow of the object rotating at high speed (e.g., 6500 rpm or 10000 rpm).

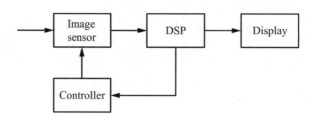

Figure 7.29 Block diagram of an optical flow system

(a)   (b)

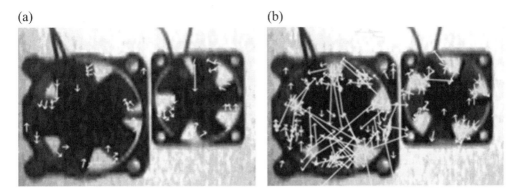

Figure 7.30 Optical flow: (a) at a capture interval of 100 μs; (b) at a capture interval of 1000 μs

However, when an image sensor can conduct continuous image capture at high speed (e.g., at a capture interval of 100 μs with an accumulation of 45 μs as in this image sensor), the optical flow of the object rotating at high speed can be accurately obtained.

Next, the effect of reduction in power consumption by the optical flow system with the image sensor fabricated in Subsection 7.3.3.2 is discussed. A conventional system with a known image sensor reported by Furuta *et al.* [15] and a DSP C6678 manufactured by Texas Instruments [21] is assumed to be able to capture a 128 × 128 -sensor pixel monochrome image with 8-bit gray scale at a frame rate of 10000 fps (corresponding to a capture interval of 100 μs) in an imaginary system. It is also assumed that power consumption scales linearly with sensor pixel count, gray-scale bit depth, and frame rate. The optical flow is assumed to be calculated at a sampling ratio of 1 fps. On the basis of the specifications of the image sensor described in [15] (number of sensor pixels: 514 × 530, ADC resolution: 12 bits, frame rate: 3500 fps, power: 1 W), the image sensor in the imaginary system is then estimated to consume 115 mW $[= 1\,W \times (128 \times 128) / (514 \times 530) \times 8\,bit / 12\,bit \times 10000\,fps / 3500\,fps]$. According to the specifications of the DSP described in [21] (number of sensor pixels: 128 × 128, throughput: 9.79 fps, power: 10 W), the DSP in the system is estimated to consume a power of 1.02 W $(= 10\,W \times 1\,fps / 9.79\,fps)$. Thus, the total power consumption of the imaginary system is estimated to be 1.135 W.

Next, the image sensor is replaced with one capable of high-speed continuous image capture without a change in system function. To hold the same system function, this image sensor captures images at a readout frame rate of 1 fps and a capture interval of 100 μs. On the basis of the result under the third condition in Table 7.6, the power consumption by the image sensor is estimated to be 7.9 μW $[= 9.2\,μW \times (128 \times 128) / (240 \times 80)]$. Supposing that the DSP consumes the same power as the above (1.02 W), the total power consumption is estimated to be 1.02 W. Thus, if the image sensor in the conventional system is replaced with the high-speed continuous-capture one, the power consumption of the entire system can be reduced by 10.1% $[= (1.135\,W - 1.02\,W) / 1.135\,W]$. Assuming that the power consumption by DSP will decrease in the future, and focusing only on the image sensor, the power consumed by the proposed image sensor decreases significantly from the conventional image sensor by 99.99% $[= (115\,mW - 7.9\,μW) / 115\,mW]$. Therefore, use of the high-speed continuous-capture image sensor in the optical flow system should effectively reduce the power consumption in future optical flow systems.

7.4   Motion Sensor

7.4.1   Overview

As described in Sections 7.2 and 7.3, the CAAC-IGZO image sensor is characterized by a storage node using a capacitor and a CAAC-IGZO FET with an extremely low off-state current. Not only can it be used for sampling images, the circuit can also be configured into an analog arithmetic circuit – i.e., a *functional image sensor* with image pre-processing that can unload the image signal processor (ISP) from such tasks.

A typical application of functional image sensors is motion sensing. Depending on the configuration, it can be applied to surveillance, security, and product inspection, and other intermittent machine vision tasks. In addition, some autonomously powered sensor network for Internet of Things (IoT) applications require microwatt-order power. Some prior studies on low-power motion sensors have proposed configurations that can detect motion by processing captured data without using external devices [8,22–24]. These are roughly classified into an event-driven type [22] and a frame-based type [23]. The event-driven type has low power consumption but needs a complicated sensor pixel configuration, making normal imaging difficult. The frame-based type can perform normal imaging but can detect motion only from a preceding frame (reference frame), and therefore considerable computing power is necessary to analyze the image differences, unless special hardware is employed. Recently, a compromise configuration of both types has also been proposed by implementing low-power motion-detection algorithms [24].

This section introduces a CAAC-IGZO image sensor with differential circuits in the sensor pixels as an example of functional image sensors. The sensor can hold the captured data of a reference frame in a storage node for a long period and generate differential data between the reference and current frames using an analog differential circuit in each sensor pixel. In addition to normal imaging, it can determine the difference between frames using a simple analog processor without an ADC. It is therefore expected to become an efficient event-driven motion sensor while maintaining the advantages of a frame-based motion sensor.

7.4.2   Configuration

7.4.2.1   System Architecture

The image sensor utilizes the good charge-retention characteristics of CAAC-IGZO FETs to achieve an extremely low-power motion sensor. Figure 7.31 shows a block diagram of such a sensor and its operating modes. By switching driving modes, the sensor pixel can conduct either normal imaging or differential imaging, the latter of which captures current frame data and calculates the differential data between the reference and current frame data through a differential circuit implemented by CAAC-IGZO FETs. Thanks to the good charge-retention characteristics, the interval between the reference and current frames can be very long, and hence detect motion in a wide range of speeds, including even a minute motion over a long period. This can be done row by row at extremely low power consumption.

The motion sensor has a motion-capture mode, a wait mode, and an imaging mode. First, in the motion-capture mode, the sensor sequentially checks the differential data of the sensor pixels in each row without turning on the column driver or ADC. Upon detection of changes

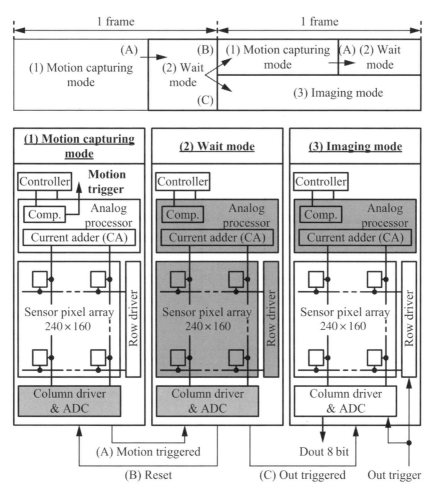

Figure 7.31 Block diagram of the motion sensor and operational states of each block at each operation mode. Gray indicates a deactivated (wait) mode. *Source*: Adapted from [8]

in the sensor pixel differential data in a particular row, the sensor stops checking the differential data for the rest of the rows. Specifically, the controller stops the analog processor and the row driver, and the sensor is switched to a wait mode automatically. Thus, the system can save the power used for the data readout of the rest of the rows. In the wait mode, all circuit blocks except the controller are deactivated. After the target frame period where motion is detected, the motion sensor is switched back to the motion-capture mode automatically, or to an imaging mode by an out-trigger from external devices. Finally, as an operation following the target frame motion capture, the motion sensor obtains the captured data in the current frame or the differential data showing a difference between the reference and current frames and then outputs the obtained data in the imaging mode.

7.4.2.2   Sensor Pixel

As described above, this motion sensor requires the function of retaining captured data from a reference frame and acquiring differential data between the target and reference frames. A sensor pixel configuration to provide this function is illustrated in Figure 7.32. Figure 7.32 shows the circuit diagram of the test sensor pixel (a), the timing diagram of differential data capture (b), and of normal image capture (c). Note that the sensor pixel circuit

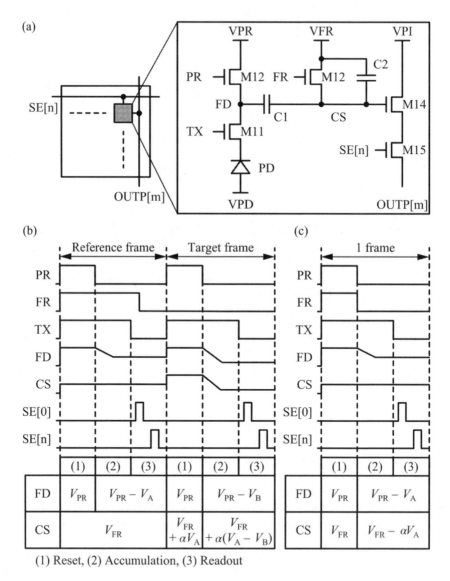

(1) Reset, (2) Accumulation, (3) Readout

Figure 7.32   Sensor pixel circuit diagram (a), timing diagram during differential data capture (b), and timing diagram during normal image capture (c). *Source*: Adapted from [8]

can be fabricated entirely by CAAC-IGZO FETs, thereby increasing the effective exposure area of the photodiode. Moreover, the extremely low off-state current of CAAC-IGZO FETs allows FD and CS nodes in the sensor pixel circuit to be non-volatile storage nodes. In other words, they can retain the charges corresponding to the captured data for a long period. The differential circuit consists of CAAC-IGZO FETs and a capacitor C1, whose electrodes are connected to FD and CS nodes.

Now, the differential data capture in the sensor pixels from Figure 7.32(b) is explained, starting with the reference frame operation. In the reset period, the electric potential of the CS node is set to the reset potential V_{FR}, and the electric potential of the FD node is set to the reset potential V_{PR}. In a subsequent accumulation period, while the CS node potential is kept at V_{FR}, an accumulation operation sets the FD node potential to the potential corresponding to the reference frame captured data $(V_{PR} - V_A)$, where V_A is the potential corresponding to the pixel value of the reference frame pixel. In the following readout period, the outputs that correspond to the reset potential of the CS node (V_{FR}) (here, the outputs are equivalent to the differential data of zero) are sequentially read out.

Next, the target frame operations are explained. In the reset period (1), the FD node potential is set to the reset potential V_{PR}. Then, the FD node potential increases by V_A, from $V_{PR} - V_A$ to V_{PR}. The capacitive coupling with capacitor C1 sets the CS node potential to $V_{FR} + \alpha V_A$, increased from V_{FR} by αV_A, where α is a constant determined by the capacitors C1 and C2, and the gate capacitance of the transistor, M14. In the accumulation period (2), the accumulation operation decreases the FD node potential from V_{PR} to the potential corresponding to the target frame captured data $(V_{PR} - V_B)$. Then, the capacitive coupling with capacitor C1 decreases the CS node potential from $V_{FR} + \alpha V_A$ to $V_{FR} + \alpha(V_A - V_B)$, which is the potential corresponding to the differential data between the reference and target frames. In the readout period (3), the outputs (equivalent to the differential data) corresponding to the CS node potential $V_{FR} + \alpha(V_A - V_B)$ are read out sequentially.

In the above, the potential of the CS and FD nodes can be maintained for a long period by setting the CAAC-IGZO FETs M11, M12, and M13 off. In other words, the sensor pixels can retain the reference frame data for a long time, and also hold the differential data between reference and target frames acquired by the differential circuit.

As shown in Figure 7.32(c), normal image capture of the current frame is achieved by setting the CS node potential to a given value (e.g., V_{FR}) and controlling FR synchronized with PR. Thus, sensor pixels can perform normal image capture as well as differential data capture.

7.4.2.3   Analog Processor

This motion sensor aims to offer low power consumption by specializing in the configuration that generates a trigger only when the differential data exceed a threshold value. Figure 7.33 shows the analog processor designed to detect whether there is a certain difference beyond a threshold value; here, this analog processor itself does not obtain the value of difference data. The drawing shows the circuit diagram and timing diagram of a motion-capture period. In this analog processor, the current when each sensor pixel has a differential data of zero is set as the reference value. The absolute values of the current corresponding to the differential data are added to the reference value, and the total value is checked. Adding the differential data as an analog value (current) without A/D conversion can simplify the analog processor.

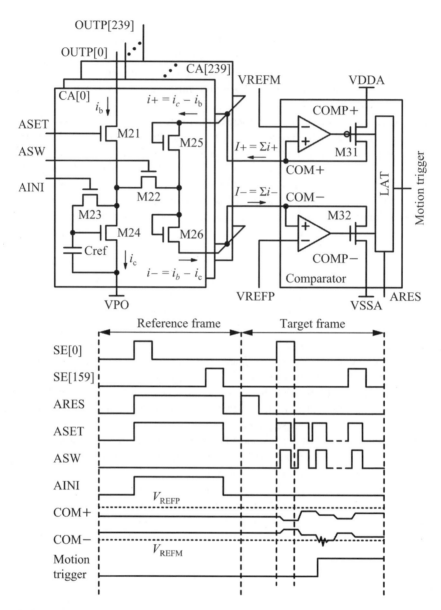

Figure 7.33 Circuit diagram of analog processor and timing diagram of event-triggered motion-capture period. *Source*: Adapted from [8]

The operations of the analog processor are as follows. First, in the readout period of the reference frame, each column capacitor Cref in a current adder (CA) is set to the potential corresponding to the current of each sensor pixel (i_c). The current i_c corresponds to the current with the CS potential in each sensor pixel being V_{FR} – i.e., the current with the difference being zero; therefore, the output from the sensor pixels connected to each row has a constant current value. The potential set to Cref at each column can be retained for a long period by switching off the

CAAC-IGZO FET M23 of the column. The transistor M24 of each column acts as a constant current source (current i_c) and enables column-variation compensation.

Next, in the target frame readout period, each column sensor pixel of the selected row outputs the current (i_b) following the CS potential $[V_{FR} + \alpha(V_A - V_B)]$ to an output line of each row. Here, the differential current $i+$ ($=i_c - i_b$ when $i_c > i_b$) and $i-$ ($=i_b - i_c$ when $i_c < i_b$) flows through M25 and M26, respectively. The COM+ and COM− node potentials vary depending on the total differential currents $I+$ and $I-$ in all columns. The two comparators, COMP+ and COMP−, compare the potentials of the COM+ and COM− nodes with a reference voltage, so that the analog processor can determine whether any image change has occurred (i.e., whether motion has been detected).

The reference voltages V_{REFM} of the comparator COMP+ and V_{REFP} of COMP− are set to the lower and upper limits of the voltage errors including sensor pixel variations and noises. The above operation allows motion capture with a given reference frame. Outputting a motion trigger lets the controller deactivate the comparator bias and the constant current source (ASET) in each column, and the sensor is automatically switched to a wait mode, thus reducing the power consumption as much as possible. In each frame, the analog processor checks the changes in the differential data of the sensor pixels sequentially selected row by row. Thus, from the output timing of the detection trigger in a frame, the row address in which motion is detected can be determined. The detection sensitivity of the analog processor is ensured by using a constant current source, which compensates for variation in the columns.

7.4.3 Prototype

7.4.3.1 Configuration

A prototype motion sensor described in this section was fabricated using a hybrid process of 0.5-μm CAAC-IGZO and 0.18-μm Si FETs [8]. The micrograph is shown in Figure 7.34.

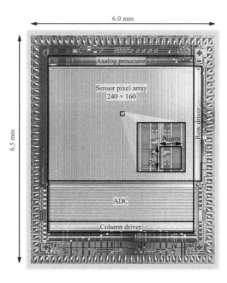

Figure 7.34 Micrograph of motion sensor. *(For color detail, please see color plate section)*

7.4.3.2   Measurement

The upper row of Figure 7.35 shows the target images with letters to be captured that are switched over time. The upper left image is captured as the reference frame. The lower row presents differential images, each of which shows different points between the reference frame and the target frame images. The differential images show that the differential data between the current frame and the reference frame can be obtained even after 60 s.

The retention characteristics of the obtained uniform, monochrome image are measured, and it was found that the pixel value of the image changes by approximately 0.39 LSB (corresponding to 2.6 mV of the analog potential at node CS) on average in 60 s. From the result, the time taken for the data to change by 1 LSB is estimated to be approximately 153.85 s.

To confirm that motion capture can be accurately conducted, a DC fan that rotates at 3150 rpm (315° per frame at 60 fps) is captured. Figure 7.36 shows the images captured frame

	Reference frame	Target frame	
		after 0 s	after 60 s
Target image	**MOTION** SENSOR	**MOTION** **After 0s**	**MOTION** **After 60s**
Captured image (Differential)		SENSOR **After 0s**	SENSOR **After 60s**

Figure 7.35   Target images to be captured and displays showing differential data. *Source*: Adapted from [8]

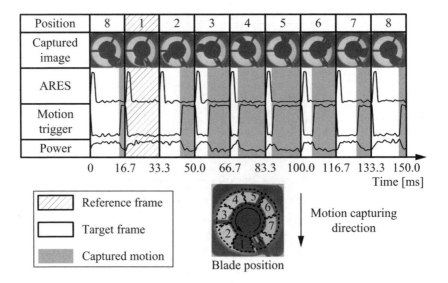

Figure 7.36   Motion-capture result (gray indicates a wait state). *Source*: Adapted from [8]

Table 7.8   Specifications of motion sensor. *Source*: Reprinted from [8], with permission of IEEE © 2015

Process	0.5-μm CAAC-IGZO/0.18-μm Si		
Die size	6.5 mm × 6.0 mm		
Number of sensor pixels	240 × 160		
Sensor pixel size	20 μm × 20 μm		
Sensor pixel configuration	5 transistors, 2 capacitors		
Fill factor	27.5%		
FPN	2.4% contrast		
Supply voltage	Nominal	1.8 V	
	Sensor pixel, row/column driver, ADC, analog processor	3.3 V	
ADC	8-bit single slope		
Power consumption/mode	Motion capture	Imaging	Wait
Power (at 60 fps)	25.3 μW	3.6 mW	1.88 μW
Figure of merit (FOM) (power/pixel·fps)	10.98 pJ	1.56 nJ	0.82 pJ

by frame of the fan and the measured waveforms of the reset signal ARES (the start of each frame), the motion trigger, and the power consumption in readout operation. The image of the fan with a blade at position "1" is used as a reference frame. Motion is detected and captured when the blade is at a different position. As can be seen, misdetection does not occur even 60 s after the blade is fixed at the position of the reference frame. The sensor can accurately detect that there is no motion even when the target is resting over as long as 60 s, because the captured data of the reference frame can be accurately retained. This means that the sensor can even detect motion at one sensor pixel in 60 s. Unlike a frame-based motion sensor, which has difficulty in capturing such slow motion, the present motion sensor can capture a slowly changing target.

The specifications of the prototype motion sensor are summarized in Table 7.8. Fixed-pattern noise (FPN) can be reduced by adding a compensation circuit such as correlated double sampling (CDS). This motion sensor consumes 25.3 μW at 60 fps in the motion-capture mode and 1.88 μW at 60 fps in the wait mode. The power consumption in the motion-capture and wait modes is 1/140 and 1/2000 of the power consumption of the display mode of 3.6 mW at 60 fps, respectively. As described in Table 7.9, the motion sensor can detect motion with the lowest energy among similar sensors.

7.4.4   Sensor Pixel Threshold-Compensation Function

7.4.4.1   Introduction

Previous subsections introduced the concept and measurement results of the motion sensor using the extremely low off-state current of CAAC-IGZO FETs. Motion sensor applications, such as surveillance and product-inspection cameras, require capturing of minute motion and differences. This subsection presents a method of increasing the motion-capture sensitivity, allowing motion capture even under low contrast.

Table 7.9 Comparison of figure of merit. *Source*: Adapted from [8, 25]

	Digital architecture		Analog architecture	
	[26]	[22]	[23]	This motion sensor
Process	90 nm	0.35 μm	0.5 μm	0.5 μm (CAAC-IGZO) 0.18 μm (Si)
Sensor pixel array	8K × 2K	128 × 128	90 × 90	240 × 160
ADC	12-bit	15-bit	6-bit	8-bit
Sampling rate	120 fps	free	30 fps	60 fps
FOM Motion cap.	1.41 nJ + α	N/A	17.28 nJ	10.98 pJ
Wait	non	non	non	0.82 pJ
Imaging	1.41 nJ	N/A	17.28 nJ	1.56 nJ

As described above, the non-volatile analog memory in the sensor pixels retains the captured data from the reference frame, and the calculated difference between the reference frame data and the current frame data is generated. Motion capture is performed by comparing the differential data with certain reference levels of the analog processor. Motion-capture sensitivity depends strongly on the range of reference levels and requires narrowing of the reference level within a range that does not cause misdetection. Therefore, to increase the sensitivity, it is indispensable to suppress the variations among the sensor pixels, in other words, to improve the uniformity of the sensor pixels.

Although a column CDS is generally used to suppress the variation (in particular, the sensor pixel bias voltage), a normal column CDS cannot be used to suppress sensor pixel variations in this motion sensor because the sensor pixels of the system retain the reference frame image data. Hence, a method of increasing the uniformity among sensor pixels by mounting a threshold-compensation circuit against amplifier transistors should be employed, whereby the motion-capture sensitivity is improved [9].

7.4.4.2 Sensor Pixel

The configuration of a sensor pixel circuit with a threshold-compensation function is shown in Figure 7.37(a). In this circuit, 2Tr2C (gray area) is added to the sensor pixel circuit of the motion sensor described in the previous subsections, providing the threshold-compensation circuit.

The timing diagram in Figure 7.37(b) shows the threshold-compensation operation of the sensor pixel circuit. First, in period T1, the potential of capacitor C4 is initialized by switching on the transistors M13, M15, M16, and M17, and setting VPO to V_{POH} and the current source bias (BR) to V_{BRH}. V_{POH} is the sensor pixel output for zero differential data, in other words, the reference output level. At this time, the node AS potential V_{AS} is smaller than $V_{AG} - V_{th}$, where V_{AG} is the potential of node AG and V_{th} is the threshold voltage of M14. Subsequently, in period T2, to retain the charge corresponding to the M14 threshold voltage V_{th} in C4, M16 is switched off and the node AG discharges to the initial potential, $V_{AG0}(= V_{POH} + V_{th})$. The node CS potential is the initial value V_{CS0}. In period T3, switching off M17 sets the node AG in a floating state, and the potential corresponding to the M14 threshold voltage V_{th} is stored in capacitor C4. Finally, in period T4, the compensation operation is completed by switching

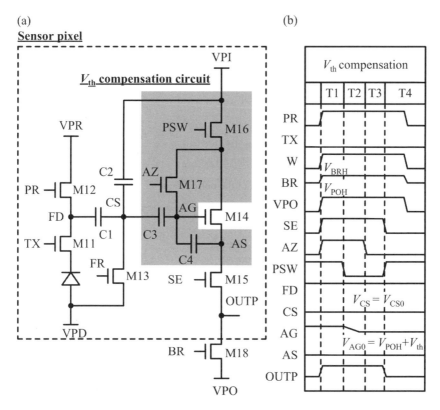

Figure 7.37 (a) Sensor pixel circuit with threshold-compensation circuit and (b) timing diagram. *Source*: Adapted from [9]. Copyright 2015 The Japan Society of Applied Physics

M16 on and M13 off, and by setting BR and VPO to low potential (V_{BRL} and V_{POL}, respectively).

The operations above allow threshold compensation as follows: assuming that the transistor M14 is a source follower, the output voltage of the sensor pixel is $V_{AS} = V_{AG} - V_{th}$. If the CS potential is V_{CS}, the AG potential is expressed as $V_{AG} = V_{AG0} + \beta(V_{CS} - V_{CS0})$, where β is a coupling coefficient. Thus, the output potential of the sensor pixel is given by $V_{AS} = V_{AG0} - V_{th} + \beta(V_{CS} - V_{CS0}) = V_{POH} + \beta(V_{CS} - V_{CS0})$, whereby the value that ignores the threshold voltage can be read out. This value is compared with the reference output level V_{POH}, and the difference can be output when the CS potential differs from the initial value. Note that CAAC-IGZO FETs in the sensor pixels can retain their threshold data for a long time, leaving threshold-compensation operations per frame unnecessary.

7.4.4.3   Prototype

On the basis of the sensor pixels in Figure 7.37(a), a prototype motion sensor is fabricated using the hybrid process of 0.5-µm CAAC-IGZO and 0.18-µm Si FETs and examined in terms of the effect of the sensor pixel threshold compensation [9]. The specifications of the motion sensor are shown in Table 7.10.

Table 7.10 Specifications of motion sensor. *Source*: Adapted from [9]. Copyright 2015 The Japan Society of Applied Physics

	w/o V_{th} compensation circuit	w/ V_{th} compensation circuit
Process	0.5-μm CAAC-IGZO/0.18-μm Si	
Die size	6.5 mm × 6.0 mm	
Number of sensor pixels	240 × 160	
Sensor pixel size	20 μm × 20 μm	
Sensor pixel configuration	5 transistors, 2 capacitors	7 transistors, 4 capacitors
Fill factor	29.75%	28.25%
FPN	1.73%	1.45%
Detection threshold voltage	200 mV at 120 of the output value	97.5 mV at 120 of the output value
Supply voltage	Nominal	1.8 V
	Sensor pixel, row/column driver, ADC, analog processor	3.3 V
ADC	8-bit single slope	

The sensor pixel uniformity is measured and summarized below. Figure 7.38(a) shows the histograms of all sensor pixel outputs for motion sensors with and without a compensation circuit. The uniformity is measured such that the transistor M13 of Figure 7.37(a) is switched on, the CS potential is set to reset voltage V_{PD}, and the analog output of all sensor pixels, OUTP, is converted into a digital value and read out. Changes in reset voltage V_{PD} cause variations in the sensor output. The sensor output was digitally converted by an 8-bit ADC, and thus varies from 0 to 255.

The standard deviation among the sensor pixel outputs is calculated from Figure 7.38(a) and shows that the variation among sensor pixels decreases – e.g., by approximately 8.3 at an output of 120. Thus, it can be seen that the motion sensor with a compensation circuit improves the uniformity at each output. Although the output-gain variation in the motion sensor with a compensation circuit is 37.8% larger than that of the motion sensor without one, the output-offset variation in the former is 22.2% smaller than that in the latter. The increased gain variation is considered to be caused by the increase in the number of transistors in the sensor pixels and by capacitance variation; nevertheless, the overall variation is suppressed. FPN, calculated from the variation among the sensor pixels, also decreases by 16.2% because of the compensation circuit.

The detection accuracy of the analog processor in a prototype with a threshold-compensation circuit is compared with the accuracy of that without one. Figure 7.38(b) represents ΔV_{REF} of $V_{REFP} - V_{REFM}$, the smallest reference voltage range to keep the motion trigger (the output signal of the analog processor) in a non-active state. In each circuit, ΔV_{REF} is calculated in the following manner: the transistor M13 is switched on, the CS potential is set to the reset voltage V_{PD}, and the sensor pixel output OUTP is set to a certain output level; this output level is sent to the analog processor, and then ΔV_{REF} is calculated. The smaller ΔV_{REF}, the more precisely the sensor can detect even small changes and reduce misdetection. Figure 7.38(b) shows that the threshold-compensation circuit reduced ΔV_{REF} from 200 mV to 97.5 mV at an output of 120. In other words, the detectable reference voltage range decreased to 1/2.05, which means that the circuit with threshold compensation has a detection threshold contrast 2.05 times higher than one without threshold compensation.

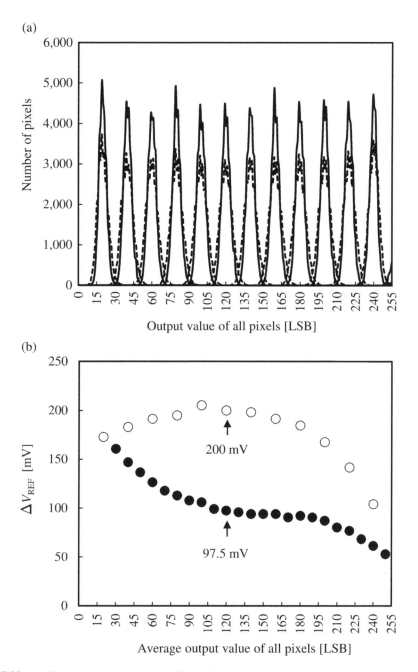

Figure 7.38 (a) Output variations (dashed line: without V_{th} compensation circuit, solid line: with V_{th} compensation circuit) and (b) threshold-compensation effect of motion-capture range ΔV_{REF} (unfilled circles: without V_{th} compensation circuit, filled circles: with V_{th} compensation circuit). *Source*: Adapted from [9]. Copyright 2015 The Japan Society of Applied Physics

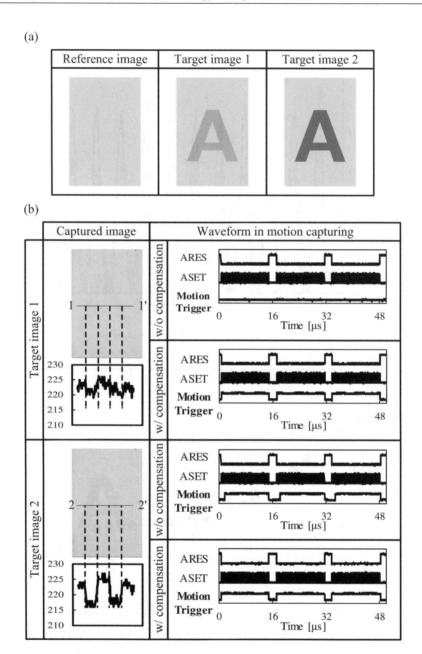

Figure 7.39 Comparison of motion-capture results with or without threshold compensation. *Source*: Adapted from [9]

An actual motion-capture operation is measured to confirm that the above-mentioned threshold-compensation circuit improves the sensor pixel uniformity and detection accuracy. A reference frame image of a transparent film and target frame images of several films on which a letter "A" was printed at different optical densities are shown in Figure 7.39(a).

Figure 7.39(b) illustrates the images of the letter "A" with various contrasts captured by the motion sensor and the waveforms during motion capture (motion-capture mode and wait mode). The upper row shows the motion-capture result with a low-contrast letter (1) and the lower row shows the result with a high-contrast letter (2). Each row shows the motion-capture difference between the sensors with and without a threshold-compensation circuit. The leftmost columns show the normally captured images of letter "A" with different optical densities; however, since it is difficult to see the density difference on paper, the pixel value distribution on lines 1–1' and 2–2' is shown in Figure 7.39(b). The motion sensor without a compensation circuit detects the difference for a dark printed "A" and generates a motion trigger, but cannot detect the difference for light "A." In contrast, the motion sensor with a compensation circuit detects the differences in both cases and generates motion triggers. Therefore, the threshold compensation operation improves the motion-capture sensitivity, allowing motion detection at low contrast.

The variation among the sensor pixels is approximately 8% one hour after the threshold-compensation operation. Therefore, the effect of threshold correction can be retained by threshold compensation with a frequency as low as once in several hours, because CAAC-IGZO FETs can maintain the potential of capacitor C4.

References

[1] Fossum, E. R. (1993) "Active pixel sensors: Are CCDs dinosaurs?," *Proc. SPIE*, 1900, 2.

[2] Aw, C. H. and Wooley, B. A. (1996) "A 128 × 128-pixel standard-CMOS image sensor with electronic shutter," *IEEE J. Solid-State Circuits*, **31**, 1922.

[3] Aoki, T., Ikeda, M., Kozuma, M., Tamura, H., Kurokawa, Y., Ikeda, T., *et al.* (2011) "Electronic global shutter CMOS image sensor using oxide semiconductor FET with extremely low off-state current," *IEEE Symp. VLSI Circuits Dig. Tech. Pap.*, 174.

[4] Boyle, W. S. and Smith, G. E. (1970) "Charge coupled semiconductor devices," *Bell Syst. Tech. J.*, **49**, 587.

[5] Geurts, T., Cools, T., Esquenet, C., Sankhe, R., Prathipati, A., Syam, M. R. E., *et al.* (2015) "A 25 Mpixel, 80 fps, CMOS imager with an in-pixel-CDS global shutter pixel," *International Image Sensor Workshop*, Vaals, The Netherlands.

[6] Yoneda, S., Okamoto, Y., Nakagawa, T., Maeda, S., Aoki, T., Kozuma, M., *et al.* (2014) "300 μs Short interval continuous capturing image sensor with *c*-axis-aligned crystalline oxide semiconductor FET/p-channel silicon FET stacked CMOS structure," *Ext. Abstr. Solid State Dev. Mater.*, 988.

[7] Lucas, B. D. and Kanade, T. (1981) "An iterative image registration technique with an application to stereo vision," *Proc. DARPA Image Understanding Workshop*, 121.

[8] Ohmaru, T., Nakagawa, T., Maeda, S., Okamoto, Y., Kozuma, M., Yoneda, S., *et al.* (2015) "25.3 μW at 60 fps 240 × 160-Pixel vision sensor for motion capturing with in-pixel non-volatile analog memory using crystalline oxide semiconductor FET," *IEEE Int. Solid-State Circuits Conf. Dig. Tech. Pap.*, 118.

[9] Maeda, S., Ohmaru, T., Inoue, H., Nakagawa, T., Kurokawa, Y., Ikeda, T., *et al.* (2015) "Low contrast motion capturing vision sensor with crystalline oxide semiconductor FET-based in-pixel threshold voltage compensation circuit," *Ext. Abstr. Solid State Dev. Mater.*, 810.

[10] Dickinson, A., Ackland, B., Eid, E.-S., Inglis, D., and Fossum, E. R. (1995) "A 256 × 256 CMOS active pixel image sensor with motion detection," *IEEE Int. Solid-State Circuits Conf. Dig. Tech. Pap.*, 226.

[11] Nitta, Y., Muramatsu, Y., Amano, K., Toyama, T., Yamamoto, J., Mishina, K., *et al.* (2006) "High-speed digital double sampling with analog CDS on column parallel ADC architecture for low-noise active pixel sensor," *IEEE Int. Solid-State Circuits Conf. Dig. Tech. Pap.*, 500.

[12] Yasutomi, K., Itoh, S., and Kawahito, S. (2010) "A 2.7e- temporal noise 99.7% shutter efficiency 92 dB dynamic range CMOS image sensor with dual global shutter pixels," *IEEE Int. Solid-State Circuits Conf. Dig. Tech. Pap.*, 398.

[13] Funaki, M., Shimizu, T., Orihara, S., Kawanaka, H., Kurihara, M., Sato, H., *et al.* (2008) "New global shutter CMOS imager with 2 transistors per pixel," *IEEE Symp. VLSI Circuits Dig. Tech. Pap.*, 197.

[14] Fowler, B., Gamal, A. E., and Yang, D. X. D. (1994) "A CMOS area image sensor with pixel-level A/D conversion," *IEEE Int. Solid-State Circuits Conf. Dig. Tech. Pap.*, 226.

[15] Furuta, M., Nishikawa, Y., Inoue, T., and Kawahito, S. (2007) "A high-speed, high-sensitivity digital CMOS image sensor with a global shutter and 12-bit column-parallel cyclic A/D converters," *IEEE J. Solid-State Circuits*, **42**, 766.

[16] Okura, S., Nishikido, O., Sadanaga, Y., Kosaka, Y., Araki, N., Ueda, K., *et al.* (2014) "A 3.7 M-pixel 1300-fps CMOS image sensor with 5.0 G-pixel/s high-speed readout circuit," *IEEE Symp. VLSI Circuits Dig. Tech. Pap.*, 149.

[17] Tochigi, Y., Hanzawa, K., Kato, Y., Kuroda, R., Mutoh, H., Hirose, R., *et al.* (2013) "A global-shutter CMOS image sensor with readout speed of 1-Tpixel/s burst and 780-Mpixel/s continuous," *IEEE J. Solid-State Circuits*, **48**, 329.

[18] Etoh, T. G., Poggemann, D., Kreider, G., Mutoh, H., Theuwissen, A. J. P., Ruckelshausen, A., *et al.* (2003) "An image sensor which captures 100 consecutive frames at 1000000 frames/s," *IEEE Trans. Electron Devices*, **50**, 144.

[19] Etoh, T. G., Son, D. V. T., Yamada, T., and Charbon, E. (2013) "Toward one giga frames per second – evolution of *in situ* storage image sensors," *Sensors*, **13**, 4640.

[20] Ishii, I., Taniguchi, T., Yamamoto, K., and Takaki, T. (2012) "High-frame-rate optical flow system," *IEEE Trans. Circuits Syst. Video Technol.*, **22**, 105.

[21] Igual, F. D., Botella, G., García, C., Prieto, M., and Tirado, F. (2013) "Robust motion estimation on a low-power multi-core DSP," *EURASIP J. Appl. Signal Process.*, 2013, 99.

[22] Lichtsteiner, P., Posch, C., and Delbruck, T. (2008) "A 128 × 128 120 dB 15 μs latency asynchronous temporal contrast vision sensor," *IEEE J. Solid-State Circuits*, **43**, 566.

[23] Chi, Y. M., Mallik, U., Clapp, M. A., Choi, E., Cauwenberghs, G., and Etienne-Chummings, R. (2007) "CMOS camera with in-pixel temporal change detection and ADC," *IEEE J. Solid-State Circuits*, **42**, 2187.

[24] Park, S., Cho, J., Lee, K., and Yoon, E. (2014) "243.3 pJ/pixel Bio-inspired time-stamp-based 2D optic flow sensor for artificial compound eyes," *IEEE Int. Solid-State Circuits Conf. Dig. Tech. Pap.*, 126.

[25] Ohmaru, T., Nakagawa, T., Maeda, S., Okamoto, Y., Kozuma, M., Yoneda, S., *et al.* (2015) "25.3 μW at 60 fps 240 × 160 pixel vision sensor for motion capturing with in-pixel non-volatility analog memory using crystalline oxide semiconductor FET," *IEEE International Solid-State Circuits Conference*, San Francisco, CA.

[26] Toyama, T., Mishina, K., Tsuchiya, H., Ichikawa, T., Iwaki, H., Gendai, Y., *et al.* (2011) "A 17.7 Mpixel 120 fps CMOS image sensor with 34.8 Gb/s readout," *IEEE Int. Solid-State Circuits Conf. Dig. Tech. Pap.*, 420.

8

Future Applications/Developments

8.1 Introduction

In previous chapters we described typical LSI applications configured by c-axis-aligned crystalline indium–gallium–zinc oxide (CAAC-IGZO) field-effect transistors (FETs), such as memory devices, CPUs, FPGAs, and image sensors. These LSIs exhibit more appealing characteristics than LSIs based solely on Si FETs. LSIs based on CAAC-IGZOs will be developed further in a wide range of fields, as covered in this chapter.

Section 8.2 introduces radio frequency (RF) devices employing CAAC-IGZO LSIs. Low power consumption is advantageous for RF devices, which must operate under restricted power supply. In addition, the high-temperature properties of CAAC-IGZO FETs ensure their operation in demanding environments. For both of these reasons, RF devices using CAAC-IGZO FETs are promising.

Section 8.3 describes an application of additional image sensors, namely X-ray detectors. The applicable range of image sensors can be extended by replacing the visible-light sensor element (Si photodiode) with a sensor that detects invisible light (e.g., infrared, ultraviolet, and X-rays). The CAAC-IGZO X-ray detectors are expected to be exploited increasingly in the medical and industrial fields.

Section 8.4 introduces an image coder–decoder (CODEC), an application that exploits various CAAC-IGZO LSIs, including FPGAs. Advanced encoding of video signals is required for broadcasting digital television (TV), and CODECs are therefore of major importance. Because of the reconfigurability and intermittent nature of the video streams, CAAC-IGZO FETs are well positioned to be deployed in such applications.

As mentioned in previous chapters, CAAC-IGZO LSIs can be utilized in memory devices with superior charge-retention characteristics. This characteristic enables an ideal analog memory that can retain both analog and digital data. DC–DC converters, analog programmable

Physics and Technology of Crystalline Oxide Semiconductor CAAC-IGZO: Application to LSI, First Edition.
Edited by Shunpei Yamazaki and Masahiro Fujita.
© 2017 John Wiley & Sons, Ltd. Published 2017 by John Wiley & Sons, Ltd.

devices, and neural networks are presented as analog memory applications in Sections 8.5, 8.6, and 8.7, respectively. These possible analog circuit applications may to some extent reverse the current transition from analog to digital circuits, especially since CAAC-IGZO FETs can be monolithically integrated with conventional high-speed Si circuitry. Further, Section 8.8 explains memory-based computing, and Section 8.9 outlines backtracking programs with power gating. The authors hope that the reader will appreciate the potential of new LSI applications made possible with CAAC-IGZO.

8.2 RF Devices

8.2.1 Overview

A typical RF device is a wireless IC tag, which wirelessly identifies an object without physical contact. RF technology is emerging in various fields, such as logistics, product management, electronic money management, and security systems [1].

Wireless IC tags are roughly classified as passive or active. A passive wireless tag requires no battery and generates power by harvesting the energy of the radio-frequency (RF) field from a reader–writer (RW). The RF signal is demodulated and processed and then modulated again for communication back to the RW. The RW provides the power needed to operate the passive wireless IC tag. Accordingly, to improve the convenience (e.g., increase the communication distance) of a passive RF device, low power consumption is required. A combination of a wireless interface and a CAAC-IGZO LSI with low power consumption should therefore provide an attractive solution for a passive wireless IC tag.

In contrast, active wireless IC tags are battery-driven devices. Using the power from a battery, the active wireless tag demodulates a wireless signal from the RW, executes signal processing, and then modulates the wireless signal or generates another signal for communication. Thus, the limitation on the power consumption is less severe than in a passive wireless tag. However, an active wireless IC tag cannot have a large-capacity battery because of its required shape and size (usually the form factor of a credit card). To improve the convenience of the tag (e.g., reduce the need for battery replacement), the power consumption must be reduced similarly to that of a passive wireless tag. Therefore, CAAC-IGZO LSIs offer advantages for active IC tags also.

As a specific example of a CAAC-IGZO LSI application, this section introduces the passive wireless IC tag equipped with non-volatile oxide semiconductor random access memory (NOSRAM) – i.e., a NOSRAM passive wireless IC tag. This can be employed (for example) in medical instrument management. By adding a sensor function, the wireless IC tag becomes applicable to monitoring and management of structures such as tunnels and bridges.

8.2.2 NOSRAM Wireless IC Tag

8.2.2.1 Overview

First, the structure of a NOSRAM passive wireless IC tag is described. The CAAC-IGZO FET used in the NOSRAM has a good low off-state current even at elevated temperatures. Furthermore, it enables writing data at high speed and low power consumption, as well as long-term data retention.

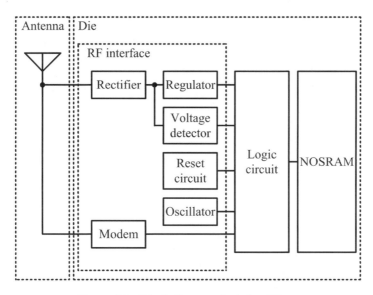

Figure 8.1 Block diagram of wireless IC tag

The structure of a typical NOSRAM-equipped wireless IC tag is shown in Figure 8.1. The rectifier rectifies the wireless signal transmitted from the RW and received by the antenna, generating a direct-current (DC) voltage. The DC voltage is then adjusted to the appropriate power supply voltage by the regulator. The modem demodulates the wireless signal, generating a command based on the signal transmitted from the RW (received signal). In response to the data to be transmitted to the RW (transmitted signal), it then modulates the electromagnetic (EM) wave. When the RW receives the EM wave, it establishes communication with the wireless IC tag. If the generated DC voltage is too low, the voltage detector outputs a signal that limits access of the logic circuit to the NOSRAM. This scheme prevents incorrect reading and writing owing to RF signal strength (i.e., insufficient power supply voltage). The reset circuit and the oscillator generate a reset signal and a clock signal, respectively, for the logic circuit. The logic circuit is based on a wireless communication standard such as ISO/IEC18000-63 [2]. To fully utilize the features of the NOSRAM, extended commands must be defined as described below.

8.2.2.2 Fabrication Example

A NOSRAM wireless IC tag has been fabricated to verify its characteristics performance experimentally. A photograph and specifications of the die of a wireless IC tag are shown in Figure 8.2 and Table 8.1, respectively. The 5.0 mm × 5.0 mm wireless IC tag was fabricated by a hybrid process using 0.8-μm CAAC-IGZO and 0.35-μm Si FETs. The carrier frequency is 920 MHz and the communication protocol is based on the ISO/IEC18000-63 standard. Access to the NOSRAM is accomplished by executing mandatory read and write commands according to the standard.

The operation and reliability of the NOSRAM wireless IC tag have been evaluated by a data-retention test. The test involved data writing to the wireless tag, data retention, and data reading.

(a)

(b)

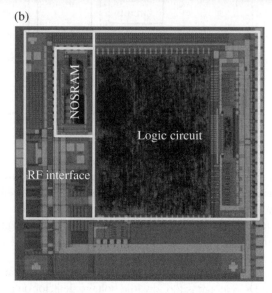

Figure 8.2 Photograph of NOSRAM wireless IC tag: (a) whole area and (b) die

Table 8.1 Specifications of NOSRAM wireless IC tag

Tag	Carrier frequency		920 MHz
	Protocol		ISO/IEC18000-63
	Die size		5.0×5.0 mm^2
NOSRAM	Technology	CAAC-IGZO FET	0.8 μm
		Si FET	0.35 μm
	Voltage	CAAC-IGZO FET	3.3 V
		Si FET	1.8 V/1.2 V
	Module area		1.1×0.5 mm^2
	Cell area		8.0×8.2 μm^2
	Capacitance		20.6 fF
	Number of bits		1024 bits
Logic circuit	Technology	Si FET	0.35 μm
	Voltage		1.2 V
	Area		3.4×3.15 mm^2

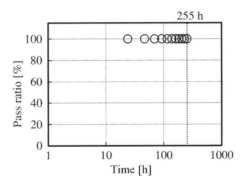

Figure 8.3 Data-retention characteristics of wireless IC tag at 130°C

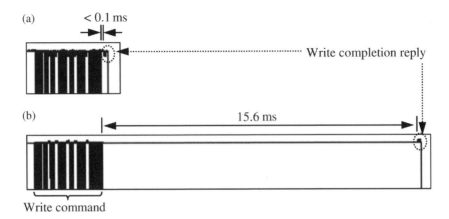

Figure 8.4 Writing times of (a) NOSRAM-based and (b) EEPROM-based wireless IC tags

In the data-writing task, the data were written to the NOSRAM by transmitting a write command from the RW to the wireless IC tag. The data-retention task assessed the data-retention capability of the NOSRAM. In the data-reading task, data were read out from the NOSRAM by transmitting a read command from the RW. The wireless tag was kept in an oven at 130°C during the data-retention evaluation, whereas the data-reading and writing tasks were carried out at room temperature (because RF interfaces and logic circuits with Si FETs are not operated at 130°C). The results are shown in Figure 8.3. As shown, the data written to the NOSRAM were retained even after 255 h at 130°C.

Next, a data-writing time test was performed. This test measured the interval between the signals transmitted from the RW and the wireless IC tag, corresponding to a write command and the completion of writing to the NOSRAM, respectively. The measurement was referenced to a wireless signal waveform obtained at the time of data writing to the wireless IC tag (when the write command from the RW was received by the wireless IC tag and the data were written to the NOSRAM). For comparison, a tag of "Belt, NXP UCode G2iL" manufactured by SMARTRAC and employing an electrically erasable programmable read-only memory (EEPROM) [3] was tested in a similar manner. The measurement results are shown in Figure 8.4. The writing time of the fabricated wireless IC tag is 0.1 ms or shorter, compared with 15.6 ms for the general-use IC tag. Accordingly, the fabricated wireless IC tag is capable of writing at an extremely high speed compared with the EEPROM-based IC tag.

The two results – good data retention at high temperature and data writing within a short time – reflect the features of NOSRAM, implying that CAAC-IGZO LSIs are potentially applicable to RF devices.

8.2.3   Application Examples of NOSRAM Wireless IC Tags

8.2.3.1   Management of Medical Equipment

NOSRAM wireless IC tags can be used directly in fields such as medical instrument management. Checking a large number of medical utensils before and after surgery, for example, is complicated and risky by manual management [4]. Wireless IC tags are expected to facilitate medical utensil management through their RW operations by affixing medical instruments with wireless IC tags and writing their relevant data (such as usage history) to the tags. However, normal wireless IC tags are unusable, because their data-retention characteristics decrease under the high-temperature sterilization of medical utensils at 130°C (although high reliability of the data retention is mandatory).

As mentioned in the previous subsection, NOSRAM wireless IC tags retain their written data even after 255 h at 130°C. This duration is equivalent to 510 sterilizations in an autoclave (assuming that each autoclave treatment runs for 30 min at 130°C). This durability is sufficient for practical use in medical utensil management. Therefore, on account of its long-term retention capability at high temperature, NOSRAM wireless IC tags can be exploited.

8.2.3.2   Building Management

NOSRAM wireless IC tags are also applicable to building management. Tunnels, bridges, and other structures built in the past are now ageing [5]. Rather than reconstruct these old buildings at enormous cost, authorities monitor and repair them as needed, ensuring their long-term viability. However, the management is extremely complicated and is required for many buildings, incurring large cost. Moreover, many parts of the buildings are difficult to monitor by manpower alone. Thus, wireless IC tags are expected to offer efficient data monitoring.

High-Speed Monitoring from Running Trains
Because NOSRAM wireless IC tags can write to the NOSRAM at high speed and with low power consumption, they can be applied efficiently to high-speed monitoring of buildings from running trains.

Figure 8.5 shows the structure of a wireless IC tag suitable for this type of management. The structure of Figure 8.1 is supplemented with a strain sensor and an analog-to-digital converter (ADC). The wireless IC tag supports an extension command for sensing (sensing command). Specifically, the sensing command from the RW converts the output from the distortion sensor into digital data, which are then written to the NOSRAM and can be read out by a read command.

A system based on the sensor-equipped wireless IC tags for checking the degree of deterioration of a tunnel construction is illustrated in Figures 8.6 (schematic) and 8.7 (flow diagram).

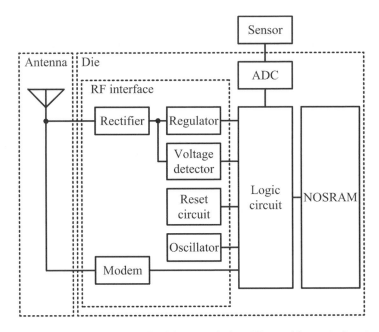

Figure 8.5 Block diagram of NOSRAM wireless IC tag with sensor function

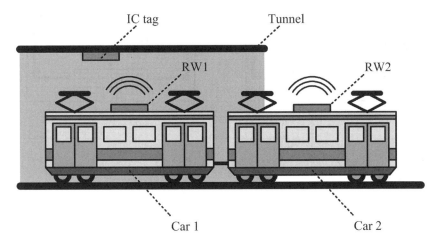

Figure 8.6 System for checking degree of deterioration in train tunnel. System employs NOSRAM wireless IC tag with sensor function

As shown in Figure 8.6, a wireless IC tag equipped with a distortion sensor is attached to the inner wall of the tunnel. The strain sensor detects changes such as cracks in the inner wall. RW1 and RW2 are attached to the roofs of the first and second cars, respectively. As the train runs through the tunnel, the first car approaching the wireless IC tag sends a sensing command from RW1 to that tag [Figure 8.7(a)]. Upon reception of the sensing command, the wireless IC tag

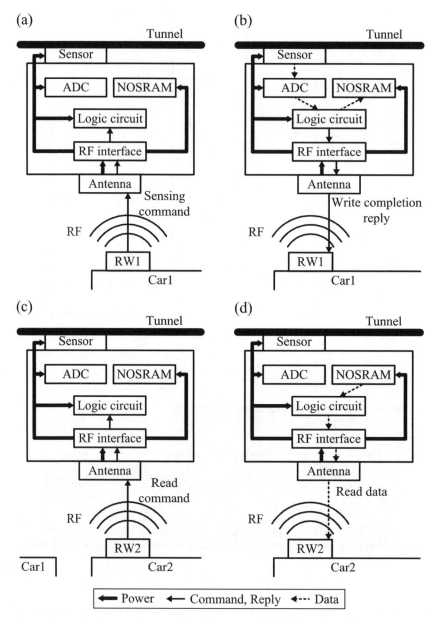

Figure 8.7   Flow diagram of system illustrated in Figure 8.6

converts the output of the distortion sensor into digital data through its ADC, writes the data to the NOSRAM, and transmits a response signal notifying the completion of writing [Figure 8.7(b)]. Next, when the second car approaches the wireless IC tag, RW2 transmits a read command [Figure 8.7(c)]. Upon receipt of the read command, the wireless IC tag reads and transmits the output data of the strain sensor from the NOSRAM [Figure 8.7(d)].

As described above, the structure separating sensing and data transmission ensures time for sensing and storage, which is particularly effective in detecting a weak signal. Although a fast-moving train enables only short-time communication, the degree of deterioration can easily be checked on a regular basis. In similar systems based on wireless IC tags with volatile memory, such as SRAM, the data retention needs a power supply. If non-volatile memory such as EEPROM or flash memory is used, it is difficult to supply the required power and to secure the time needed to write data to the memory. In contrast, NOSRAM – which is a type of non-volatile memory capable of writing at high speed and with low power consumption – can be employed efficiently in construction management.

Continuous Batch Monitoring of Long-Term Data

The above system is not applicable in constructions which cannot easily be accessed by trains, cars, trucks, etc. Such constructions need to be monitored by other methods, such as manual monitoring. In the case where manual monitoring cannot be carried out frequently, continuous batch monitoring by the above-described wireless IC tags is effective for construction management and maintenance. Specifically, the low-power data writing and long-term data retention of NOSRAM allow sensing over a long period and readout in a batch by a single operation.

Figure 8.8 shows a NOSRAM wireless IC tag suitable for this type of construction management. In this design, the system of Figure 8.1 is supplemented by a secondary battery and a power controller. The wireless IC tag receives its operating power from the wireless RW signals and the secondary battery, which supplies power when the wireless RW signal is missing or too weak to drive the circuit operation. The power supply by the secondary battery is mediated by the power controller. When the wireless signal provides excess power, the power controller charges the secondary battery. When no wireless signal is sent from the RW, the power from

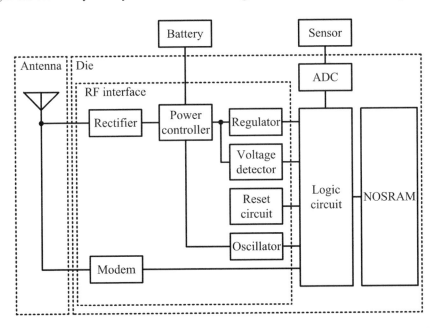

Figure 8.8   Block diagram of NOSRAM wireless IC tag with sensor function and secondary battery

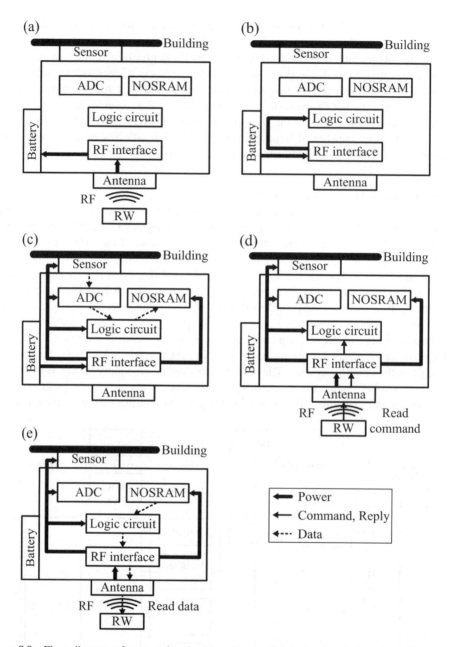

Figure 8.9 Flow diagram of system for checking degree of deterioration in buildings. System uses NOSRAM wireless IC tag with sensor function and secondary battery

the secondary battery is expended in periodic sensing by the distortion sensor. During this routine monitoring, the sensor output is converted to digital data through the ADC, which are written to and retained in the NOSRAM.

A system based on the wireless IC tag is illustrated in Figure 8.9 (flow diagram). The wireless IC tag is mounted with a strain sensor and attached to a construction in advance, and the strain sensor monitors the changing cracks continuously. When a building manager brings the RW closer to the wireless IC tag, the RW transmits a wireless signal that charges the secondary battery of the wireless IC tag [Figure 8.9(a)]. After a sufficient charge time, the RW stops the wireless signal. The wireless IC tag counts the time and samples the sensor even without the RW signal [Figure 8.9(b)]. After a certain period, the wireless IC tag converts the output of the strain sensor to digital data through its ADC, drawing power from the secondary battery, and writes these data to the NOSRAM [Figure 8.9(c)]. The elapsed time recording and sensing are then repeated [Figure 8.9(b) and (c)]. Throughout this process, the wireless IC tag records the strain changes in the NOSRAM, even when no wireless signal is transmitted from the RW. That is, it retains a long-term history of the strain change. When a building manager brings the RW close to the wireless IC tag for monitoring purposes, the RW sends a read command [Figure 8.9(d)] and the wireless IC tag successively reads and transmits a batch of output data from the sensor in the NOSRAM [Figure 8.9(e)].

The above system performs a long-term batch monitoring of data relating to crack development in a construction. Thus, it can reduce the cost of monitoring parts that are difficult to monitor manually, improving the efficiency of building management. A similar system based on wireless IC tags incorporating volatile memory, such as SRAM, requires power for data retention and a large-capacity secondary battery. Non-volatile memories such as EEPROM or flash memory also require a high-capacity secondary battery and power for the memory-writing process. A large capacity may not be viable in wireless IC tags, because the size and shape of the tags must meet certain requirements, such as thinness and easy installation. Since the NOSRAM consumes low power in data writing and retaining, reducing the size of the battery is possible. NOSRAM wireless IC tags can acquire sensor data for a longer time and at lower charge frequencies.

8.3   X-Ray Detector

8.3.1   Outline

The image sensor mentioned in Chapter 7 is intended for use in the visible-light region. By replacing the visible-light detector of the image sensor (i.e., the photodiode) with an element sensitive to non-visible radiation, a wider range of image capture is possible (e.g., diagnostics using X-rays).

X-ray detectors capture the X-ray image of an object and convert it into electrical signals. They are currently used in medical apparatus for diagnosis, inspection apparatus at airports, and various other devices. Like image sensors for visible light, X-ray detectors accumulate the charges generated by X-ray irradiation on sensor elements in the capacitors of the sensor pixels and release the charges as electrical signals.

This section describes the operational principle, fabrication, and evaluation of an X-ray detector employing CAAC-IGZO FETs.

The backplane of the general X-ray detector comprises amorphous silicon FETs but in this case they are replaced with CAAC-IGZO FETs, which have higher mobility. The resulting X-ray detector therefore offers superior image-capturing ability, and can capture moving images.

8.3.2   X-Ray Detection Principle

In general, X-ray detection modes are classified as direct or indirect conversion (Figure 8.10) [6]. In the direct-conversion mode, the irradiating X-rays generate hole–electron pairs in a photoconductor layer, similar to visible light in a light-sensing Si photodiode [7,8]. A

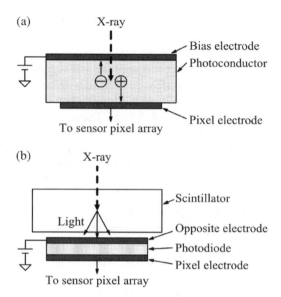

Figure 8.10 Mechanisms of X-ray detection in (a) direct-conversion mode and (b) indirect-conversion mode

high-voltage bias generates an electric field, by which the charges move and accumulate at the sensor pixels. The X-ray-sensitive material in the photoconductor is usually amorphous Se.

The sensor element of the indirect-conversion detector consists of a wavelength converter and a photoconductor [9]. Irradiating X-rays are converted into visible light in a so-called scintillator. The visible light is then converted into charges in the photoconductor (photodiode) layer, a process similar to general image sensors. Common scintillator materials are CsI and Gd_2O_2S.

Both direct and indirect-conversion X-ray detection modes have their merits and demerits. In the direct-conversion mode, charge moves to the pixel electrode under a strong electric field applied to the photoconductor layer. The direct-conversion mode achieves relatively high spatial resolution but demands a high voltage, which increases the cost and introduces a large noise. Conversely, the indirect-conversion mode requires many components to convert X-rays into visible light and the resolution is negatively affected by light scattering, as shown in Figure 8.10(b). Also, the scintillation efficiency is rather low and generates short-wavelength visible light, so the overall sensitivity suffers. Both factors deteriorate the spatial resolution of this mode. However, by applying technologies from visible-image sensors, the voltage required by indirect X-ray conversion can easily be lowered, and the noise can effectively be removed by existing techniques. Exploiting these features of X-ray detectors, researchers have developed various kinds of X-ray detectors.

8.3.3   CAAC-IGZO X-Ray Detector

8.3.3.1   Sensor Pixel

This subsection introduces direct and indirect-conversion X-ray detectors constructed from the CAAC-IGZO image sensors mentioned in Chapter 7. By combining these detectors, we can expect to realize the advantages of CAAC-IGZO sensors in an X-ray detector (CAAC-IGZO

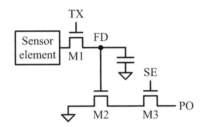

Figure 8.11 Circuit configuration of sensor pixel

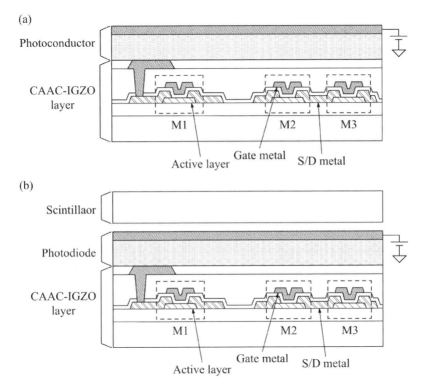

Figure 8.12 Device structure of a CAAC-IGZO X-ray detector in (a) direct-conversion mode and (b) indirect-conversion mode

X-ray detector). A specific example is shown in Figure 8.11. In this circuit, the X-ray-induced charge generated in the sensor element is accumulated (stored) in the retention node FD, which has a high electrical insulating property conferred by the CAAC-IGZO FET. Note that the sensor pixel and the CAAC-IGZO image sensor introduced in Chapter 7 have similar circuit configurations. In other words, the sensor pixel of the X-ray detector can operate under the same driving mode as the CAAC-IGZO image sensor.

Figure 8.12 shows the device structure of a sensor pixel in the CAAC-IGZO X-ray detector. The sensor element is stacked on a CAAC-IGZO layer composed of M1, M2, and M3. In the direct and indirect-conversion modes, the sensor element is a photoconductor and a stacked scintillator/photodiode, respectively.

8.3.3.2 Block Diagram and Operation

Here, the whole structure and operation of the CAAC-IGZO X-ray detector are described. CAAC-IGZO X-ray detectors should be designed for larger-area applications than the image

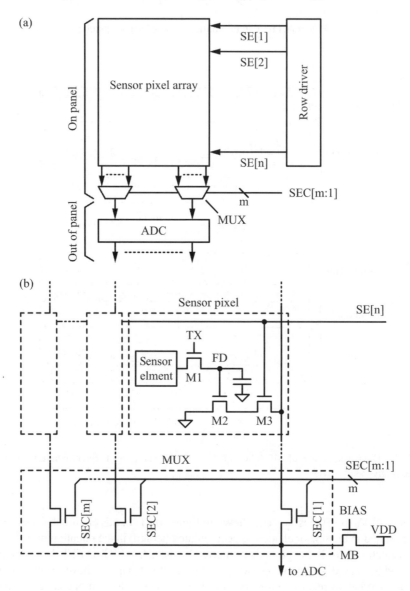

Figure 8.13   (a) Block diagram of X-ray detector and (b) circuit diagram of pixel configuration and MUX

sensor in Chapter 7. Therefore, the sensor pixels should be formed from CAAC-IGZO FETs on a glass substrate, as in the backplane of a display. A difference from displays is that the peripheral circuits (including the driver) are implemented in a Si LSI which is mounted on the glass, a so-called chip on glass (COG).

Figure 8.13(a) and (b) shows a block diagram of the CAAC-IGZO X-ray detector and the connections between the multiplexer (MUX) and the sensor pixel employing CAAC-IGZO FETs, respectively. In the CAAC-IGZO X-ray detector, the pixel array, row driver, and MUX are fabricated on the glass substrate. The row driver selects a particular row (SE[n]) from the pixel array. Each sensor pixel in the selected row outputs an analog signal (sensor output) corresponding to the data retained in the node FD. The outputs of the m sensor pixels in the selected row are input to the MUX. Among the m sensor outputs, the MUX selects a single output and relays it to an ADC located off the panel.

The voltage of the sensor output depends on the channel resistance of transistor M2 in the sensor pixel (selected by the selection signals SE and SEC) and the channel resistance of transistor MB connected to the output portion. As the potential at the charge-retention node FD in each pixel increases, the channel resistance of transistor M2 decreases, reducing the voltage of the sensor output. The voltage of the sensor output can be controlled by the potential of BIAS on the gate of transistor MB. In this way, the sensor output of the sensor pixel can be amplified. Even a large-area X-ray detector constructed with CAAC-IGZO FETs on a glass substrate can read data at a relatively high speed.

The CAAC-IGZO X-ray sensor operates similarly to its visible-light counterpart. In the sensor pixel, the charge accumulation node (FD) is reset, charge accumulates under light exposure, and the sensor pixel data are read out if the pixel is selected. In the X-ray detector, all of the sensor pixels are simultaneously subjected to a sequential reset and charge-accumulation operation (see the global shutter system of Section 7.2). In the subsequent reading operation, the selection signals SE[1] to SE[n] are sequentially selected by the row driver, as shown in Figure 8.14. In the selection period of the row-by-row reading operation, the MUX sequentially obeys the column-selection signals SEC[1] to SEC[m], and sequentially outputs m signals. The selection signal SE[n] is selected at the end of one frame.

The reset operation can be constructed in several ways. For instance, the reset transistor for resetting the FD potential can be connected to the FD as in visible-light image sensors. Alternatively, the reset operation can be triggered by applying a forward bias to the photodiode, as mentioned in Subsection 7.4.2. The latter is preferred for indirect X-ray sensing with amorphous Si photodiode sensor elements, which requires relatively few transistors.

The above-described global shutter system in the CAAC-IGZO X-ray detector offers the following advantages:

- moving objects can be imaged without distortion;
- the X-ray dose is significantly reduced by the one-time light exposure, dramatically reducing the administered dose.

These advantages should widen the applications of CAAC-IGZO X-ray detectors.

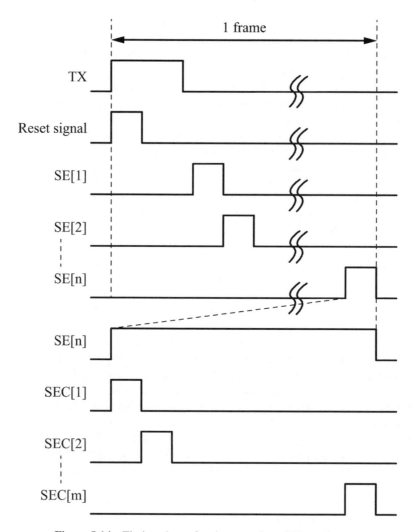

Figure 8.14 Timing chart of entire operation of X-ray detector

8.3.4 Fabrication Example and Evaluation

Figure 8.15 and Table 8.2 show a photograph and the specifications of a prototype X-ray detector, respectively. This detector has been designed to operate in indirect-conversion mode with a scintillator and a photodiode array made from Gd_2O_2S:Tb and amorphous Si, respectively. All FETs in the sensor pixel circuit, the row driver, and the MUX ($m = 12$) have been fabricated on a glass substrate by a 3.0-μm CAAC-IGZO FET process. Flexible printed circuits (FPCs) have been used to connect the sensor with the external ADC via the input and output terminals of the panel.

Figure 8.16 shows an image captured by the fabricated CAAC-IGZO X-ray detector. The objects are a clock, a copper coin, and lead boards. As shown in the figure, the gray level

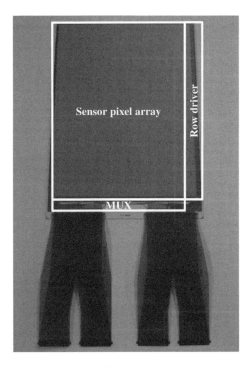

Figure 8.15 Photograph of a CAAC-IGZO X-ray detector panel (scintillator not shown)

Table 8.2 Specifications of the CAAC-IGZO X-ray detector

Process technology	3.0 μm CAAC-IGZO
Panel size	100.5 mm × 139 mm
Number of pixels	384 × 512
Pixel size	120 μm × 120 μm
Resolution	106 ppi
Pixel configuration	3 transistors, 1 capacitor, 1 photodiode
Photodiode	amorphous Si
Scintillator	$Gd_2O_2S{:}Tb$

depends on the material, because the X-ray transmittance depends on the material and thickness. In Figure 8.16, since the clock was highly transparent to X-rays, the detector captured both its surface and its interior.

The above-described CAAC-IGZO X-ray detector performs the basic operation of X-ray image capturing and offers additional advantages when combined with a global shutter drive. The CAAC-IGZO X-ray detector will be developed further in the future.

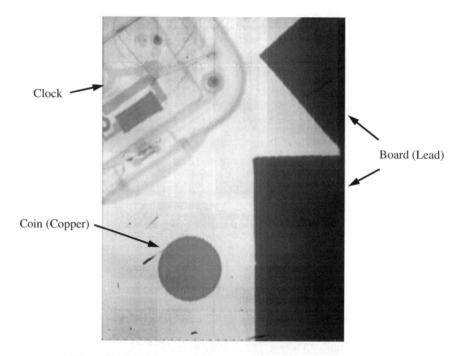

Clock

Board (Lead)

Coin (Copper)

Figure 8.16   Objects captured by the CAAC-IGZO X-ray detector

8.4 CODEC

8.4.1 Introduction

This section discusses the future application of CAAC-IGZO FETs to a CODEC, an LSI that encodes and decodes data for, e.g., satellite broadcasting. Although the application introduced here is intended for 8K satellite broadcasting, it is also effective for other compression techniques than that of 8K satellite broadcasting.

8K broadcasting, proposed as the next-generation TV broadcasting standard, has attracted much attention in recent years, and related technologies are rapidly emerging. An uncompressed 8K video format consists of quadruple data compared with 4K video and a 16-fold increase compared with FHD. In addition, a doubling of the frame rate to 120 Hz and an enlarging of the color depth from 8 to 12 bits/color is being proposed. To transmit and store 8K video data to homes, an efficient compression system is therefore required. The current, widely used compression system is ITU-T H.265∥ISO/IEC 23008-2 [hereafter referred to as H.265/HEVC (high-efficiency video coding)] [10,11]. A system suitable for 8K broadcasting is depicted in Figure 8.17. The 8K broadcasting system consists of a data transmission side [Figure 8.17(a)] and an 8K TV system on the data reception side [Figure 8.17(b)]. On the data transmission side, the imagery captured by an 8K video camera is compressed in an encoder IC. The compressed signal is transmitted from the transmitting antenna and conveyed to homes via a broadcasting satellite (BS). On the data reception side, the 8K TV system receives the analog signal transmitted from the broadcasting satellite at the receiving antenna and front-end module, and converts it to a digital signal. The decoder IC decodes the digital signal and displays the decompressed video data on an 8K display.

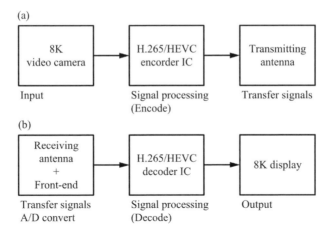

Figure 8.17   8K Broadcasting system: (a) data transmission and (b) data reception (8K TV system)

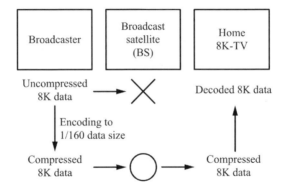

Figure 8.18   Conceptual diagram of a satellite broadcasting system based on data compression–extension by H.265/HEVC

8.4.2   Encoder/Decoder

Figure 8.18 is a conceptual diagram of a satellite broadcasting system using H.265/HEVC for data compression and decompression. The 8K data are compressed (encoded) at the signal transmission side and decompressed (decoded) at the signal reception side. The H.265/HEVC compresses the enormous uncompressed 8K data to a size that can be transmitted/received via BS. The compression rate is approximately 1/160, approximately twice that of H.264/AVC. Such high compression is essential for the enormous amount of signal transmission in 8K broadcasting.

Figure 8.19(a) and (b) depict an H.265/HEVC encoder and decoder, respectively. The H.265/HEVC encoder mainly compresses video data (raw data) and generates transmission video data (encoded data), whereas the H.265/HEVC decoder mainly plays the opposite role of signal processing. It performs a discrete cosine transform (DCT) on uncompressed video data (raw data), and the transformed data are quantized and entropy coded to generate coded data. Specifically, the H.265/HEVC decoder decompresses the compressed video data (encoded data) and produces video data (decoded data). The encoded data are entropy decoded

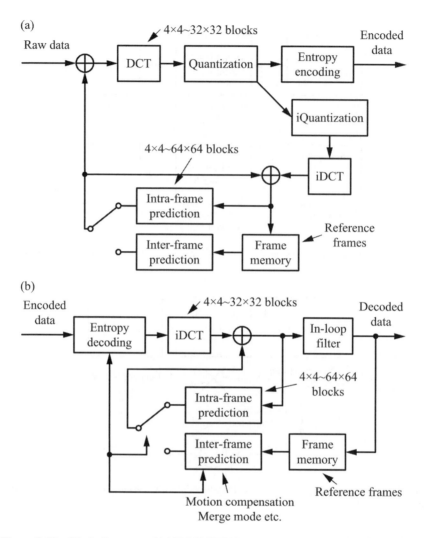

Figure 8.19   Block diagrams of (a) H.265/HEVC encoder and (b) H.265/HEVC decoder

and an inverse-discrete cosine transform (iDCT) is performed; the transformed data are decoded via an in-loop filter.

Data compression by H.265/HEVC is roughly divided into one spatial compression and two temporal compressions. These two classes are described below.

1. When the spatial difference is small between adjacent pixels, the data can be compressed and decompressed by estimating the data of a target pixel from the data of the adjacent pixels; that is, by intra-frame prediction.
2. When the video data between frames are very similar, they can be compressed and decompressed by estimating the video data of a target frame from the video data of the previous and

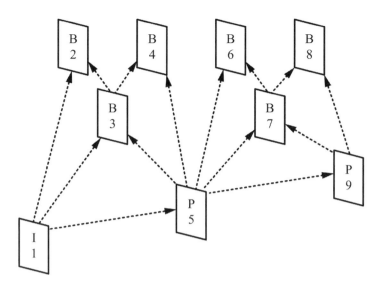

Figure 8.20 Data reference relationships among frames (I, P, and B denote I picture, P picture, and B picture, respectively. Numbers indicate order in which data are displayed)

next frames; that is, by inter-frame prediction. This prediction requires a frame memory for storing the previous and next frames.

The encoded data of each frame are classified by their prediction processes. Intra-predicted picture (I picture) data are encoded only within a single frame by intra-frame prediction, whereas predicted picture (P picture) data are encoded from an earlier I picture by inter-frame prediction. A bi-directional predicted picture (B picture) is predicted either forward or backward in time from I and/or P pictures. The encoded data are transmitted sequentially, as shown in Figure 8.20.

Figure 8.19(b) illustrates the decoding process of H.265/HEVC. Block sizes of iDCT or intra-frame prediction in the decoding process are variable. Accordingly, the contents of the digital processing and the processing scale vary with block size.

8.4.3 CAAC-IGZO CODEC

As described in Subsection 8.4.2, the block size to be processed depends on the received data. Specifically, when the content of signal processing varies, the FPGA hardware based on CAAC-IGZO FETs (see Section 6.4) can be adjusted so that unnecessary circuit blocks are not used, improving the efficiency of signal processing.

Figure 8.20 shows the data relationships in a reference frame represented by an I picture. Here the video data should be stored over multiple frames over a long period. In this case, large-scale video data of a reference frame is stored in a general DRAM, which needs to be refreshed frequently, usually every 16 ms. If the video data of the reference frame instead are stored in a dynamic oxide semiconductor random access memory (DOSRAM), power can be reduced significantly, since it does not require frequent refreshes (refer to Chapter 4).

The above discussion highlights the promising applicability of CAAC-IGZO FETs to a decoder for 8K satellite broadcasting; for instance, an H.265/HEVC decoder. The decoder is expected to combine LSIs based on CAAC-IGZO FETs such as DOSRAM and FPGA with existing CAAC-IGZO FET technologies.

The ultimate target is a TV system that combines a decoder constructed from CAAC-IGZO FETs with a display. The 8K TV system is advantageous for several reasons. For instance, if a frame and its succeeding frame are nearly identical, the initial video data stored in the frame memory of a decode LSI can be read continuously, which is effective for power saving. In this case, in combination with the decoder LSI, the display is driven by idling stop (IDS) driving, which leads to further power saving. In IDS driving, the image signals can be retained because the pixels of the display are configured by CAAC-IGZO FETs, which have a low leak current. This results in reduction in the display refresh rate when displaying still images. Details are provided in *Physics and Technology of Crystalline Oxide Semiconductor CAAC-IGZO: Application to Displays* [12]. In addition, because an 8K TV system that combines an H.265/HEVC decoder with an 8K display has to process a large amount of data, the use of CAAC-IGZO FETs is expected to contribute significantly to power saving.

Accordingly, the 8K TV system is believed to be an important low-power technology for 8K TV in households. Furthermore, decoders using CAAC-IGZO FETs can be applied to TV systems for tablets and smartphones.

8.5   DC–DC Converters

8.5.1   Introduction

So far we have discussed LSIs with memories based on CAAC-IGZO FETs, which effectively retain charge [13–15]. In particular, NOSRAM exhibits extraordinary charge-retention characteristics and can also be used as a multi-level memory. As such, it retains both digital and analog data – i.e., it could also be used as an analog voltage memory, such as the CAAC-IGZO TFTs in displays with slow refresh (IDS driving).

This section introduces DC–DC converters as an analog memory application of CAAC-IGZO FETs. DC–DC converters, employed as power supply circuits in various devices, obtain the desired voltages by boosting or bucking the input DC voltage. Power-conversion efficiency is an essential parameter of DC–DC converters. In particular, extending the battery runtime of modern smartphones and wearable devices requires DC–DC converters with low power consumption and high power-conversion efficiency. When a DC–DC converter converts power close to the consumed power of circuits included in the DC–DC converter, the current consumption of an internal analog circuit is one of the main factors in reducing the power-conversion efficiency. In other words, the power-conversion efficiency is largely reduced by the power consumption in the internal circuit.

The DC–DC converter described below significantly reduces the power consumption of the conversion and hence improves the efficiency. When CAAC-IGZO FETs with an analog memory are used as sampling circuits, the bias and input reference voltages of the CLK generator and hysteresis comparator (i.e., the analog voltage) can be retained for a long time. Consequently, the power is reduced in the voltage reference circuits, including the bias circuit and bias generator.

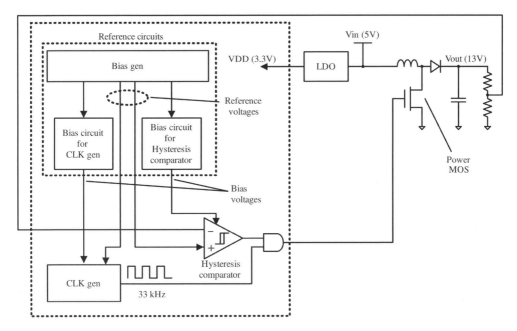

Figure 8.21   Circuit diagram of non-hybrid DC–DC converter fabricated with a 0.35-µm CMOS process technology (converter A). This hysteretic-controlled boost DC–DC converter is designed for low power consumption

8.5.2   *Non-hybrid DC–DC Converter*

Figure 8.21 shows a circuit diagram of a non-hybrid DC–DC converter fabricated with a 0.35-µm Si CMOS process technology (hereafter referred to as converter A). The power consumption is reduced by the hysteresis-controlled boost converter, which is configured from reference circuits, a hysteresis comparator, a CLK generator, and an AND circuit. To achieve the desired output voltage, the hysteresis comparator compares the output of the converter and one of the reference voltages, and the AND circuit performs an AND calculation between the comparison result and the CLK from the CLK generator. Subsequently, the AND result is input to the power MOS gate.

During DC–DC converter operations, the hysteresis comparator and the CLK generator must also operate. The reference circuits, supplementary circuits for operating the hysteresis comparator and CLK generator, consume a constant amount of current during operation.

8.5.3   *Fabricated CAAC-IGZO Bias Voltage Sampling Circuit with Amplifier*

This subsection proposes a DC–DC converter to which the technology of a CAAC-IGZO bias voltage sampling circuit with amplifier is applied. The DC–DC converter proposed in this section (Figure 8.22) resolves one problem in non-hybrid DC–DC converters fabricated with a 0.35-µm CMOS process technology (converters A): the losses in power-conversion efficiency at low-power input/output caused by the constant power of the reference circuit. The

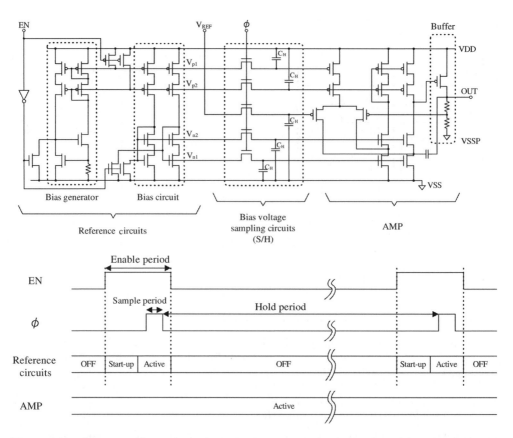

Figure 8.22 Fabricated CAAC-IGZO bias voltage sampling circuit with an amplifier and its timing diagram

amplifier and the reference circuit (bias generator and bias circuit) are complemented with a sampling circuit implemented by CAAC-IGZO FETs. Figure 8.22 also shows the timing diagram of the proposed DC–DC converter.

The bias generator is a threshold-referenced circuit which, together with the reference circuit, generates the reference voltage of the amplifier and the bias voltage of the current source. The amplifier receives the reference voltage V_{REF} and outputs the same V_{REF}.

The bias voltage sampling circuit is based on CAAC-IGZO FETs. This circuit works as an analog memory, sampling and retaining V_{REF} supplied from the bias circuit and the bias voltage (analog potential). At the retention period after sampling of the reference and bias voltages by the bias voltage sampling circuit, the power supply can be shut off. Replacing the amplifier in the circuit of Figure 8.22 with a comparator or CLK generator would provide the same effect.

Here, the set time is defined as the period in which the circuit can retain the reference and bias voltages needed for stable operations of the amplifier. After the set time passes, the power supply is restarted to the reference circuit and the bias voltage sampling circuit samples the reference and bias voltages. Repeating this operation reduces the average current consumed by the reference circuit.

8.5.4 Evaluation Results of Fabricated CAAC-IGZO Bias Voltage Sampling Circuit with Amplifier

To discuss the effect of the CAAC-IGZO bias voltage sampling circuit described above, the circuit shown in Figure 8.22 was fabricated by a hybrid process: 0.35-μm process for Si FETs and 0.8-μm process for CAAC-IGZO FETs. A micrograph of the circuit is presented in Figure 8.23.

In this prototype, the reference voltage was input externally to the amplifier. The circuit was configured to output 1.8 V at a reference voltage of 1.25 V. The bias voltage sampling circuit was configured from a 10-pF hold capacitor (CH) and CAAC-IGZO FETs of width 3 μm and length 20 μm. The transistors in the other circuits were Si FETs with $L = 3$ μm.

First, the stability of the amplifier output voltage in the test-fabricated circuit was measured. The power supply voltage was 2.3 V, and the bias and reference voltages were sampled by the bias voltage sample circuit. During the subsequent hold period, the power supply to the bias generator and the bias circuit was stopped, and the output voltage and current consumption of the amplifier were measured. The measured results are plotted in Figure 8.24, which shows that the output voltage and the consumed current decrease by 8.95 μV/s and 0.0083%/s, respectively. The variation in the retention voltage was calculated to be 6.22 μV/s, since the prototype was designed so that the output voltage of the amplifier was 1.44 times as large as the retention voltage. A reference voltage (retention voltage) variation of 6.22 μV/s can be ignored in an internal circuit not requiring high precision, such as a DC–DC converter.

Subsequently, the average current consumption was measured as a function of the hold period. The measured results are plotted in Figure 8.25. Considering the required wake-up period of a general reference circuit (the transitional time from quiescence to an active state), the enable period was set to 2 ms. At a hold period of 220 ms, the average current consumed by the reference circuit was 0.059 μA. During constant operation, the consumed current by the reference circuit was 6.27 μA. Therefore, by limiting the circuit operation to the sampling period, the current consumption was reduced to below 1/100 of the original consumption.

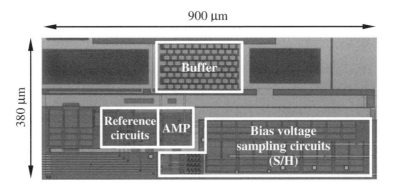

Figure 8.23 Micrograph of CAAC-IGZO bias voltage sampling circuit with an amplifier fabricated by the hybrid process

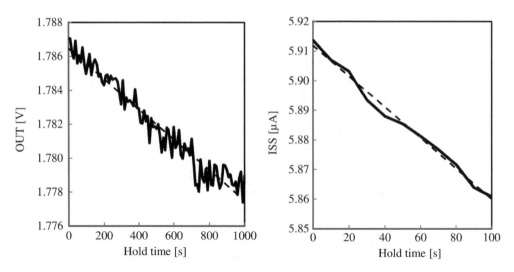

Figure 8.24 Measured output voltage (left) and current consumption (right) of the circuit fabricated by the hybrid process

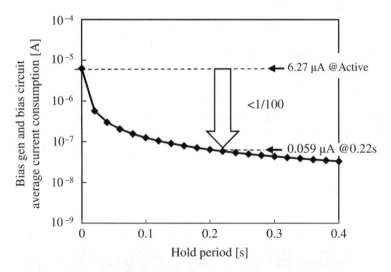

Figure 8.25 Measured average current consumption of the bias generator and bias circuit in the circuit fabricated by the hybrid process

8.5.5 Proposed DC–DC Converter

Although the CAAC-IGZO bias voltage sampling circuit (S/H) is applied to the amplifier, the S/H can also be applied to the hysteresis comparator and CLK generator in Figure 8.21, which offers the same effect. Figure 8.26 shows the comparator fabricated by the S/H, Figure 8.27 shows the hysteresis comparator configured with the comparator shown in Figure 8.26, and Figure 8.28 shows a CLK generator to which the S/H is applied.

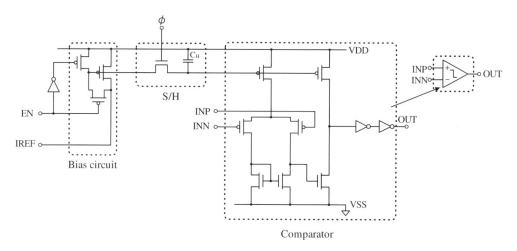

Figure 8.26 Comparator using CAAC-IGZO bias voltage sampling circuit (S/H)

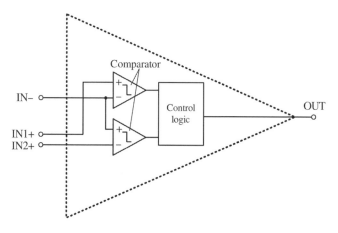

Figure 8.27 Hysteresis comparator configured by the comparator in Figure 8.26

Figure 8.29 depicts a DC–DC converter where the circuits shown in Figures 8.26–8.28 are used and a timer circuit is added. The timer circuit monitors the set time and generates control signals for the sampling of the reference and bias voltages in the bias voltage sampling circuit, and a signal for controlling the operations of the reference circuit. As the CAAC-IGZO FETs in the bias voltage sampling circuit have good retention characteristics, the circuit can achieve low power consumption even when the set time monitoring is imprecise. In other words, the timer circuit can be configured with asynchronous counters to reduce its current consumption.

Table 8.3 lists the simulated current consumed by the analog circuits in the DC–DC converter fabricated only with a 0.35-μm CMOS process technology (converter A) and the proposed DC–DC converter in quiescent states. The current consumption of the reference

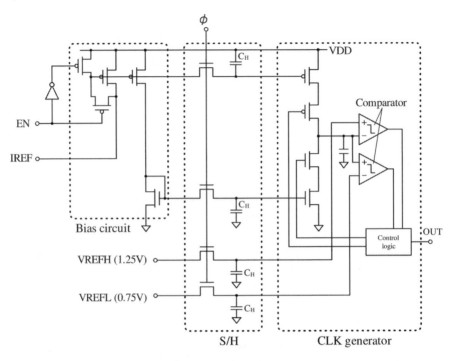

Figure 8.28   CLK generator fabricated by the CAAC-IGZO bias voltage sample and hold circuit (S/H)

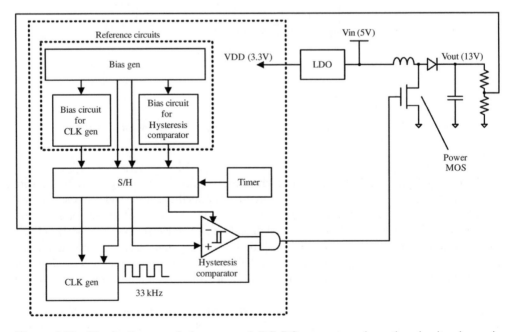

Figure 8.29   Circuit diagram of the proposed DC–DC converter where the circuits shown in Figures 8.26–8.28 are used in the reference circuit, and a timer circuit is added to converter A

Table 8.3 Simulated current consumption in quiescent state by analog circuits in ICs of converter A and proposed DC–DC converter

Technology		0.35-μm CMOS	0.35-μm CMOS +0.8-μm CAAC-IGZO (proposal)
Power supply voltage		3.3 V	
Clock frequency		33 kHz	
Enable period		—	2 ms
Hold period		—	220 ms
Quiescent current	Hysteresis comparator	10 μA	
	CLK generator	9.5 μA	
	Timer	—	0.07 μA
	BGR	14 μA	0.128 μA
	VREF generator	3.8 μA	0.036 μA
	Bias generator	11.4 μA	0.104 μA
	Bias circuit of hysteresis comparator	1 μA	0.009 μA
	Bias circuit of CLK generator	4.5 μA	0.041 μA
	Total of reference circuits	34.7 μA	0.319 μA
	Total	54.2 μA	19.8 μA

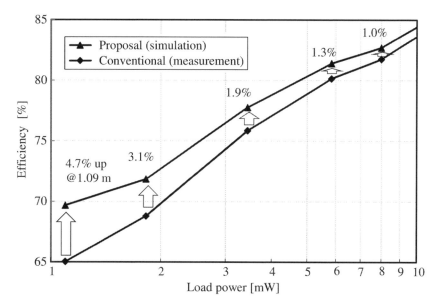

Figure 8.30 Measured power-conversion efficiency of converter A and simulated power efficiency of the proposed DC–DC converter

circuit is over 100 times smaller in the proposed DC–DC converter than in converter A (0.319 µA vs. 34.7 µA). The proposed converter reduces the quiescent current consumption of the entire analog circuit by 63%, from 54.2 µA to 19.8 µA. At V_{DD} = 3.3 V and an oscillation circuit frequency of 33 kHz, the simulated current consumed by the timer circuit is 0.07 µA. Figure 8.30 plots the measured power-conversion efficiency of converter A and the simulated power efficiency of the proposed converter. Here, V_{in} = 5 V, V_{DD} = 3.3 V, and V_{out} = 13 V. Under a load power of 1.09 mW, the efficiency is 4.7% higher in the proposed converter than in converter A.

Clearly, the power efficiency of the DC–DC converter is improved by mounting a bias voltage sampling circuit constructed from CAAC-IGZO FETs. Thus, CAAC-IGZO FETs can retain analog potentials, which is very promising and beneficial for analog applications, particularly when the load is low – i.e., DC–DC converters for liquid crystal displays or other voltage-driven devices.

8.6 Analog Programmable Devices

8.6.1 Overview

In the CAAC-IGZO FPGA introduced in Chapter 6, the configuration memory works as a memory element holding binary digital data. As seen with the multi-level NOSRAM memory, however, CAAC-IGZO FETs also have the potential to work as a non-volatile analog programmable device.

As an application example of an analog programmable device, this section introduces a voltage-controlled oscillator (VCO) using an analog programmable element composed of CAAC-IGZO FETs [16]. The signals output by the VCO oscillate at the intended frequency according to the analog voltage data (AVD) stored in the analog programmable element. The VCO can switch between oscillation frequencies and achieve stable oscillation in a short time after power-on.

8.6.2 Design

A circuit diagram of the proposed VCO with an analog programmable element is presented in Figure 8.31. The VCO is a ring-oscillator-type circuit composed of odd-numbered-stage inverters and voltage-controlled switches (VCSs, corresponding to analog programmable elements). The VCSs control the connections between the inverters. Borrowing from the multicontext method in the CAAC-IGZO FPGA, each VCS is constructed from k-stage voltage-controlled elements (VCEs). The inverter connections are made by the VCE selected by a signal line CL. The intended AVD (V_{DATA}) is stored by the VCE on node SN and controls the channel resistance of transistor MG. In other words, the VCS resistance is controlled by the AVD programmed in the selected VCE, enabling control of the VCO oscillation frequency. When transistor MW is a CAAC-IGZO FET and is turned off, node SN becomes an electrically isolated floating node that retains the AVD for a long time. The transistor MW has a back gate and its subthreshold voltage can be controlled by changing the back-gate voltage.

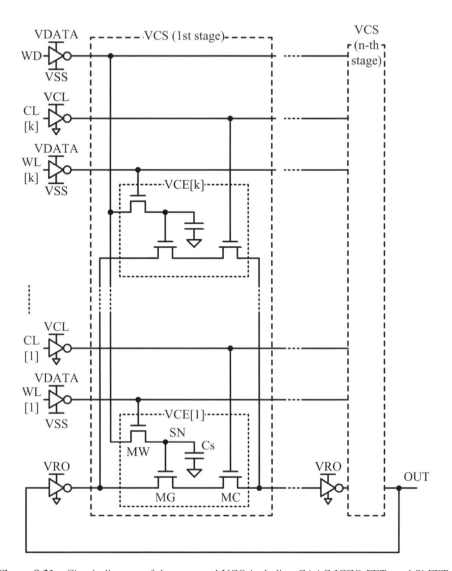

Figure 8.31   Circuit diagram of the proposed VCO including CAAC-IGZO FETs and Si FETs

Multiple VCEs offer the advantage of variable AVDs. Therefore, the oscillation frequency of the VCO can be changed immediately by selecting a VCE, depending on the AVD programmed therein.

8.6.3   Prototype

8.6.3.1   Whole Structure

The proposed VCO has been fabricated by a hybrid process using 1.0-μm CAAC-IGZO and 0.5-μm Si FETs [16–19]. The circuit design of the prototype VCO is presented in Figure 8.32. The prototype consisted of 101 stage inverters and VCSs and had two contexts ($k = 2$). The

(a)

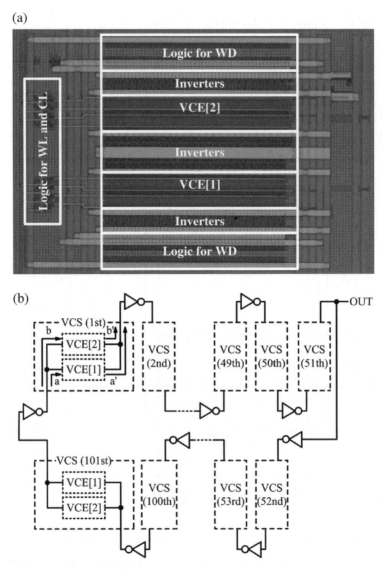

(b)

Figure 8.32 Fabricated layout of VCO with two VCEs ($k = 2$): (a) die photograph and (b) simplified diagram

transistor sizes (channel widths W) of transistors MW, MG, and MC are 4 μm, 16 μm, and 16 μm, respectively. The gate capacitance of transistor MG and the storage capacitance are 16 fF and 2 fF, respectively. Consequently, their combined capacitance at node SN is 18 fF.

Figure 8.32(a) shows a photograph of the fabricated VCO. The VCO includes logic for WL and CL, logic for WD, inverters, VCE[1], and VCE[2]. The logic for WL and CL consists of signal lines WL and CL and a signal supply buffer. The logic for WD consists of a signal line WD and a signal supply buffer. The 101 stage inverters and VCSs are folded at the 52nd stage. In this layout, the outputs from the inverters on the lower side are input to the two VCEs (VCE [1] and VCE[2]) immediately above, and the outputs from the two VCEs are input to the

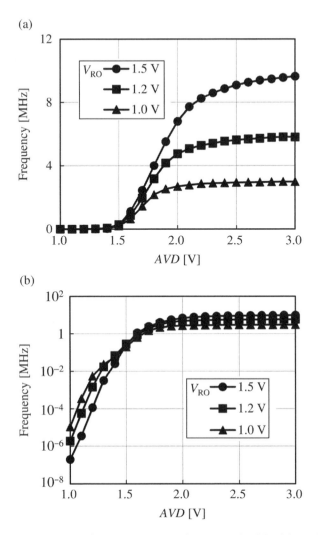

Figure 8.33   Relationship between frequency and AVD in proposed VCO with two VCEs on (a) linear scale and (b) log scale

inverters on the upper side. That is, when VCE[1] is selected, the inverters are connected via wirings a and a′; similarly, when VCE[2] is selected, the inverters are connected via wirings b and b′. The layout ensured no difference in wiring length between the inverters through a selected VCE [see Figure 8.32(b)], thus avoiding the effects of wiring length on the VCO performance.

8.6.3.2   Measurements and Results

AVD Dependence of Oscillation Frequency
The VCO controlled the oscillation frequency by referring to the AVD in the analog program-mable element. Figure 8.33 plots the relationship between oscillation frequency and the AVD programmed in each VCE, measured on a test-element group (TEG) with a 2-VCE VCS (that

is, for each VCS, the number n of VCEs is 2). For measurements, VCE[1] is selected. The supply voltage V_{RO} of the inverter is varied as 1.0, 1.2, and 1.5 V. The voltage V_{CL} applied to the gate of the transistor MC is fixed at 3.0 V. The back-gate voltage V_{BG} of transistor MW is 0 V, and V_{SS} was −0.2 V. The write time is 1.0 ms.

As shown in Figure 8.33, the oscillation frequency could be controlled by changing the AVD. For example, at V_{RO} = 1.5 V and AVD ranging from 1.0 to 3.0 V, the oscillation frequencies ranged from 197 mHz to 9.65 MHz. In other words, the oscillation frequency of the VCO changed over seven orders of magnitude. The change ratio of the oscillation frequencies depended on the AVD. For example, in the AVD range from 1.0 to 1.5 V, the change ratio is 1.24 decades/100 mV, dramatically reducing to 0.06 decades/100 mV in the AVD range from 2.5 to 3.0 V. This result is attributable to the high conductivity of transistor MG and the large inverter delay (relative to the transistor) in the higher AVD range. Consequently, the delay in inverters is larger than in the transistor MG. In contrast, in the lower AVD range, the oscillation frequency changed at 0.82 decades/100 mV, 1.10 decades/100 mV, and 1.24 decades/100 mV at V_{RO} of 1.0 V, 1.2V, and 1.5 V, respectively. This result can be explained by the relatively low conductivity of transistor MG at low analog voltage data, and the long transistor delay (relative to the inverter). Consequently, the oscillation frequency depends strongly on the AVD.

The oscillation frequency of the VCO also depends on V_{RO}. However, in the high-AVD range, the inverter delay dominates as described above, and the oscillation frequency changes largely with V_{RO}. Accordingly, V_{RO} may be changed to meet intended VCO applications; for example, a high driving voltage could achieve a wide frequency range, whereas a low driving voltage could enable fine tuning in a narrow frequency range.

Oscillation Frequency Retention Characteristics

Since the analog programmable memory is composed of CAAC-IGZO FETs, it retains its stored AVD for a long time. When V_{RO} was set to 1.5 V and an AVD of 2.5 V stored in VCE[1], the oscillation frequency of the VCO changed over time (see Figure 8.34). To obtain

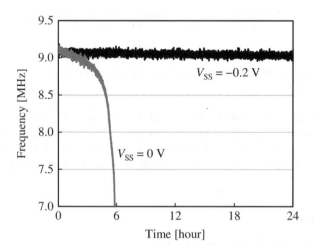

Figure 8.34 Data-retention characteristics of VCO (n = 2) with different V_{SS} values (AVD = 2.5 V, V_{RO} = 1.5 V)

Figure 8.34, V_{SS} was set to 0 and −0.2 V. Initially, the oscillation frequency is 9.10 MHz regardless of V_{SS}. When $V_{SS} = 0$ V, the frequency decreases by 7.7% after 5 h. This reduction of the oscillation frequency improves significantly when V_{SS} is negative. Specifically, when $V_{SS} = -0.2$ V, the oscillation frequency is largely stable over time, decreasing only to 9.02 MHz after 24 h. In this case, the oscillation frequency decreases only by 0.87% a day, indicating that the retention characteristics of the analog memory are improved by lowering the V_{SS}. Referring to Figure 8.33 and assuming that the AVD decreases at a constant rate, the AVD should decrease to approximately 30 mV after 24 h. Denoting the time, storage capacitance, and voltage change by t (s), C (F), and ΔV (V), respectively, the leakage current I_{leak} is expressed as follows:

$$I_{leak} = \frac{\Delta V C}{t}. \tag{8.1}$$

Inserting $t = 86,400$ s (24 h), $C = 18$ fF, and $\Delta V = 0.03$ V into Equation (8.1), the leakage current I_{leak} can be estimated to be 6×10^{-21} A.

Figure 8.35 shows the spectral change in the oscillation frequency when $V_{RO} = 1.5$ V and the VCO oscillates at $V_{SS} = -0.2$ V. In panels (a) and (b) of this figure, AVD is 2.5 V and 2.0 V, respectively. In Figure 8.35(a), the central frequencies of the spectra at $t = 0$ min and $t = 180$ min are 9.10 MHz and 9.07 MHz, respectively. That is, the oscillation frequency decreases by 0.34% over 3 h. When the ADV is 2.0 V [Figure 8.35(b)], the central frequencies at $t = 0$ min and $t = 180$ min are 6.63 MHz and 6.58 MHz, respectively, representing a 0.74% reduction in oscillation frequency. These results suggest that when V_{SS} is set appropriately ($V_{SS} = -0.2$ V) the oscillation frequency barely changes over time – i.e., the data-retention characteristics are excellent for any AVD.

From the spectra in Figure 8.35, we can determine the figure of merit (FOM) at $t = 0$ min by [20]:

$$FOM = Phn - 20\log\left(\frac{F_C}{\Delta F}\right) + 10\log(1000P). \tag{8.2}$$

Here, Phn, F_C, and P denote the phase noise, central frequency, and power consumption, respectively. For an AVD of 2.5 V, the FOM can be estimated to be −151.8 dBc/Hz.

Quick Start-up

Owing to its non-volatile analog programmable memory, the VCO maintains the AVD over a long time, even during its quiescent period. In other words, the AVD corresponding to the set oscillation frequency is retained in the analog programmable element, even when the supply voltage is switched off. The restored supply voltage is followed rapidly by an output signal at the desired oscillation frequency.

Figure 8.36 shows the waveforms of the output OUT. To obtain these measurements, the ring oscillator was rebooted 1 h after powering off the supply voltage V_{RO}. V_{RO}, AVD, V_{SS}, and V_{BG} were fixed at 1.5 V, 2.5 V, −0.2 V, and 0 V, respectively. The output OUT resumed oscillating immediately after the restart of V_{RO}.

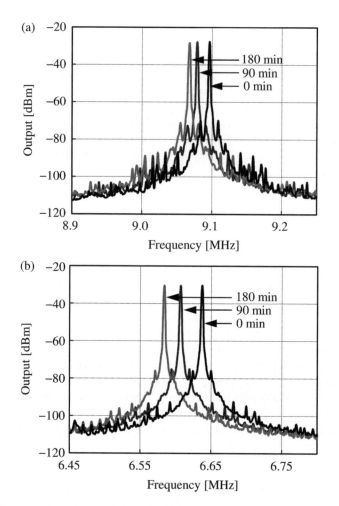

Figure 8.35 Data-retention characteristics of VCO ($n = 2$) with different AVDs ($V_{RO} = 1.5$ V, $V_{SS} = -0.2$ V): (a) AVD = 2.5 V and (b) AVD = 2.0 V

Quick Frequency Switching

The multicontext architecture of the VCO enables instant switching among the VCEs in the VCS. That is, after programming the AVDs in the VCS, we can instantly change the oscillation frequency of the VCO by selecting a new VCE.

Figure 8.37 shows the output waveforms of the TEG with $n = 2$. The analog programmable memories of VCE[1] and VCE[2] was programmed with AVDs of 1.8 V and 2.5 V, respectively, and the selected VCE was switched during VCO operation. During the first 1.0 μs, VCE[1] was selected, and the VCO oscillated at 4.0 MHz in response to the 1.8 V AVD. At $t = 1.0$ μs, the selected VCE was switched from VCE[1] to VCE[2], and the oscillation frequency of the output OUT increased instantly to 9.1 MHz.

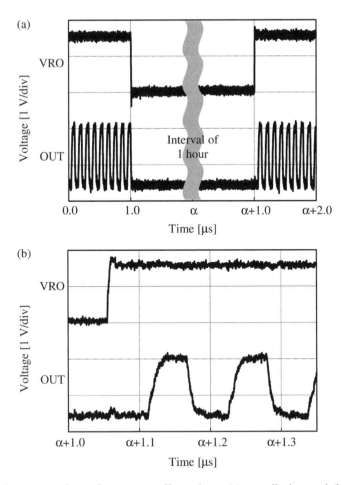

Figure 8.36 Output waveforms from power-off to reboot: (a) overall view and (b) enlarged view immediately after reboot

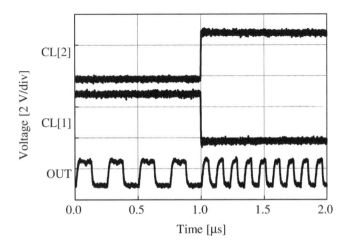

Figure 8.37 Output waveforms during switching between selected VCEs ($V_{RO} = 1.5$ V)

8.6.4   Possible Application to Phase-Locked Loop

In a general phase-locked loop (PLL), the VCO is built into a feedback loop that is forced to operate even after the oscillation frequency is locked. In contrast, the proposed VCO retains the AVD in the analog programmable element over long periods. Consequently, once the desired oscillation is set, the power supply to the circuits forming the feedback loop can be stopped, and power can be supplied only to the VCO, largely reducing the power consumption of the PLL. For instance, in a PLL operating at approximately 20 GHz and 1.5 V, circuits other than the VCO account for 60% of the total current consumption [21]. When the PLL incorporates the proposed VCO, circuits other than the VCO can be powered on only during the infrequent refresh periods, which are required to maintain the oscillation frequency. In this way, the power consumption can be reduced drastically. Furthermore, because the programmable elements hold the AVD corresponding to the previous oscillation frequency, high-speed rebooting of the PLL is expected.

When installing the proposed VCO in an actual PLL, we should thoroughly examine the operation stability at the frequency lock and the AVD-dependent frequency changes in the whole PLL. Nevertheless, the VCO-based PLL is a promising application of analog programmable devices based on CAAC-IGZO FETs.

8.7   Neural Networks

8.7.1   Introduction

As a future application of CAAC-IGZO technology, this section proposes a neural network based on CAAC-IGZO FETs. As discussed above, CAAC-IGZO FETs can be configured as an analog memory with high-speed writing capability and long-term data retention. Such an analog memory is suitable as a weight memory in neural networks. Below, a neural network and its potential construction from CAAC-IGZO FETs are briefly described.

8.7.2   Neural Networks

Neural networks attempt to model the neuron structure of the human brain. The models are described by neurons that are connected to other neurons through synapses, each of which is weighted.

The functions of neurons and synapses can be expressed most simply by the linear threshold model [see Figure 8.38 and Equations (8.3) and (8.4)]. Here, $x_1, \ldots, x_n$ are the input signals from n neurons, and $w_1, \ldots, w_n$ denote the weights of the respective n synapses. The weighted sum of the neurons is represented by Equation (8.3), where net denotes the membrane potential in a neuron. Equation (8.4) gives the output signal y of the neuron, where $f(x)$ is an activation function such as a step function (Figure 8.39). A neuron fires when net exceeds the threshold of neuron firing θ. The output signals y of firing and quiescent neurons are 1 and 0, respectively:

$$net = w_1x_1 + w_2x_2 + w_3x_3 + \cdots + w_nx_n \tag{8.3}$$

$$y = f(net - \theta). \tag{8.4}$$

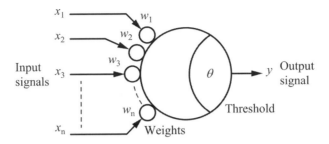

Figure 8.38   Linear threshold model of neurons and synapses

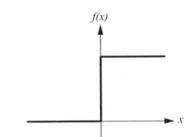

Figure 8.39   Step function

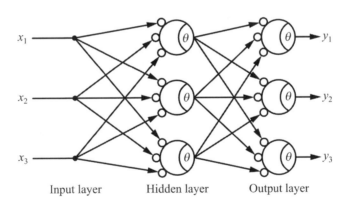

Input layer   Hidden layer   Output layer

Figure 8.40   Multilayer perceptron neural network

Neural network structures are roughly classified as multilayer perceptron or recurrent neural networks. In multilayer perceptron networks (Figure 8.40) [22], the neurons are layered and signals are unidirectionally transferred from the input side to the output side. The output data are uniquely determined from the input data. In other words, the output information cannot be changed by updating the weights. Thus, back propagation or similar is used to update the weight to change output information and perform learning.

Figure 8.41 shows an example of a recurrent neural network [23], in which the neurons are connected to each other. The signals are not confined to certain directions, and output signals

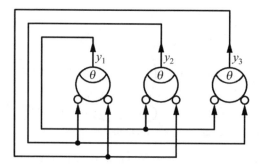

Figure 8.41   Recurrent neural network

can directly become input signals; in other words, the network has feedback. Hopfield networks and Boltzmann machine networks are included in this category.

8.7.3   CAAC-IGZO Neural Network

In a hardwired circuit construction of the above neural networks, neurons must receive signals from other neurons, calculate their weighted sum, and send signals to other neurons. The synapses store the weights, indicating the bond strengths between neurons, and the weighting function that multiplies each weight by the corresponding neuron's output. Therefore, each neuron needs an adder or similar functionality to calculate the sum of signals from other neurons, and each synapse needs a memory to store the weight and multiplier.

When the adder, memory, and multiplier are configured from digital circuits, a multi-bit adder, high-capacity memory, and multi-bit multiplier are required. The resulting network would be a large-scale IC requiring microfabrication. Furthermore, if the number of layers and number of neurons in each layer increase, the number of wirings increases exponentially.

In contrast, when the adder, memory, and multiplier are configured from analog circuits, the number of circuits is reduced from the digital case, but analog memories are very difficult to construct. A DRAM-type analog memory enables very short-time data retention. Although we could employ a high-capacity capacitor and perform regular data recovery by a refresh operation, these implementations would increase the die size and power consumption. Weights can also be stored in a flash memory, but it would increase the programming time, thereby extending the convergence time of learning with frequent weight updates.

The above problems could be solved by a neural network with an analog memory composed of CAAC-IGZO FETs and a capacitor. This network would enable high-speed writing, improving the learning speed and reducing the power consumption (as no refresh operation would be required). Moreover, the analog circuitry would require less die size and wirings than the digital circuit.

The extremely low off-state current of the CAAC-IGZO FET enables easy construction of a high-speed writing analog memory. A neural network with this type of memory would offer several advantages. First, the CAAC-IGZO FETs memorize the weight over a longer time than DRAM; second, the learning could be performed more frequently than is possible with flash memory; third, if appropriate, the network could be unlearned by changing the threshold

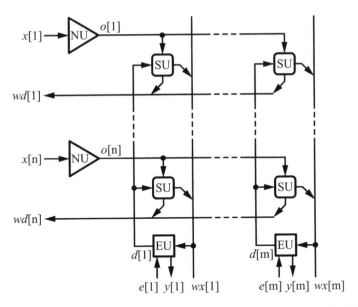

Figure 8.42   Unit layer of multilayer perceptron neural network constructed by CAAC-IGZO FETs

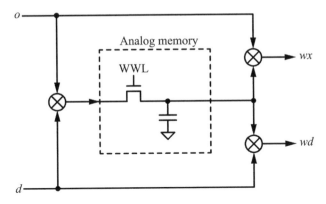

Figure 8.43   Synapse unit in multilayer perceptron neural network constructed from CAAC-IGZO FETs; ⊗ denotes multiplier

through the back gate of the CAAC-IGZO FET. Multilayer perceptron and recurrent neural networks constructed from CAAC-IGZO FETs are described below.

Figure 8.42 schematizes a multilayer perceptron neural network built from CAAC-IGZO FETs. Each layer is composed of n neuron units NU, m error units EU, and $m \times n$ synapse units SU. The two unit layers form a multilayer perceptron neural network with three layers. Each NU has a selector and is selected to be an input neuron or a hidden neuron. Each EU has a selector and is selected to be an error-signal generator or an output neuron. The SU is composed of multipliers and an analog memory (see Figure 8.43). The analog memory consists of one CAAC-IGZO FET and one capacitor that together can store one weight. The weights are

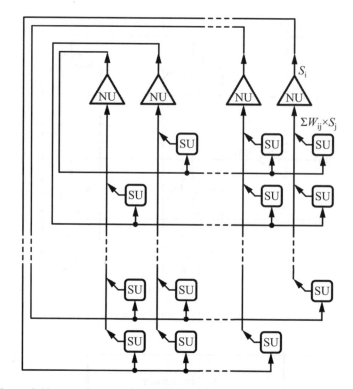

Figure 8.44 Recurrent neural network constructed from CAAC-IGZO FETs

updated by accumulating charges in the capacitor via the CAAC-IGZO FET. This operation occurs at much higher speed than is possible by flash memory, and the weight can be retained for a longer time than is possible in DRAM.

Figure 8.44 schematizes a recurrent neural network built from CAAC-IGZO FETs. The network consists of neuron units NU and synapse units SU. Each SU has a weight modifier that writes and stores a weight (see Figure 8.45). A weight is written by decreasing the voltage of node NA using a step-down charge pump composed of M1, M2, and C1, while increasing the voltage of NA using a step-up charge pump composed of M3, M4, and C2. The weight is stored by holding the NA voltage in C3. Transistors M1–M4 are CAAC-IGZO FETs, enabling long-term retention of the weight.

8.7.4   Conclusion

As a future application, the application of CAAC-IGZO technology to neural networks is proposed. The CAAC-IGZO FETs provide analog memories with various structures, which can be assembled into the circuit components of a neural network. Such an analog memory would empower the neural network with high-speed writing (consequently, with high-speed learning), low power consumption (by removing the need for refresh operations), and a simplified circuit (by replacing the digital circuitry with analog circuitry).

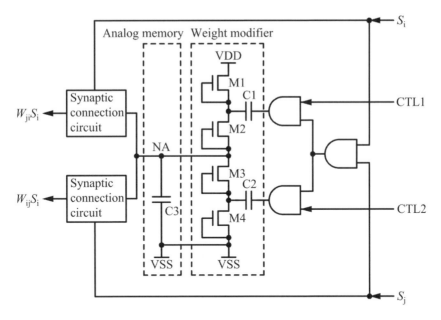

Figure 8.45 Synapse unit of recurrent neural network constructed from CAAC-IGZO FETs

8.8 Memory-Based Computing

Programmable circuits such as FPGA are now widely used. They typically include both pro-grammable wiring (PW, a switched network) and a programmable function (PF), as shown in Figure 8.46(a). Although programmable routing can be used to realize any topology and can accommodate any circuits to be implemented, its delay may not be predictable. Consequently, resolving all timing problems may not be easy; significant logical and layout restructuring may be required to fix timing closure problems, which may take a long time. This is a serious prob-lem when attempting to implement large circuits, which may include millions of gates. Things become much worse if the circuit is implemented with multiple chips, as the delays among chips are much different from those inside chips.

One way to minimize delay issues is to use look-up tables (LUT), which are con-nected through a predetermined "fixed" topology, as shown in Figure 8.46(b). When the routing is fixed, delays can be estimated accurately, and the design process can con-centrate on logical correctness. When non-volatile memories are used for implementing LUTs, the entire circuit becomes completely non-volatile, and the power can be turned on or off at any time.

Figure 8.47 shows an example. The circuit (a) has a fixed network topology; nine program-mable functions with two inputs are interconnected. If each PF is programmed as shown in (b), it becomes a one-bit full adder. In contrast, if each PF is programmed as shown in (c), it becomes a one-bit full subtractor. In all cases, the topology is fixed.

Now an important question is "how many logic functions appearing in practical designs can be implemented on top of predetermined fixed-topology LUT networks?" We performed a pre-liminary experiment as shown below.

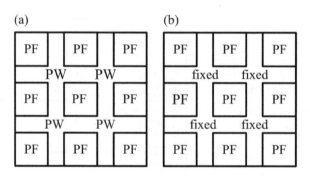

Figure 8.46   Programmable circuit: (a) with programmable wiring; (b) with fixed routing

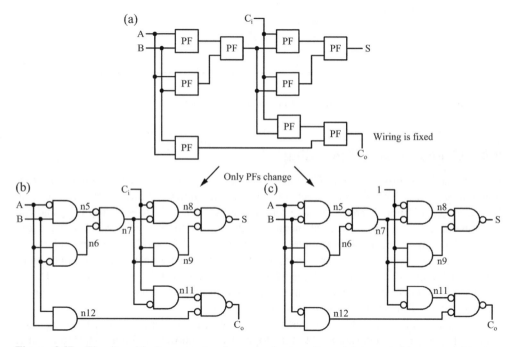

Figure 8.47   Circuits with the same topology can be transformed into an adder or a subtractor: (a) programmable functions set for one-bit full adder; (b) programmable functions set for one-bit full subtractor

Suppose we attempt to implement a specific 12-input single-output logic function. Overall, there are 2^{4096} different functions with 12 inputs. If we use a single 12-input LUT, as shown in Figure 8.48(a), we can realize all of them with 4096 bits of memory. In contrast, if, for example, we use a network of 4-input LUTs, as shown in Figure 8.48(b), we only need $17 \times 16 = 272$ bits (each 4-input LUT needs 16 bits), but it can implement only a very small subset of all possible logic functions of 12 inputs because $2^{272} \ll 2^{4096}$.

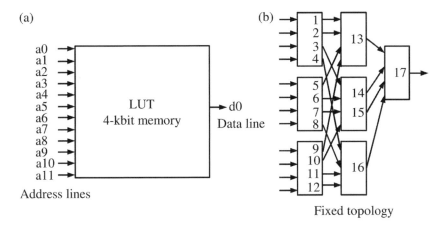

Figure 8.48   Implementation of 12-input logic functions with LUTs: (a) schematic diagram indicating that all 12-input logic functions can be implemented with a LUT having 4096 bits of storage; (b) example of a 12-input circuit having a set of 4-input LUTs and their connections. The circuit in (b) needs only 272 bits of storage for all LUTs

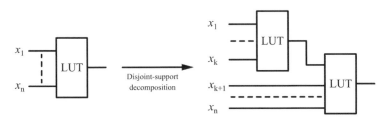

Figure 8.49   Disjoint support decomposition of logic functions

Therefore, the practical question is "how many logic functions appearing in real-life (benchmark) circuits can be implemented with fixed-topology programmable circuits such as the one presented in Figure 8.48(b)?" For such an evaluation, 300,000 12-input single-output logic functions are extracted from various benchmark circuits. Then, we check whether the programmable circuits of fixed topologies can implement them. The extracted circuits are classified as disjoint support decomposable (DSD), partially DSD, and non-DSD [24]. In a DSD circuit, if a given logic function of $x1$, ..., xn is decomposed as shown in Figure 8.49, the inputs to the two LUTs are disjoint.

In contrast, in a non-DSD circuit, all input variables must appear as inputs to both LUTs. In general, DSD functions are relatively simpler than non-DSD functions. A partially DSD circuit means that some of the inputs must appear in both LUTs.

We use several different fixed 3-input and 4-input LUT networks (Figure 8.50). The LUTs are programmable but the wiring is completely fixed regardless of the logic functions to be implemented.

The results are shown in Figure 8.51. For each topology of Figure 8.50, the percentage of the implementable functions out of the extracted 12-input logic functions discussed above is

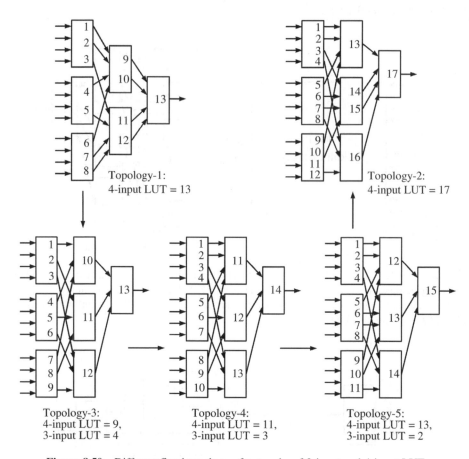

Figure 8.50 Different fixed topology of networks of 3-input and 4-input LUTs

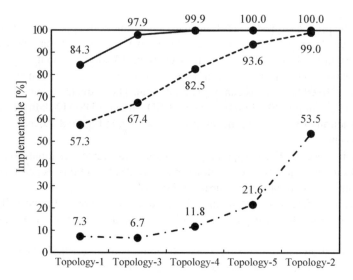

Figure 8.51 Percentages of implementable functions out of the extracted 12-input logic functions (solid line: DSD; dashed line: partial DSD; chain line: non-DSD)

reported. DSD functions are easier to implement than non-DSD functions. The topologies shown in Figure 8.50 can be used to implement most of the DSD functions. Even for non-DSD functions, half of them can be implemented in topology 2. Consequently, many logic functions encountered in real hardware designs can be implemented with fixed-topology programmable circuits. With non-volatile memory, the entire computation can be non-volatile, that is, the power can be shut off and on at any time.

8.9 Backtracking Programs with Power Gating

In this section, we present an application of power-gating mechanisms to efficient programming. Power-off operations in power gating preserve the internal state in the non-volatile area, and power-on operations restore the internal state. This means that when some programs are executed after a power-gating operation and a power-on operation is performed during execution without a power-off operation, the program state immediately goes back to the state when the previous power-off operation was performed. This is an instant backtrack and can be used effectively in backtrack-based programming as well as fault-tolerant computation. With CAAC-IGZO FET-based power-gating mechanisms, only a few cycles for power on and off are required. Consequently, there is almost no penalty to performance due to power gating.

An example of a power-gating execution sequence is presented in Figure 8.52. In this example, after operation 2 is executed, the processor enters into power gating and its power is cut off. Here, it is assumed that even without power, all of the storage elements in the processor – such as memories, register files, and flip-flops – can retain their values by utilizing non-volatile memory, such as that using a CAAC-IGZO FET as discussed in this book. After a certain amount of time, the processor is powered on, and it simply resumes its computation, as if there was no power-off period. Therefore, only a few cycles of powering on and off are

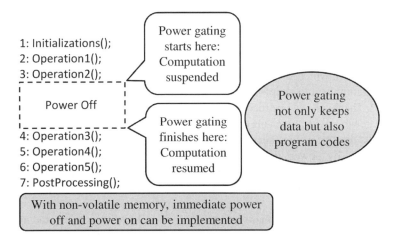

Figure 8.52 Program that includes power-gating operations. *Source*: Reprinted from [25] by permission of Association for Computing Machinery, Inc. © 2014. http://doi.acm.org/10.1145/2660859.2660959

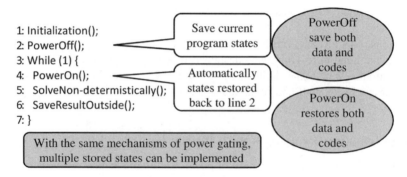

Figure 8.53   Use of power gating for quick and efficient backtracking in programming. *Source*: Reprinted from [25] by permission of Association for Computing Machinery, Inc. © 2014. http://doi. acm.org/10.1145/2660859.2660959

required, which means that the processors can utilize power gating very frequently to save energy.

From the viewpoint of programs running on processors, power-off operations save all information required for resuming the computations to some non-volatile areas, and power-on operations load them back to the processors. This means that in each power-gating sequence, all states of the programs running on the processor are kept in non-volatile areas, even after the power is turned on again. Therefore, if power-on operations are repeated without any additional power-off operations, the states of the programs running on the processor automatically return back to the point where the last power-off operations were performed. As long as we do not perform the power-off operations again, each time a power-on operation is performed, the states of the programs will automatically return to the state at the last power-off operation point. This is a very efficient way of implementing backtracking, which may be a frequent operation of search and optimization programs.

Figure 8.53 shows an example of such power-gating use. After the computation in line 1, the program proceeds to the operations corresponding to a power-off (but the power may not necessarily be turned off), which save the current states of the programs running on the processor. Then, the program falls into an infinite loop, where non-deterministic search and optimization are performed. The results can be recorded indirectly because, when the entire program returns back to the state where the power-off operations were last performed, even the search and optimization results disappear. In order not to lose the results, they must be sent to external locations, not affected by power gating (line 6). After that, we need to backtrack to the state corresponding to line 2, which is automatically performed by executing the power-on operations. Mechanisms for power gating can be used as very efficient ways to perform automatic backtracking.

This can be extended further if there are multiple non-volatile areas to be used in conjunction with power gating, in a similar manner as the current implementations of quick power gating [25]. That is, when we are performing power-off operations, we can choose which of the available non-volatile areas to use to save the states of the programs running on the processor. Depending on the power-gating architectures, there can be minimal increases in area for such storages. If we implement such mechanisms to select the saving areas, multiple states of the programs can be chosen as backtrack targets, which makes it possible to program much more complicated non-deterministic algorithms.

Apart from using power-gating mechanisms for efficient realization of non-deterministic algorithms, implementation of high-speed complete backtracking to multiple targets may enable efficient executions of interruption-based programming [25]. As there are only a few additional cycles required, owing to efficient realization of power gating with non-volatile memory, power gating could be applied in various programming settings, which could be an interesting research topic.

References

[1] Finkenzeller, K. (2010) *RFID Handbook: Fundamentals and Applications in Contactless Smart Cards, Radio Frequency Identification and Near-Field Communication*, 3rd edn. Wiley: London.

[2] *Information technology – Radio frequency identification for item management – Part 63: Parameters for air interface communications at 860 MHz to 960 MHz Type C*. ISO/IEC 18000-63, 2013.

[3] NXP Semiconductors N.V., "SL3S1203_1213 UCODE G2iL and G2iL+." Product data sheet available at: www.nxp.com/documents/data_sheet/SL3S1203_1213.pdf (2014) [accessed February 12, 2016].

[4] Gawande, A. A., Studdert, D. M., Orav, E. J., Brennan, T. A., and Zinner, M. J. (2003) "Risk factors for retained instruments and sponges after surgery," *N. Engl. J. Med.*, **348**, 229.

[5] The Road Committee of the Panel on Infrastructure Development, "Recommendations for Full-Scale Maintenance of Aging Roads." Available at: www.mlit.go.jp/road/road_e/pdf/recommendation.pdf (2014) [accessed February 12, 2016].

[6] Seibert, J. A. (2006) "Flat-panel detectors: How much better are they?," *Pediatr. Radiol.*, **36**, 173.

[7] Kasap, S., Frey, J. B., Belev, G., Tousignant, O., Mani, H., Greenspan, J., *et al.* (2011) "Amorphous and poly-crystalline photoconductors for direct conversion flat panel X-ray image sensors," *Sensors*, **11**, 5112.

[8] Wronski, M. M. and Rowlands, J. A. (2008) "Direct-conversion flat-panel imager with avalanche gain: Feasibility investigation for HARP-AMFPI," *Medical Phys.*, **35**, 5207.

[9] Li, D. and Xhao, W. (2008) "SAPHIRE (scintillator avalanche photoconductor with high resolution emitter read-out) for low dose X-ray imaging: Spatial resolution," *Medical Phys.*, **35**, 3151.

[10] *ITU-T Recommendation Series H Audiovisual and multimedia systems. 265: High Efficiency Video Coding*. International Telecommunication Union, Geneva, 2015.

[11] Sullivan, G. J., Ohm, J.-R., Han, W.-J., and Wiegand, T. (2012) "Overview of the high efficiency video coding (HEVC) standard," *IEEE Trans. Circuits Syst. Video Technol.*, **22**, 1649.

[12] Yamazaki, S. and Kimizuka, N. (in press) *Physics and Technology of Crystalline Oxide Semiconductor CAAC-IGZO: Application to Displays*. Wiley: Chichester.

[13] Yamazaki, S., Koyama, J., Yamamoto, Y., and Okamoto, K. (2012) "Research, development, and application of crystalline oxide semiconductor," *SID Symp. Dig. Tech. Pap.*, **43**, 183.

[14] Ohmaru, T., Yoneda, S., Nishijima, T., Endo, M., Dembo, H., Fujita, M., *et al.* (2012) "Eight-bit CPU with non-volatile registers capable of holding data for 40 days at 85°C using crystalline In–Ga–Zn oxide thin film transistors," *Ext. Abstr. Solid State Dev. Mater.*, 1144.

[15] Inoue, H., Matshuzaki, T., Nagatsuka, S., Okazaki, Y., Sasaki, T., Noda, K., *et al.* (2012) "Nonvolatile memory with extremely low-leakage indium–gallium–zinc-oxide thin-film transistor," *J. Solid-State Circuits*, **47**, 2258.

[16] Okamoto, Y., Nakagawa, T., Aoki, T., Kozuma, M., Kurokawa, Y., Ikeda, T., *et al.* (2014) "CAAC-OS-based nonvolatile programmable analog device: Voltage controlled oscillator realizing instant frequency switching," *Ext. Abstr. Solid State Dev. Mater.*, 452.

[17] Kozuma, M., Okamoto, Y., Nakagawa, T., Aoki, T., Ikeda, M., Osada, T., *et al.* (2014) "Crystalline In–Ga–Zn–O FET-based configuration memory for multi-context field-programmable gate array realizing fine-grained power gating," *Jpn. J. Appl. Phys.*, **53**, 14EE12.

[18] Aoki, T., Okamoto, Y., Nakagawa, T., Ikeda, M., Kozuma, M., Osada, T., *et al.* (2014) "Normally-off computing with crystalline InGaZnO-based FPGA," *IEEE Int. Solid-State Circuits Conf. Dig. Tech. Pap.*, 502.

[19] Okamoto, Y., Nakagawa, T., Aoki, T., Ikeda, M., Kozuma, M., Osada, T., *et al.* (2015) "A boosting pass gate with improved switching characteristics and no overdriving for programmable routing switch based on crystalline In–Ga–Zn–O technology," *IEEE Trans. Very Large Scale Integr. (VLSI) Syst.*, 422.

[20] Gao, X., Klumperink, E. A. M., Geraedts, P. F. J., and Nauta, B. (2009) "Jitter analysis and a benchmarking figure-of-merit for phase-locked loops," *IEEE Trans. Circuits Syst.*, **56**, 117.

[21] Ding, Y. and Kenneth, K. O. (2007) "A 21-GHz 8-modulus prescaler and a 20-GHz phase-locked loop fabricated in 130-nm CMOS," *IEEE J. Solid-State Circuits*, **42**, 1240.

[22] Morie, T. and Amemiya, Y. (1994) "An all-analog expandable neural network LSI with on-chip backpropagation learning," *IEEE J. Solid-State Circuits*, **29**, 1086.

[23] Arima, Y., Murasaki, M., Yamada, T., Maeda, A., and Shinohara, H. (1992) "A refreshable analog VLSI neural network chip with 400 neurons and 40K synapses," *IEEE J. Solid-State Circuits* **27**, 1854.

[24] Mishchenko, A. (2014) "Enumeration of irredundant circuit structures," *Proc. Int. Workshop Logic Synthesis*, **1**.

[25] Fujita, M. (2014) "Highly-pipelined and energy-saved computing with arrays of non-volatile memories," *Proc. ICONIAAC'14*, 46.

Appendix

FET Symbols

Symbol	Description
D G S	CAAC-IGZO FET
D G BG S	CAAC-IGZO FET with back gate
S G D	Pch-Si FET
D G S	Nch-Si FET

S: source
D: drain
G: gate
BG: back gate

Physics and Technology of Crystalline Oxide Semiconductor CAAC-IGZO: Application to LSI, First Edition.
Edited by Shunpei Yamazaki and Masahiro Fujita.
© 2017 John Wiley & Sons, Ltd. Published 2017 by John Wiley & Sons, Ltd.

Unit Prefixes

Multiple	Prefix	Symbol	Multiple	Prefix	Symbol
10^{24}	yotta	Y	10^{-1}	deci	d
10^{21}	zetta	Z	10^{-2}	centi	c
10^{18}	exa	E	10^{-3}	milli	m
10^{15}	peta	P	10^{-6}	micro	μ
10^{12}	tera	T	10^{-9}	nano	n
10^{9}	giga	G	10^{-12}	pico	p
10^{6}	mega	M	10^{-15}	femto	f
10^{3}	kilo	k	10^{-18}	atto	a
10^{2}	hecto	h	10^{-21}	zepto	z
10^{1}	deka	da	10^{-24}	yocto	y

Index